Classical Theory of Gauge Fields

Valery Rubakov

Translated by Stephen S Wilson

PRINCETON UNIVERSITY PRESS

PRINCETON AND OXFORD

Original title: *Klassicheskie Kalibrovochnye Polia*
©V.A. Rubakov, 1999   ©Éditorial URSS, 1999

Published by Princeton University Press, 41 William Street, Princeton, New Jersey 08540
In the United Kingdom: Princeton University Press, 3 Market Place, Woodstock, Oxfordshire OX20 1SY

Library of Congress Control Number 2002102049
ISBN 0-691-05927-6 (hardcover : alk. paper)

The publisher would like to acknowledge the translator of this volume for providing the camera-ready copy from which this book was printed

Printed on acid-free paper. ∞
www.pupress.princeton.edu
Printed in the United States of America
10 9 8 7 6 5 4 3 2 1

Contents

Preface

This book is based on a lecture course taught over a number of years in the Department of Quantum Statistics and Field Theory of the Moscow State University, Faculty of Physics, to students in the third and fourth years, specializing in theoretical physics.

Traditionally, the theory of gauge fields is covered in courses on *quantum field theory*. However, many concepts and results of gauge theories are already in evidence at the level of classical field theory, which makes it possible and useful to study them in parallel with a study of quantum mechanics. Accordingly, the reading of the first ten chapters of this book does not require a knowledge of quantum mechanics, Chapters 11–13 employ the concepts and methods conventionally presented at the start of a course on quantum mechanics, and a thorough knowledge of quantum mechanics, including the Dirac equation, is only required in order to read the subsequent chapters. A detailed acquaintance with quantum field theory is not mandatory in order to read this text. At the same time, at the very start, it is assumed that the reader has a knowledge of classical mechanics, special relativity and classical electrodynamics.

Part I of the book contains a study of the basic ideas of the theory of gauge fields, the construction of gauge-invariant Lagrangians and an analysis of spectra of linear perturbations, including perturbations about a non-trivial ground state. Part II is devoted to the construction and interpretation of solutions, whose existence is entirely due to the nonlinearity of the field equations, namely, solitons, bounces and instantons. In Part III, we consider certain interesting effects, arising in the interactions of fermions with topological scalar and gauge fields.

The book has an Appendix, which briefly discusses the role of instantons as saddle points of the Euclidean functional integral in quantum field theory and several related topics. The purpose of the Appendix is to give an initial idea about this rather complex aspect of quantum field theory; the presentation there is schematic and makes no claims to completeness (for example, we do not touch at all upon the important questions relating to supersymmetric gauge theories). To read the Appendix, one needs to be acquainted with the quantum theory of gauge fields.

Of course, the majority of questions touched upon in this book are considered in one way or another in other existing books and reviews on quantum field theory, a far from complete list of which is given at the end of the book. In a sense, this book may serve as an introduction to the subject.

The book contains two mathematical digressions, where elements of the theory of Lie groups and algebras and of homotopy theory are studied, without any claim to completeness or mathematical rigor. This should make it possible to read the book without constantly resorting to the specialized literature in these areas.

We should like to thank our colleagues F.L. Bezrukov, D.Yu. Grigoriev, M.V. Libanov, D.V. Semikoz, D.T. Son, P.G. Tinyakov and S.V. Troitsky for their great help in the preparation and teaching of the lecture course, attentive reading of the manuscript and preparation for its publication.

Classical Theory of Gauge Fields

Part I

Chapter 1

Gauge Principle in Electrodynamics

1.1 Electromagnetic-field action in vacuum

The electromagnetic field in a vacuum is described by two spatial vectors $\mathbf{E}(x)$ and $\mathbf{H}(x)$, the electric and the magnetic fields. They form the antisymmetric field strength tensor $F_{\mu\nu}$:

$$F_{i0} = -F_{0i} = E_i$$

$$F_{ij} = \varepsilon_{ijk} H_k$$

(where $H_i = \frac{1}{2}\varepsilon_{ijk} F_{jk}$).

Here and throughout the book, Greek indices refer to space–time and take values $\mu, \nu, \lambda, \rho, \ldots = 0, 1, 2, 3$ (in d-dimensional space–time $\mu, \nu, \lambda, \rho, \ldots = 0, 1, 2, \ldots, (d-1)$). Latin indices refer to space and take values $i, j, k, \ldots = 1, 2, 3$ ($i, j, k, \ldots = 1, 2, \ldots, (d-1)$ in d-dimensional space–time). Repeated Greek indices are assumed to denote summation with the Minkowski metric tensor, $\eta_{\nu\mu} = \mathrm{diag}\,(+1, -1, -1, -1)$, where, we shall not usually distinguish between superscripts and subscripts (for example, $F_{\mu\nu} F_{\mu\nu}$ will denote $F_{\mu\nu} F_{\lambda\rho} \eta^{\mu\lambda} \eta^{\nu\rho}$). Repeated Latin indices will be assumed to denote summation with the Euclidean spatial metric (thus, $F_{\mu\nu} F_{\mu\nu}$ denotes

$$-F_{0i} F_{0i} - F_{i0} F_{i0} + F_{ij} F_{ij} = -2(\mathbf{E}^2 - \mathbf{H}^2),$$

and the scalar product of two vectors k^μ and x^μ is $k^\mu x^\mu = k^0 x^0 - \mathbf{kx}$).

The electromagnetic-field action in vacuum has the form

$$S = -\frac{1}{4} \int d^4x F_{\mu\nu} F_{\mu\nu}. \tag{1.1}$$

The dynamical variables here are vector potentials A_μ, such that

$$F_{\mu\nu} = \partial_\mu A_\nu - \partial_\nu A_\mu. \tag{1.2}$$

Here and throughout the book, $\partial_\mu \equiv \partial/\partial x^\mu$.

The variational principle for the action (1.1) leads to the Maxwell equations in vacuum:

$$\partial_\mu F_{\mu\nu} = 0. \tag{1.3}$$

Problem 1. *Show that the Maxwell equations (1.3) are indeed extremality conditions for the action with respect to variations of the field $A_\mu(x)$ with fixed values of A_μ at the boundary of the space–time volume in which the system is located.*

Problem 2. *Show that equations (1.3) are equivalent to the pair of equations*

$$\begin{aligned}
\mathrm{div}\,\mathbf{E} &= 0 \\
-\mathrm{curl}\,\mathbf{H} + \frac{\partial \mathbf{E}}{\partial t} &= 0,
\end{aligned}$$

and that definition (1.2) is equivalent to the equations

$$\begin{aligned}
\mathrm{div}\,\mathbf{H} &= 0 \\
\mathrm{curl}\,\mathbf{E} + \frac{\partial \mathbf{H}}{\partial t} &= 0.
\end{aligned} \tag{1.4}$$

Show that the pair of equations (1.4) can be written in the Lorentz covariant form

$$\varepsilon_{\mu\nu\lambda\rho} \partial_\nu F_{\lambda\rho} = 0.$$

The last equation (consequence of definition (1.2)) is called the Bianchi identity for the electromagnetic-field strength tensor.

The energy of an electromagnetic field is expressed in terms of the electric and magnetic fields,

$$E = \frac{1}{2} \int (\mathbf{E}^2 + \mathbf{H}^2) d^3x. \tag{1.5}$$

Other dynamical quantities (for example, the field momentum) are also expressed in terms of $\mathbf{E}$ and $\mathbf{H}$, i.e. in terms of the field tensor $F_{\mu\nu}$.

1.2   Gauge invariance

The strength tensor $F_{\mu\nu}$, and the associated action, the energy and the field equations are invariant under the transformations

$$A_\mu(x) \to A'_\mu(x) = A_\mu(x) + \partial_\mu \alpha(x), \qquad (1.6)$$

where $\alpha(x)$ is an arbitrary function of the space–time coordinates. For example, for the strength tensor we have

$$
\begin{aligned}
F_{\mu\nu} \to F'_{\mu\nu} &= \partial_\mu(A_\nu + \partial_\nu \alpha) - \partial_\nu(A_\mu + \partial_\mu \alpha) \\
&= \partial_\mu A_\nu - \partial_\nu A_\mu = F_{\mu\nu}.
\end{aligned}
$$

Field transformations, whose parameters depend in an arbitrary manner on the space–time points are called gauge transformations (as opposed to global transformations, whose parameters are constant, i.e. do not depend on x^μ; we shall discuss global transformations a little later). Thus, the transformation (1.6) is the simplest gauge transformation.

The main postulate of the theory of gauge fields is the requirement that all physical (observable) quantities, and also actions and equations of motion, should be gauge invariant, i.e. invariant under the corresponding gauge transformations. This principle is satisfied in electrodynamics; thus, electrodynamics is the simplest example of a gauge theory.

In electrodynamics, one can construct the gauge-invariant quantity

$$\oint A_\mu dx^\mu, \qquad (1.7)$$

where the integration is performed along some closed contour C in space–time. In simply connected space–time (without holes), this quantity reduces to $\mathbf{E}$ and $\mathbf{H}$; for example, for a purely spatial contour C, we can write

$$\oint_C A_i dx^i = -\int_\Sigma \mathbf{H} d\Sigma,$$

where Σ is a surface in space, encircled by the contour C. If the space–time is not simply connected, the gauge-invariant quantity (1.7) cannot be expressed in terms of $\mathbf{E}$ and $\mathbf{H}$. For example, in $(2 + 1)$-dimensional space–time, where the space has a hole (or in $(3 + 1)$-dimensional space–time, whose space contains an infinitely long incised cylinder), it may be the case that $\mathbf{E} = \mathbf{H} = 0$ everywhere, but

$$\oint_C \mathbf{A} d\mathbf{x} \neq 0,$$

where the contour C winds around the hole. This quantity, or more precisely, the phase factor

$$\exp\left[ie \oint_C \mathbf{A}d\mathbf{x}\right] \tag{1.8}$$

(e is the electron charge) turns out to be observable in quantum theory (Aharonov–Bohm effect). Physically, such a situation is realized as follows: one can take a long thin solenoid in which a magnetic field is concentrated and surround it by an impenetrable (for all particles and fields) wall. The electric and magnetic fields outside the wall are equal to zero; however, the scattering of electrons off this wall depends upon the value of the phase factor (1.8).

1.3   General solution of Maxwell's equations in vacuum

We shall find the solution of Maxwell's equations (1.3). It is convenient to perform the Fourier transformation

$$A_\mu = \int_{k^0 \geq 0} \left[e^{ikx} a_\mu(k) + \text{c.c.}\right] d^4 k.$$

Here and throughout the book, 'c.c.' denotes the complex conjugate quantity (thus, in this formula, 'c.c'$= e^{-ikx} a_\mu^*(k)$). The $a_\mu(k)$ are complex functions of the four-vector k. We note that

$$kx = k_\mu x_\mu = k_0 x_0 - \mathbf{kx},$$

where $\mathbf{k}$ is a three-dimensional wave vector and k_0 is the plane wave frequency.

Equations (1.3) are equivalent to the following equations for Fourier transforms:

$$k^\mu k_\mu a_\nu - k^\mu k_\nu a_\mu = 0,$$

or

$$k^2 a_\nu - k_\nu (ka) = 0, \tag{1.9}$$

where $k^2 = k^\mu k_\mu$; $ka = k^\mu a_\mu$.

Let us first consider the case $k^2 \neq 0$. Then it follows from (1.9) that the vector $a_\nu(k)$ is directed along k_ν,

$$a_\nu(k) = k_\nu c(k). \tag{1.10}$$

Substituting (1.10) in (1.9), we see that there are no constraints on $c(k)$; equations (1.9) are turned into an identity. Thus, for $k^2 \neq 0$, the general solution is of the form (1.10).

Let us now consider the case $k^2 = 0$, i.e. $k^0 = |\mathbf{k}|$. Equations (1.9) then reduce to the four-dimensional orthogonality condition:

$$k^\mu a_\mu = 0.$$

There are three linearly independent vectors orthogonal (in the four-dimensional sense) to k^μ. One of these is k^μ itself, the other two, $e_\mu^{(\alpha)}$, $\alpha = 1, 2$, can be chosen to be purely spatial vectors which are orthogonal in the three-dimensional sense to the vector $\mathbf{k}$ and to each other:

$$
\begin{aligned}
e_0^{(\alpha)} &= 0 \\
e_i^{(\alpha)} k_i &= 0 \\
e_i^{(\alpha)} e_i^{(\beta)} &= \delta^{\alpha\beta}
\end{aligned}
\tag{1.11}
$$

(the last condition also fixes the normalization of $\mathbf{e}^{(\alpha)}$). Thus, for $k^2 = 0$, the general solution has the form

$$a_\mu(k) = k_\mu c(k) + e_\mu^{(\alpha)}(k) b_\alpha(k),$$

where $c(k)$ and $b_\alpha(k)$ are arbitrary complex functions of the three-vector $\mathbf{k}$ (since $k^0 = |\mathbf{k}|$).

Combining the cases $k^2 \neq 0$ and $k^2 = 0$, we obtain that the general solution of the Maxwell equations has the form

$$A_\mu(x) = A_\mu^\perp(x) + A_\mu^\|(x),\tag{1.12}$$

where

$$
\begin{aligned}
A_\mu^\perp &= \int d^3k \left[e^{ikx} e_\mu^{(\alpha)}(\mathbf{k}) b_\alpha(\mathbf{k}) + \text{c.c.} \right]_{k^0 = |\mathbf{k}|} \\
A_\mu^\| &= \int d^4k \left[e^{ikx} k_\mu c(k) + \text{c.c.} \right].
\end{aligned}
\tag{1.13}
$$

The field $A_\mu^\|(x)$ is a pure gauge,

$$A_\mu^\|(x) = \partial_\mu \alpha(x),$$

where

$$\alpha(x) = \int d^4k \left[\frac{1}{i} e^{ikx} c(k) + \text{c.c.} \right].$$

As one might expect, the general solution contains an arbitrary scalar function (if $A_\mu(x)$ is a solution of Maxwell's equations, then $A_\mu'(x) = A_\mu(x) + \partial_\mu \alpha(x)$ is also a solution). The non-trivial part of the solution, $A_\mu^\perp(x)$ is the collection of plane waves, moving with the speed of light (with frequency k^0 equal to the modulus of the wave vector $\mathbf{k}$).

Problem 3. *Assuming that the vectors $e_\mu^{(1)}$ and $e_\mu^{(2)}$ (see (1.11)) are real (linear polarization), show that in a plane wave of the form*

$$A_\mu(x) = e^{ikx} e_\mu^{(1)} b_1$$

the magnetic field is orthogonal (in the three-dimensional sense) to the electric field at every point x, and the electric and the magnetic fields are orthogonal to the wave vector **k**.

We note that (1.13) is not the only possible way of writing the general solution of the Maxwell equations. In some problems, instead of the plane-wave decomposition, it is convenient to use a decomposition in terms of other sets of functions, for example, in terms of spherical harmonics. We shall not dwell on these decompositions, since the important properties of solutions as far as we are concerned, the propagation of waves with the speed of light and the presence of two independent amplitudes b_α for each wave vector **k**, are most conveniently seen in a plane-wave basis.

1.4   Choice of gauge

The gauge invariance of the Maxwell equations means that their general solution contains an arbitrary scalar function of space–time coordinates x^μ. We showed this explicitly in Section 1.3 for the example of an electromagnetic field without sources. This property of the non-uniqueness of a solution is inconvenient in more complex cases when sources exist (and also presents a difficulty for field quantization). Thus, a supplementary condition is often imposed on the field A_μ, in order to decrease this arbitrariness or, so to speak, fix the gauge. The following gauge conditions are often used.

a) The Coulomb gauge

$$\operatorname{div} \mathbf{A} \equiv \partial_i A_i = 0. \tag{1.14}$$

This condition, like all the others, is not invariant under gauge transformations; if $A_\mu(x)$ satisfies this condition, then $A'_\mu(x) = A_\mu(x) + \partial_\mu \alpha(x)$ also satisfies it, if and only if

$$\partial_i \partial_i \alpha \equiv \Delta \alpha = 0,$$

i.e. if and only if $\alpha(x)$ is spatially constant or increases in space. If we restrict ourselves to fields decreasing at spatial infinity (we implicitly assumed this in Section 1.3) then the vector potential is uniquely fixed. It is trivial to check this fact for the solution (1.12): condition (1.14) gives

$$\mathbf{k}^2 c(k) = 0,$$

i.e. $A_\mu^\parallel(x) = 0$. Thus, the solution of the Maxwell equations in the Coulomb gauge has the form

$$A_\mu(x) = A_\mu^\perp(x),$$

where $A_\mu^\perp$ is given by expression (1.13).

b) The Lorentz gauge

$$\partial_\mu A_\mu = 0.$$

Unlike the Coulomb condition, this condition is invariant under residual gauge transformations

$$A_\mu \to A_\mu' = A_\mu + \partial_\mu \alpha,$$

where $\alpha(x)$ satisfies the d'Alembert condition

$$\partial_\mu \partial_\mu \alpha \equiv \Box \alpha = 0.$$

Thus, the Lorentz gauge fixes the solution modulo a sum of longitudinal waves $\partial_\mu \alpha(x)$, propagating with the speed of light. In the language of the solution (1.12), this means that $c(k) = 0$ for $k^2 \neq 0$, while for $k^2 = 0$ the function $c(\mathbf{k})$ is arbitrary.

c) The gauge $A_0 = 0$. Residual gauge invariance is characterized by gauge functions α, independent of time, $\partial_0 \alpha = 0$. The general solution of the Maxwell equations has the form

$$A_\mu(x) = A_\mu^\perp(x) + B_\mu(\mathbf{x})$$

(we note that $A_0^\perp(x) = 0$), where $B_0 = 0$ and

$$B_i(\mathbf{x}) = \partial_i \alpha(\mathbf{x}).$$

This corresponds to $c(k) \neq 0$, provided $k^0 = 0$.

Problem 4. *Find the residual gauge transformations and the general solution of the Maxwell equations in the axial gauge*

$$\mathbf{n}\mathbf{A} = 0,$$

where $\mathbf{n}$ is some fixed unit three-vector, which is constant in space–time.

Chapter 2

Scalar and Vector Fields

2.1 System of units $\hbar = c = 1$

Throughout the book, we shall use the system of units $\hbar = c = 1$. The only non-trivial dimension then is the dimension of mass. Length and time have dimension $1/M$. Indeed

$$[c] = \frac{L}{T}, \quad [\hbar] = M\frac{L^2}{T} \tag{2.1}$$

(the last equation is evident, for example, from the relation $E = \hbar\omega$). From (2.1) and $\hbar = c = 1$ it follows that

$$L = T = \frac{1}{M}.$$

Physically, $1/M$ is the Compton wavelength of a particle of mass M.

Problem 1. *Find, in conventional units, the length and time corresponding to $1/gram$.*

Problem 2. *a) Find the dimension of $\mathbf{E}$ and $\mathbf{H}$ in the system of units $\hbar = c = 1$. Show that the electromagnetic field action is dimensionless.*
b) The same thing in space–time of dimension d.

Problem 3. *a) Find the dimension of the electric charge in the system of units $\hbar = c = 1$.*
b) Find the numerical value of the quantity $e^2/(4\pi)$ (fine structure constant), where e is the electron charge.
c) The same as in a), but in d-dimensional space–time.

11

Problem 4.   *The electron-Volt (eV) is the energy obtained by an electron in passing through a potential difference of 1 V. Find the relationship between 1 gram and 1 eV in the system $\hbar = c = 1$. Find, in conventional units, the length and time corresponding to $1/GeV$, where $1\ GeV = 10^9\ eV$. Find 1 Gauss and 1 V/cm (units of the magnetic and electric fields) in the system $\hbar = c = 1$ and unit of mass 1 GeV.*

2.2   Scalar field action

The electric field is the best known, but not the simplest, example of a field. The simplest case arises if, instead of four functions of space–time coordinates $A_\mu(x)$ (forming a four-vector under Lorentz transformations) one considers a single real function $\varphi(x)$ (Lorentz scalar). Another example arises when the scalar field is complex, in other words, when the field has two real components $\operatorname{Re}\varphi(x)$ and $\operatorname{Im}\varphi(x)$.

In quantum theory, there is a correspondence between fields and particles. Thus, the electromagnetic field describes the photon; π^0-mesons, η-mesons and certain other particles are described by real scalar fields; $\pi^{\pm}$-mesons correspond to a single complex scalar field. Vector bosons $W^{\pm}$ and Z are described by (massive) vector fields, distinct from the electromagnetic field. One important class of fields which will not be considered until Part III, since it does not have a classical c-number analogue, describes fermions, particles with a half-integer spin (electrons, neutrinos, etc.).

Let us consider a real scalar field $\varphi(x)$ and construct the simplest Lagrangian for it. We require that, as a result of the variation of the action, we obtain second-order differential equations, and so the derivatives in the Lagrangian should occur no more than quadratically. We shall also require the Lagrangian to be a Lorentz scalar. In addition to these two quite general requirements we temporarily impose one further requirement, namely linearity of the field equations. This implies that the action is quadratic with respect to the field. These requirements lead to the action

$$S = \int d^4x \left[\frac{1}{2}(\partial_\mu \varphi)^2 - \frac{1}{2}m^2\varphi^2 \right], \qquad (2.2)$$

where m^2 is the only arbitrary parameter, and $(\partial_\mu \varphi)^2 = \partial_\mu \varphi \partial_\mu \varphi$. The signs in (2.2) are determined by the requirement of non-negative energy; the field energy will be considered below.

Problem 5.   *Generally speaking, there are more than two invariants satisfying the requirements listed above. The Lagrangian could be chosen in* *the form*

$$\mathcal{L} = a(\partial_\mu\varphi)^2 + b\partial_\mu\partial_\mu\varphi + c\varphi\partial_\mu\partial_\mu\varphi + \frac{m^2}{2}\varphi^2 + d\varphi, \qquad (2.3)$$

where $a, \ldots, d$ are arbitrary constants. Assuming $m^2 \neq 0$, show that the theory with the more general Lagrangian (2.3) is equivalent to the theory (2.2).

Variation of the action (2.2) leads to the Klein–Gordon–Fock equation

$$\partial_\mu\partial_\mu\varphi + m^2\varphi = 0. \qquad (2.4)$$

Let us derive this equation from the variational principle with the action (2.2). Let $\delta\varphi(x)$ be the field variation; then

$$\delta S = \int d^4x[\partial_\mu\varphi\delta(\partial_\mu\varphi) - m^2\varphi\delta\varphi]. \qquad (2.5)$$

Now, the change in the derivative of the field is equal to the derivative of the change in the field, $\delta(\partial_\mu\varphi) = \partial_\mu(\delta\varphi)$. In the first term in (2.5), we integrate by parts

$$\delta S = \int d^4x[-\partial_\mu\partial_\mu\varphi\delta\varphi - m^2\delta\varphi] + \int d\Sigma^\mu[\partial_\mu\varphi\delta\varphi],$$

where the second integral is taken over the boundary of the volume. As usual, we suppose that $\delta\varphi = 0$ on the boundary of the four-dimensional volume, then the second integral vanishes. By requiring $\delta S = 0$ for all $\delta\varphi$, we obtain equation (2.4).

Let us find the general solution of the Klein–Gordon equation. Moving to the k-representation,

$$\varphi(x) = \int_{k^0 \geq 0} d^4k \left[e^{ikx}\tilde{\varphi}(k) + \text{c.c.}\right]$$

we obtain the equation

$$[k^2 - m^2]\tilde{\varphi}(k) = 0,$$

so that $\tilde{\varphi}$ is non-zero (and arbitrary) provided

$$k^2 \equiv (k^0)^2 - (\mathbf{k})^2 = m^2,$$

i.e. for a relationship between the frequency k^0 and the wave vector $\mathbf{k}$ of the form

$$k^0 = \sqrt{\mathbf{k}^2 + m^2}. \qquad (2.6)$$

This equation represents the dispersion law for free waves. Equation (2.6) is reminiscent of the relativistic relationship between energy and momentum

$$E = \sqrt{\mathbf{p}^2 + m^2}.$$

This is no accident: if the field $\varphi(x)$ is understood as the wave function of some particle (not a conventional non-relativistic particle with wave function satisfying the Schrödinger equation, but some relativistic particle), then the following holds for it ($\hbar = c = 1$)

$$E = k^0, \quad \mathbf{p} = \mathbf{k},$$

where m is the mass of this particle. These considerations are made precise in quantum field theory.

We shall call the parameter m the mass of the field, and fields with $m \neq 0$ massive fields. For the photon $m = 0$; the electromagnetic field is massless.

Thus the general solution of the Klein–Gordon equation is a sum of plane waves with dispersion law (2.6)

$$\varphi(x) = \int d^3k \left[e^{ikx} \tilde{\varphi}(\mathbf{k}) + \text{c.c.} \right]_{k^0 = \sqrt{\mathbf{k}^2 + m^2}},$$

where the complex amplitudes $\tilde{\varphi}(\mathbf{k})$ are arbitrary.

For the rest, it is useful to mention the analogy between classical field theory and classical mechanics with many degrees of freedom. The dynamical coordinates $q_A(t)$ of classical mechanics are analogous to the field $\varphi(\mathbf{x}, t)$, except that the index A becomes a continuous index, namely the coordinates of a point of space (for an electromagnetic field $A_\mu(\mathbf{x}, t)$, the 'index' A, labeling the dynamical coordinates is the pair $(\mu, \mathbf{x})$). Summation over the index A (for example, in the Lagrange function $L = \sum_A \frac{1}{2} \dot{q}_A \dot{q}_A + \cdots$) is replaced by integration over d^3x. We use this analogy to construct the energy of the scalar field. In classical mechanics the energy is equal to

$$E = \sum_A \frac{\partial L}{\partial \dot{q}_A} \dot{q}_A - L.$$

The analogue of the Lagrange function L in scalar field theory is the integral of the Lagrangian over space,

$$
\begin{aligned}
L &= \int d^3x \mathcal{L} = \int d^3x \left[\frac{1}{2} (\partial_\mu \varphi)^2 - \frac{m^2}{2} \varphi^2 \right] \\
&= \int d^3x \left[\frac{1}{2} \dot{\varphi}^2 - \frac{1}{2} \partial_i \varphi \partial_i \varphi - \frac{m^2}{2} \varphi^2 \right].
\end{aligned}
$$

Using (omitting the dependence on time)

$$\frac{\delta L}{\delta \dot{\varphi}(\mathbf{x})} = \dot{\varphi}(\mathbf{x}),$$

we obtain for the energy

$$E = \int d^3x \frac{\delta L}{\delta \dot{\varphi}(\mathbf{x})} \dot{\varphi}(\mathbf{x}) - L, \tag{2.7}$$

an expression of the form

$$E = \int d^3x \left[\frac{1}{2}\dot{\varphi}^2 + \frac{1}{2}\partial_i\varphi\partial_i\varphi + \frac{m^2}{2}\varphi^2 \right]. \tag{2.8}$$

From this expression, the choice of signs in (2.2) is evident: a) a negative sign in front of $(\partial_\mu\varphi)^2$ in (2.2) would lead to a negative sign in the first two terms in (2.8) and the energy of rapidly oscillating fields would be negative and would decrease unboundedly with increasing oscillation frequency; b) a positive sign for the term with m^2 in (2.2) would lead to a negative sign for the corresponding (third) term in (2.8), and the energy of a homogeneous field would be negative and unbounded from below for large fields φ.

Problem 6. *Find the dimension of the field φ. Show that the parameter m actually has the dimension of mass.*

Problem 7. *Using an expression of type (2.7), find the energy of an electromagnetic field with action (1.1) and dynamical coordinates $A_\mu(\mathbf{x}, t)$. Show that when the field equations (1.3) are satisfied, the energy thus found agrees with (1.5).*

Problem 8. *Assuming that the field $\varphi(\mathbf{x}, t)$ decreases rapidly at spatial infinity, verify that the energy (2.8) is actually conserved, $dE/dt = 0$, if the field satisfies the Klein–Gordon equation.*

2.3 Massive vector field

A massive vector field should be described by a four-vector $B_\mu(x)$. However, if B_μ is the gradient of a scalar field $B_\mu(x) = \partial_\mu\chi(x)$, then it does not make sense to talk about the vector nature of B_μ: the system is described by the scalar field $\chi(x)$. In other words, any vector field $B_\mu(x)$ can be decomposed into a transverse component (in the four-dimensional sense) $B_\mu^\perp$ and a gradient of a scalar field,

$$B_\mu = B_\mu^\perp + \partial_\mu\chi$$

$$\partial_\mu B_\mu^\perp = 0. \tag{2.9}$$

The transverse field $B_\mu^\perp$ does not reduce to a scalar field, and it is precisely the object of interest. If the mass of the field is non-zero, the dispersion law must be of the form $k^0 = \sqrt{\mathbf{k}^2 + m^2}$, which will be achieved if each component of the field $B_\mu^\perp(x)$ satisfies the Klein–Gordon equation

$$(\partial_\mu \partial_\mu + m^2) B_\nu^\perp = 0. \tag{2.10}$$

Thus, we wish to describe the field $B_\mu^\perp$, satisfying equations (2.10) and (2.9).

The action, leading to equations (2.10) and (2.9), has the form

$$S = \int \left[-\frac{1}{4} B_{\mu\nu} B_{\mu\nu} + \frac{m^2}{2} B_\mu B_\mu \right] d^4 x, \tag{2.11}$$

where $B_{\mu\nu} = \partial_\mu B_\nu - \partial_\nu B_\mu$ is completely analogous to the strength tensor in electrodynamics. From the action (2.11), we have the field equations

$$\partial_\mu B_{\mu\nu} + m^2 B_\nu = 0. \tag{2.12}$$

For $m^2 \neq 0$, equation (2.9) is a consequence of equations (2.12): differentiating (2.12) with respect to x^ν and using the antisymmetry of the tensor $B_{\mu\nu}$, we obtain

$$\partial_\nu \partial_\mu B_{\mu\nu} + m^2 \partial_\nu B_\nu = m^2 \partial_\nu B_\nu = 0.$$

Moreover, using the definition of the tensor $B_{\mu\nu}$ and taking into account the transversality of B_ν, from (2.12) we obtain

$$\partial_\mu \partial_\mu B_\nu - \partial_\mu \partial_\nu B_\mu + m^2 B_\nu = \partial_\mu \partial_\mu B_\nu + m^2 B_\nu = 0.$$

Thus, equations (2.12) are equivalent to the system

$$\begin{aligned} \partial_\mu \partial_\mu B_\nu + m^2 B_\nu &= 0 \\ \partial_\nu B_\nu &= 0, \end{aligned} \tag{2.13}$$

as required.

Problem 9. *Find the general solution of the system (2.13).*

Problem 10. *Find the energy for the massive vector field with action (2.11). Explain the choice of signs in (2.11).*

The action (2.11) and equations (2.13) are not invariant under gauge transformations $B_\mu \to B_\mu + \partial_\mu \alpha$. In other words, the theory of a massive vector field with action (2.11) is not a gauge theory.

2.4   Complex scalar field

Another important example of a field is the complex scalar field $\varphi(x) = (\mathrm{Re}\,\varphi)(x) + i(\mathrm{Im}\,\varphi)(x)$. We choose the Lagrangian for this field in the simplest form

$$\mathcal{L} = \partial_\mu \varphi^* \partial_\mu \varphi - m^2 \varphi^* \varphi. \tag{2.14}$$

The variational principle for the action $S = \int \mathcal{L} d^4 x$ is formulated as follows: we shall formally assume that $\varphi^*(x)$ and $\varphi(x)$ are independent and require extremality of the action with respect to variations $\delta\varphi^*$ and $\delta\varphi$ separately. Varying with respect to $\delta\varphi^*$, we obtain the equation for φ:

$$\partial_\mu \partial_\mu \varphi + m^2 \varphi = 0, \tag{2.15}$$

while variation with respect to $\delta\varphi$ gives the equation for φ^*:

$$\partial_\mu \partial_\mu \varphi^* + m^2 \varphi^* = 0. \tag{2.16}$$

Instead of the complex scalar field, we can introduce a pair of real scalar fields φ_1 and φ_2, such that

$$\varphi = \frac{1}{\sqrt{2}}(\varphi_1 + i\varphi_2).$$

Then, the Lagrangian (2.14) is rewritten in the form

$$\mathcal{L} = \frac{1}{2}\partial_\mu \varphi_i \partial_\mu \varphi_i - \frac{m^2}{2}\varphi_i \varphi_i, \tag{2.17}$$

where $i = 1, 2$ and summation over i is assumed. Thus, the complex scalar field is equivalent to two real fields of equal mass.

Problem 11.   *Considering the fields φ_i as independent, find equations for these fields, corresponding to the Lagrangian (2.17), and show that they are equivalent to equations (2.15) and (2.16).*

The new element in the theory (2.14) is the presence of a current which is conserved

$$j_\mu = -i(\varphi^* \partial_\mu \varphi - \varphi \partial_\mu \varphi^*) \tag{2.18}$$

$$\partial_\mu j_\mu = 0. \tag{2.19}$$

Problem 12.   *Show that the current j_μ is actually conserved if the field equations (2.15) and (2.16) are satisfied.*

As we shall see later, conservation of this current, i.e. equation (2.19), is associated with invariance of the Lagrangian (2.14) under (global) transformations

$$\begin{aligned}
\varphi(x) &\;\rightarrow\; \varphi'(x) = e^{i\alpha}\varphi(x) \\
\varphi^*(x) &\;\rightarrow\; \varphi'^*(x) = e^{-i\alpha}\varphi^*(x),
\end{aligned}$$

where α does not depend on the space–time point (such transformations are therefore said to be global).

The conservation of the current (continuity equation (2.19)) presupposes that there exists a charge which is conserved, if the field decreases rapidly at spatial infinity. Indeed, we define

$$Q = \int j_0 d^3x,$$

where the integration is performed over all space at a fixed time x^0 (thus, Q might depend on this x^0). We have

$$\partial_0 Q = \int d^3x \partial_0 j_0 = - \int d^3x \partial_i j_i.$$

The last integral reduces to an integral over an infinitely distant surface

$$\int d^3x \partial_i j_i = \int d\Sigma_i j_i.$$

It is equal to zero if the field φ decreases sufficiently rapidly, whence

$$\partial_0 Q = 0,$$

i.e. the charge is actually conserved.

2.5 Degrees of freedom

Summarizing our discussion of free fields (i.e. fields satisfying linear equations), we note that in all the different cases, the dispersion law has the form

$$k^0 = \sqrt{\mathbf{k}^2 + m^2}$$

where m is the mass of the field (which may be equal to zero). For $k^0 > 0$ general solutions are characterized by one or more arbitrary complex functions of the three-dimensional wave vector $\mathbf{k}$ (in electrodynamics, there is another arbitrary function $c(k)$ of the four-vector k, however, that function can be eliminated by gauge transformations). The number of these arbitrary complex functions is called the number of degrees of freedom of the field. Thus, a real scalar field has one degree of freedom; a complex scalar field has two degrees of freedom, and a real massive vector field has three degrees of freedom. In the case of an electromagnetic field, there are two physical degrees of freedom, since a general solution is characterized by two arbitrary functions $b_\alpha(\mathbf{k})$ which cannot be eliminated by gauge transformations.

We note that, depending on the choice of gauge, in electrodynamics, there may or may not exist unphysical degrees of freedom (i.e. arbitrary functions) upon which (gauge-invariant) physical quantities do not depend. In the Lorentz gauge there is one unphysical degree of freedom, since, in addition to a function $b_\alpha(\mathbf{k})$, for $k^2 = 0$, the general solution in this gauge contains an arbitrary function $c(\mathbf{k})$ (see Section 1.4, point b)). In the Coulomb gauge there are no unphysical degrees of freedom.

Problem 13. *Find the number of physical degrees of freedom in electrodynamics in d-dimensional space–time. Consider the cases $d = 2$, $d = 3$ and $d \geq 4$ separately.*

2.6   Interaction of fields with external sources

In classical electrodynamics, the interaction of an electromagnetic field with charges and currents is constructed using the current four-vector $j_\mu = (-\rho, \mathbf{j})$, where $\rho(x)$ and $\mathbf{j}(x)$ are the charge density and the current density. Taking this interaction into account, the action can be written in the form

$$S = \int d^4x \left(-\frac{1}{4} F_{\mu\nu} F_{\mu\nu} - e j_\mu A_\mu \right). \tag{2.20}$$

Variation of this action with respect to A_μ leads to the Maxwell equations with a source:

$$\partial_\mu F_{\mu\nu} - e j_\nu = 0. \tag{2.21}$$

Here, e is the unit of electric charge. Equation (2.21) implies that current is conserved

$$\partial_\mu j_\mu = 0 \tag{2.22}$$

(this equation follows from (2.21) after differentiation with respect to x^ν). Current conservation in turn leads to gauge invariance of the action (2.20), if the gauge function decreases rapidly at infinity. Indeed, the action for the field

$$A'_\mu = A_\mu + \partial_\mu \alpha$$

is equal to

$$S[A'] = S[A] - \int d^4x \, e j_\mu \partial_\mu \alpha.$$

The last integral reduces, when (2.22) is taken into account, to an integral over an infinitely distant surface in four-dimensional space–time,

$$\int d^4x \, e j_\mu \partial_\mu \alpha = \int d\Sigma_\mu \, e \alpha(x) j_\mu.$$

It is equal to zero for decreasing $\alpha(x)$, so that the action is invariant, $S[A'] = S[A]$.

Problem 14. *Suppose that the current j_μ corresponds to two static point charges, $j_0(\mathbf{x}) = -q_1\delta(\mathbf{x} - \mathbf{x}_1) - q_2\delta(\mathbf{x} - \mathbf{x}_2)$. Find the solution of equation (2.21) (Coulomb law).*

Problem 15. *Suppose that the current j_μ does not depend on time. Find the energy functional corresponding to the action (2.20). Express it in terms of electric and magnetic fields $\mathbf{E}$ and $\mathbf{H}$.*

Problem 16. *Using the results obtained in the previous two problems, find the interaction energy of the two point charges. Show that electric charges of like sign are mutually repulsive, while parallel electric currents are mutually attractive.*

Problem 17. *Solve the previous three problems in d-dimensional space–time. Consider the cases $d = 2$, $d = 3$ and $d > 4$ separately.*

By analogy with electrodynamics, one can consider the interaction of an external source with a massive scalar field. The corresponding contribution to the action will be a Lorentz scalar, if the source is a scalar with respect to Lorentz transformations, $\rho(x)$. The overall action is equal to

$$S = \int d^4x \left[\frac{1}{2}(\partial_\mu\varphi)^2 - \frac{m^2}{2}\varphi^2 \right] + \int d^4x\, \rho\varphi. \qquad (2.23)$$

Analogously to formulae (2.20) and (2.23), the interaction of the current j_μ with a massive vector field B_μ is introduced.

Problem 18. *Find the field equations for the action (2.23). Solve them for a static point source $\rho(\mathbf{x}) = -q\delta(\mathbf{x})$.*

Problem 19. *Find the energy functional for the theory with the action (2.23). Calculate the interaction energy of two static point charges q_1 and q_2, separated by distance r. Do scalar charges of like sign attract or repel each other? Find an expression for the force.*

Problem 20. *Find solutions of the previous two problems in d-dimensional space–time.*

Multiplying (2.28) by $(-i\varphi^*)$ and (2.29) by $(i\varphi)$ and subtracting the equations obtained, we obtain

$$\partial_\mu j_\mu^{(0)} = 2e\partial_\mu(\varphi^*\varphi A_\mu).$$

The right-hand side of this equation is equal to zero for arbitrary fields only if $e = 0$, i.e. if there is no interaction (we note that the field φ and its *first* derivative $\partial_\mu\varphi$ are arbitrary, i.e. they cannot be reduced to anything simpler using the equations of motion (2.28) and (2.29)). Thus, the current $j_\mu^{(0)}$ is not conserved, which contradicts equation (2.26).

In principle, it would be possible to impose the condition $\partial_\mu(\varphi^*\varphi A_\mu) = 0$ and thereby guarantee conservation of current (this would be analogous to imposing the condition $\partial_\mu A_\mu = 0$ in the theory of a massive vector field). However, this condition is highly nonlinear in the fields, and it would be impossible to recover free electrodynamics and the free massive scalar field in the weak field limit.

A way out of this situation might be sought by a trial and error approach: one might try to add terms of higher order in the fields to the current $j_\mu^{(0)}$. In electrodynamics with a scalar field this procedure is quite simple to realize, by adding a term of type $A_\mu\varphi^*\varphi$ to the current $j_\mu^{(0)}$ (i.e. by assuming in equation (2.26) that $j_\mu = j_\mu^{(0)} + \text{constant} \cdot A_\mu\varphi\varphi^*$). We shall follow a completely different approach, which makes direct use of the gauge-invariance property. This procedure is notable for the fact that it plays a key role in the construction of the theory of non-Abelian gauge fields.

Let us require the invariance of the Lagrangian under gauge transformations of the field A_μ and simultaneously of the field φ.

$$
\begin{aligned}
\varphi(x) &\rightarrow \varphi'(x) = e^{i\alpha(x)}\varphi(x) \\
\varphi^*(x) &\rightarrow \varphi'^*(x) = e^{-i\alpha(x)}\varphi^*(x) \\
A_\mu(x) &\rightarrow A'_\mu(x) = A_\mu(x) + \frac{1}{e}\partial_\mu\alpha(x)
\end{aligned}
\tag{2.30}
$$

(the constant e is introduced here for convenience in what follows). The free electromagnetic field action is invariant under these transformations. The free complex scalar field action is not invariant (we have invariance only if α does not depend on the point of space–time). Indeed, consider the derivative of the field $\varphi'(x)$,

$$\partial_\mu\varphi'(x) = e^{i\alpha(x)}[\partial_\mu\varphi(x) + i\partial_\mu\alpha(x)\varphi(x)]. \tag{2.31}$$

If the second term did not exist, we would have invariance of the expression $\partial_\mu\varphi^*\partial_\mu\varphi$ and the free Lagrangian would be invariant. The idea is that, instead of the usual derivative $\partial_\mu\varphi$, one might use in the Lagrangian another object $D_\mu \varphi$ which would reduce to $\partial_\mu \varphi$ in the weak field limit, but would transform uniformly under the transformations (2.30):

$$(D_\mu \varphi)' = e^{i\alpha(x)} D_\mu \varphi. \tag{2.32}$$

From (2.31), it is clear that the term with $\partial_\mu \alpha$ can be compensated by adding a term of the type φA to $\partial_\mu \varphi$. Thus, we arrive at the expression

$$D_\mu \varphi = \partial_\mu \varphi - ieA_\mu \varphi = (\partial_\mu - ieA_\mu)\varphi. \tag{2.33}$$

This quantity is called the covariant derivative of the field φ. Let us verify equation (2.32). We have

$$
\begin{aligned}
(D_\mu \varphi)' &= \partial_\mu \varphi' - ieA'_\mu \varphi' \\
&= e^{i\alpha}\partial_\mu \varphi + e^{i\alpha}\varphi i \partial_\mu \alpha - ieA_\mu e^{i\alpha}\varphi - ie\frac{1}{e}\partial_\mu \alpha e^{i\alpha}\varphi \\
&= e^{i\alpha} D_\mu \varphi.
\end{aligned}
$$

Thus, equation (2.32) is satisfied and $(D_\mu \varphi)^* D_\mu \varphi$ is gauge invariant.

We select an action invariant under the gauge transformations (2.30) in the form

$$S = \int d^4 x \left[-\frac{1}{4}F_{\mu\nu}F_{\mu\nu} + (D_\mu \varphi)^* D_\mu \varphi - m^2 \varphi^* \varphi \right] \tag{2.34}$$

(one might also add to this the self-interaction of the scalar field $V_I(\varphi^*\varphi)$). Nonlinear (of degree greater than two in the fields) terms in this action arise due to the term $A_\mu \varphi$ in $D_\mu \varphi$ and have the structure $A_\mu \varphi^* \partial_\mu \varphi$ and $A_\mu A_\mu \varphi^* \varphi$.

By varying the action (2.34) with respect to the field A_μ, we obtain equation (2.26) with current

$$j_\mu = -i(\varphi^* \partial_\mu \varphi - \partial_\mu \varphi^* \varphi) - 2eA_\mu \varphi^* \varphi,$$

which can be written in an explicitly gauge-invariant form

$$j_\mu = -i[\varphi^* D_\mu \varphi - (D_\mu \varphi)^* \varphi]. \tag{2.35}$$

We note that, if the field φ^* is assumed to be independent, then its covariant derivative will have the form $D_\mu \varphi^* = (\partial_\mu + ieA_\mu)\varphi^*$ (the sign in front of ieA_μ is dictated by the requirement $(D_\mu \varphi^*)' = e^{-i\alpha} D_\mu \varphi^*$ and by the sign of the transformation (2.30)), which coincides with $(D_\mu \varphi)^*$. In what follows, we shall not distinguish between $(D_\mu \varphi)^*$ and $D_\mu \varphi^*$ (since they are the same).

Let us now find the scalar field equation. Taking variations, as usual, with respect to φ^*, we have

$$D_\mu D_\mu \varphi + m^2 \varphi = 0, \tag{2.36}$$

where, of course, $D_\mu D_\mu$ is defined completely analogously to (2.33):

$$D_\mu D_\mu \varphi = (\partial_\mu - ieA_\mu)(\partial_\mu - ieA_\mu)\varphi.$$

Let us check that, taking into account equation (2.36), the current (2.35) is conserved. We have

$$\partial_\mu j_\mu = -i[\partial_\mu \varphi^* D_\mu \varphi + \varphi^* \partial_\mu D_\mu \varphi - (D_\mu \varphi)^* \partial_\mu \varphi - \partial_\mu D_\mu \varphi^* \varphi].$$

Furthermore

$$\begin{aligned}
&\partial_\mu \varphi^* D_\mu \varphi + \varphi^* \partial_\mu D_\mu \varphi \\
=\ & \partial_\mu \varphi^* D_\mu \varphi + ieA_\mu \varphi^* D_\mu \varphi + \varphi^* \partial_\mu D_\mu \varphi - ie\varphi^* A_\mu D_\mu \varphi \\
=\ & D_\mu \varphi^* D_\mu \varphi + \varphi^* D_\mu D_\mu \varphi.
\end{aligned}$$

This and an analogous equation give

$$\partial_\mu j_\mu = -i(\varphi^* D_\mu D_\mu \varphi - D_\mu D_\mu \varphi^* \varphi).$$

The expression on the right-hand side is equal to zero, by virtue of the field equation (2.36) and its conjugate equation. The current j_μ is conserved, and the system of equations

$$\begin{aligned}
\partial_\mu F_{\mu\nu} &= j_\nu & (2.37) \\
D_\mu D_\mu \varphi + m^2 \varphi &= 0 & (2.38)
\end{aligned}$$

is consistent.

Thus, the requirement of invariance of the Lagrangian under gauge transformations (2.30) leads to replacement of the conventional derivatives $\partial_\mu \varphi$ by the covariant derivatives $D_\mu \varphi$ in the action. It turns out that the current j_μ, appearing in the field equations (2.37), is automatically gauge invariant and is conserved, when one takes the field equations (2.38) into account.

We note that it is instructive to write the transformations (2.30) in the form

$$\begin{aligned}
\varphi(x) &\rightarrow \varphi'(x) = g(x)\varphi(x) & (2.39) \\
\mathcal{A}_\mu(x) &\rightarrow \mathcal{A}'_\mu(x) = \mathcal{A}_\mu(x) + g(x)\partial_\mu g^{-1}(x), & (2.40)
\end{aligned}$$

where $g(x) = e^{i\alpha(x)}$, $\mathcal{A}_\mu = -ieA_\mu$. One advantage of this notation is that $g(x)$ at any point x can be interpreted as an element of the group $U(1)$, the (multiplicative) group of complex numbers of unit modulus. The transformation (2.39) looks like a transformation under the action of a fundamental group representation, whereas $g(x)\partial_\mu g^{-1}(x)$ at any point x is an element of the Lie algebra of this group. Generalization of the transformations (2.39) and (2.40) to cases of other (non-Abelian) Lie groups leads to non-Abelian gauge fields, which are the main topic of Chapter 4 and subsequent chapters of the book.

In conclusion, we stress that in the theory of interacting fields, the Lagrangians contain both terms quadratic in the fields and terms of degree three, four and higher in the fields. The quadratic terms lead to linear terms in the field equations, while higher-order terms lead to nonlinear terms. It is usually impossible to find general solutions of nonlinear field equations (integrable models form an exception); some physically interesting solutions, whose existence is mainly due to the nonlinearity of the equations, will be considered in Part II. The situation is markedly simplified if one considers waves with a small amplitude at any point of space–time. In that case the nonlinear terms in the field equations are small in comparison with the linear terms and may be neglected. Solving linearized equations is not very difficult, as we have seen in previous sections. In this part, we shall construct full Lagrangians, and discuss small excitations. Their properties can be found from the structure of the linear terms in the field equations or, equivalently, from the quadratic part of the field Lagrangian.

In quantum theory, small field excitations correspond to (elementary) particles. Thus, the study of classical fields of small amplitude in a specific model is at the same time the study of the spectrum of the elementary particles in that model.

2.8   Noether's theorem

In the previous sections we have met examples of global invariance of the Lagrangian (translation invariance in space–time, phase transformations in complex scalar field theory, etc.). We pointed out that every such symmetry gives rise to a conserved quantity. This assertion is called Noether's theorem. In this section, we present a derivation of Noether's theorem for the two most important cases as far as we are concerned: field transformations which do not affect space–time coordinates, and translations in space–time. The first type of symmetry corresponds to conservation of the current vector j_μ^a (the index a does not relate to space–time and takes as many values as there are independent global transformations)

$$\partial_\mu j_\mu^a = 0, \tag{2.41}$$

while translations correspond to conservation of the energy–momentum tensor $T_{\mu\nu}$,

$$\partial_\mu T_{\mu\nu} = 0. \tag{2.42}$$

If the fields decrease (rapidly) at spatial infinity, (2.41) leads to charge conservation (by complete analogy with Section 2.4)

$$Q^a = \int d^3x j_0^a(x),$$

while (2.42) leads to conservation of energy–momentum

$$P^\mu = \int d^3x T^{0\mu}(x).$$

We note that Q^a are scalars, while P^μ is a vector with respect to Lorentz transformations.

Before moving directly to Noether's theorem, let us discuss the field equations in quite general form. Let Φ^I be the full set of fields in the theory; the index I denotes the whole set of indices, both 'space–time' and 'internal' indices. (For example, in scalar electrodynamics Φ^I is the collective notation for A_μ, $\operatorname{Re}\varphi$, $\operatorname{Im}\varphi$). We shall consider Lagrangians containing only the first derivatives of the fields

$$\mathcal{L} = \mathcal{L}(\Phi^I, \partial_\mu \Phi^I). \tag{2.43}$$

In what follows, it will be convenient to write

$$\partial_\mu \Phi^I = \Phi^I_{,\mu}.$$

In order to write down the field equations, let us consider the variation of the action

$$S = \int \mathcal{L}(\Phi^I(x), \partial_\mu \Phi^I(x)) d^4x.$$

We have

$$\delta S = \int d^4x \left[\frac{\partial \mathcal{L}}{\partial \Phi^I} \delta \Phi^I + \frac{\partial \mathcal{L}}{\partial \Phi^I_{,\mu}} \partial_\mu (\delta \Phi^I) \right].$$

Here and in what follows, repetition of the index I will be taken to denote summation.

Integrating the second term by parts and assuming, as always, that the variations of the field are equal to zero on the boundary of the space–time volume, we obtain

$$\delta S = \int d^4x \left[\frac{\partial \mathcal{L}}{\partial \Phi^I} - \partial_\mu \left(\frac{\partial \mathcal{L}}{\partial \Phi^I_{,\mu}} \right) \right] \delta \Phi^I.$$

From the requirement $\delta S = 0$ we have the field equations

$$-\partial_\mu \left(\frac{\partial \mathcal{L}}{\partial \Phi^I_{,\mu}} \right) + \frac{\partial \mathcal{L}}{\partial \Phi^I} = 0. \tag{2.44}$$

The number of these equations is the same as the number of independent fields in the theory (i.e. the same as the number of values taken by the index I).

Let us now move to Noether's theorem for global transformations which do not affect space–time coordinates. We shall restrict ourselves to the class of phase-type transformations:

$$\Phi^I \to \Phi'^I = (\delta^{IJ} + \varepsilon^a t_a^{IJ})\Phi^J, \tag{2.45}$$

where we consider infinitesimal transformations, ε^a are infinitesimal parameters of the transformations (independent of the space–time point), the index a, over which the summation is performed, takes as many values as there are independent symmetries, and t_a^{IJ} are numerical constants. The invariance of the Lagrangian under the transformations (2.45) means that

$$\delta \mathcal{L} = \mathcal{L}(\Phi + \delta\Phi, \Phi_{,\mu} + \delta\Phi_{,\mu}) - \mathcal{L}(\Phi, \Phi_{,\mu}) = 0, \tag{2.46}$$

where

$$\begin{aligned}
\delta\Phi^I &= \varepsilon^a t_a^{IJ} \Phi^J \\
\delta\Phi_{,\mu}^I &= \varepsilon^a t_a^{IJ} \partial_\mu \Phi^J.
\end{aligned}$$

Equation (2.46) gives

$$\frac{\partial \mathcal{L}}{\partial \Phi^I} \varepsilon^a t_a^{IJ} \Phi^J + \frac{\partial \mathcal{L}}{\partial \Phi_{,\mu}^I} \varepsilon^a t_a^{IJ} \partial_\mu \Phi^J = 0.$$

Since the infinitesimal parameters ε^a are independent, we have

$$\frac{\partial \mathcal{L}}{\partial \Phi^I} t_a^{IJ} \Phi^J + \frac{\partial \mathcal{L}}{\partial \Phi_{,\mu}^I} t_a^{IJ} \partial_\mu \Phi^J = 0. \tag{2.47}$$

We now use the field equation (2.44) to write

$$\partial_\mu \left(\frac{\partial \mathcal{L}}{\Phi_{,\mu}^I} \right) t_a^{IJ} \Phi^J$$

instead of the first term of (2.47). Thus, two terms in (2.47) are combined into the total derivative, i.e. we have conservation of current

$$\partial_\mu j_\mu^a = 0,$$

where

$$j_\mu^a = \frac{\partial \mathcal{L}}{\partial \Phi_{,\mu}^I} t_a^{IJ} \Phi^J. \tag{2.48}$$

We note that current is conserved only if the fields satisfy the dynamical equations (2.44); we also note that the proof of Noether's theorem at the same time leads to explicit construction of the current (2.48).

As an example, let us consider a complex scalar field with Lagrangian

$$\mathcal{L} = \partial_\mu \varphi^* \partial_\mu \varphi - V(\varphi^* \varphi), \tag{2.49}$$

where V is some function of the square of the modulus of the field. Let us formally assume that φ and φ^* are independent, and denote $(\varphi, \varphi^*) \rightarrow \Phi^I$, $I = 1, 2$. The Lagrangian (2.49) is invariant under the phase transformations

$$\varphi \quad \rightarrow \quad \varphi' = e^{i\alpha} \varphi$$
$$\varphi^* \quad \rightarrow \quad \varphi^{*\prime} = e^{-i\alpha} \varphi^*,$$

or in infinitesimal form

$$\varphi' \quad = \quad (1 + i\alpha)\varphi \tag{2.50}$$
$$\varphi'^* \quad = \quad (1 - i\alpha)\varphi^*.$$

Comparing with (2.48) we have (here we have a single parameter α which plays the role of ε^a)

$$t^{11} = i, \quad t^{22} = -i, \quad t^{12} = t^{21} = 0.$$

Hence the conserved current is immediately constructed

$$j_\mu = \frac{\partial \mathcal{L}}{\partial \varphi_{,\mu}} i\varphi + \frac{\partial \mathcal{L}}{\partial \varphi^*_{,\mu}} (-i)\varphi^*.$$

Explicitly, we have

$$j_\mu = i(\partial_\mu \varphi^* \varphi - \partial_\mu \varphi \varphi^*), \tag{2.51}$$

which coincides with the current discussed in Section 2.4.

Problem 22. *Let us consider the theory of a complex scalar field with Lagrangian (2.49). Let us introduce two real fields φ_1 and φ_2, such that*

$$\varphi \quad = \quad \frac{1}{\sqrt{2}} (\varphi_1 + i\varphi_2)$$

$$\varphi^* \quad = \quad \frac{1}{\sqrt{2}} (\varphi_1 - i\varphi_2).$$

1) Write the Lagrangian in terms of the fields φ_1, φ_2. 2) Write the transformations (2.50) in terms of the fields φ_1, φ_2. Find corresponding constants of the type t^{IJ} ($I, J = 1, 2$). 3) Construct the current (2.48) in terms of the fields φ_1, φ_2 and show that it agrees with (2.51).

Problem 23.   *Construct the Noether current in scalar electrodynamics with Lagrangian (2.34), using the transformations (2.30), where α does not depend on the coordinates.*

Next let us consider translations

$$\Phi^I(x^\mu) \to \Phi'^I(x^\mu) = \Phi^I(x^\mu + \varepsilon^\mu),$$

where ε^μ are four independent parameters of the translations. The Lagrangian (2.43) does not depend explicitly on the space–time coordinates, hence

$$\mathcal{L}(\Phi', \Phi'_{,\mu}) = \mathcal{L}(x^\mu + \varepsilon^\mu), \tag{2.52}$$

where, on the right-hand side, the Lagrangian for the field $\Phi^I(x)$ is calculated at the point $x^\mu + \varepsilon^\mu$. To the lowest order in ε^μ we have

$$\mathcal{L}(\Phi', \Phi'_{,\mu}) = \mathcal{L}(\Phi, \Phi_{,\mu}) + \frac{\partial \mathcal{L}}{\partial \Phi^I} \partial_\nu \Phi^I \varepsilon^\nu + \frac{\partial \mathcal{L}}{\partial \Phi^I_{,\mu}} \partial_\nu \partial_\mu \Phi^I \varepsilon^\nu$$

and

$$\mathcal{L}(x^\mu + \varepsilon^\mu) = \mathcal{L}(x) + \partial_\nu \mathcal{L} \varepsilon^\nu.$$

By virtue of the arbitrariness of ε^ν, from (2.52) we have

$$\frac{\partial \mathcal{L}}{\partial \Phi^I} \partial_\nu \Phi^I + \frac{\partial \mathcal{L}}{\partial \Phi^I_{,\mu}} \partial_\mu \partial_\nu \Phi^I = \partial_\nu \mathcal{L}.$$

Again using the field equations (2.44) in order to reduce the left-hand side to the total derivative, we obtain

$$\partial_\mu \left(\frac{\partial \mathcal{L}}{\partial \Phi^I_{,\mu}} \partial_\nu \Phi^I \right) = \partial_\nu \mathcal{L},$$

which is equivalent to the equation

$$\partial_\mu \left(\frac{\partial \mathcal{L}}{\partial \Phi^I_{,\mu}} \partial_\nu \Phi^I - \delta^\mu_\nu \mathcal{L} \right) = 0.$$

Thus the energy–momentum tensor can be defined as follows

$$T^\mu_\nu = \frac{\partial \mathcal{L}}{\partial \Phi^I_{,\mu}} \partial_\nu \Phi^I - \delta^\mu_\nu \mathcal{L}. \tag{2.53}$$

It is conserved,

$$\partial_\mu T^\mu_\nu = 0.$$

The integral conservation laws are the law of conservation of the field energy

$$E = \int T_0^0 d^3x = \int \left(\frac{\partial \mathcal{L}}{\partial \dot{\Phi}^I} \dot{\Phi}^I - \mathcal{L} \right) d^3x \qquad (2.54)$$

and the law of conservation of the field momentum

$$P_i = \int T_i^0 d^3x = \int \left(\frac{\partial \mathcal{L}}{\partial \dot{\Phi}^I} \partial_i \Phi^I \right) d^3x.$$

We note that expression (2.54) coincides with that used earlier.

It is useful to make the following remark. Instead of the Noether current j_μ^a, one could use an expression of the form

$$\bar{j}_\mu^a = j_\mu^a + \partial_\nu f_{\mu\nu}^a,$$

where $f_{\mu\nu}^a$ is an arbitrary antisymmetric tensor (depending on the fields). Indeed, by virtue of the identity $\partial_\mu \partial_\nu f_{\mu\nu}^a = 0$, the current $\bar{j}_\mu^a$ is also conserved. Analogously, instead of the Noether energy–momentum tensor (2.53) one could use an expression

$$\bar{T}_\nu^\mu = T_\nu^\mu + \partial_\lambda \Omega_\nu^{\mu\lambda}, \qquad (2.55)$$

where $\Omega_\nu^{\mu\lambda}$ is an antisymmetric tensor with respect to the indices μ, λ.

Generally speaking, the tensor (2.53) is not symmetric in the indices μ, ν; $T^{\mu\nu} \neq T^{\nu\mu}$. However, using the above remark, $\Omega_\nu^{\mu\lambda}$ can be chosen in such a way that the conserved tensor $\bar{T}^{\mu\nu}$ is symmetric in the indices μ, ν.

Problem 24. *Find the Noether and the symmetric energy–momentum tensors in the theory of a free electromagnetic field. Which of these can be expressed in terms of the field strength $F_{\mu\nu}$?*

Problem 25. *Using the invariance of the Lagrangian of the scalar field under spatial rotations, find an expression for conserved angular momentum of the field.*

Problem 26. *The previous problem, but for a free electromagnetic field.*
A symmetric energy–momentum tensor can be obtained using the following technique. Let us introduce the space–time metric $g^{\mu\nu}$ into the action, assuming it to be arbitrary, but slightly different from the Minkowski metric. Let us write the action, invariant under general coordinate transformations, in the form

$$S = \int d^4x \sqrt{-g} \mathcal{L}(g^{\mu\nu}, \partial_\mu \Phi^I, \Phi^I),$$

where $g = \det g_{\mu\nu}$, $g_{\mu\alpha}g^{\alpha\nu} = \delta^\nu_\mu$. For example, the action of a scalar field with potential $V(\varphi)$ in space–time with the metric $g^{\mu\nu}$ has the form

$$S = \int d^4x\sqrt{-g}\left(\frac{1}{2}g^{\mu\nu}\partial_\mu\varphi\partial_\nu\varphi - V(\varphi)\right), \tag{2.56}$$

while the action of a free electromagnetic field is equal to

$$S = -\frac{1}{4}\int d^4x\sqrt{-g}(g^{\mu\lambda}g^{\nu\rho}F_{\mu\nu}F_{\lambda\rho}), \tag{2.57}$$

where $F_{\mu\nu} = \partial_\mu A_\nu - \partial_\nu A_\mu$. We note that here we distinguish between superscripts and subscripts and use only the derivative

$$\partial_\mu \equiv \frac{\partial}{\partial x^\mu},$$

while for the independent variables of the electromagnetism we choose the potentials A_μ. In the general case the energy–momentum tensor is obtained by differentiating the density of the action with respect to the metric:

$$\bar{T}_{\mu\nu} = \frac{\partial(\sqrt{-g}\mathcal{L})}{\partial g^{\mu\nu}},$$

where the metric $g^{\mu\nu}$ is set equal to the Minkowski metric after differentiation (the fields Φ^I, of course are not differentiated with respect to the metric). The tensor $\bar{T}_{\mu\nu}$ is clearly symmetric with respect to the indices μ, ν. More precisely, the above relation is written in the form (taking into account the symmetry of the tensor $g^{\mu\nu}$)

$$\delta S = \frac{1}{2}\int \sqrt{-g}\bar{T}_{\mu\nu}\delta g^{\mu\nu}d^4x, \tag{2.58}$$

where $\delta g^{\mu\nu}$ are small deviations from the Minkowski metric.

Problem 27. *Obtain an explicit expression for $\bar{T}_{\mu\nu}$ from (2.58) for the theory of a scalar field with the action (2.56). Show that in this case the metric energy–momentum tensor coincides with that of Noether.*

Problem 28. *Obtain an explicit expression for the metric energy–momentum tensor (2.58) in electrodynamics with the action (2.57). Compare with known expressions.*

Problem 29. *Show, in general form, that the metric energy–momentum tensor (2.58) is conserved.*

Chapter 3

Elements of the Theory of Lie Groups and Algebras

3.1 Groups

A *group* is a set G in which a multiplication operation with the following properties is defined:

1. associativity: for all $a, b, c \in G$, $(ab)c = a(bc)$;

2. existence of a unit element $e \in G$, such that for all $a \in G$, $ae = ea = a$;

3. existence of an inverse element $a^{-1} \in G$ for each $a \in G$ such that $a^{-1}a = aa^{-1} = e$.

If the multiplication operation is commutative (i.e. $ab = ba$ for all $a, b \in G$), the group is said to be *Abelian*, otherwise it is *non-Abelian*.

Groups G_1 and G_2 are *isomorphic* if there exists a bijective mapping $f : G_1 \rightarrow G_2$ consistent with the multiplication operations

$$f(g_1 g_2) = f(g_1)f(g_2), \quad f(g^{-1}) = [f(g)]^{-1}.$$

In what follows, we shall write group isomorphisms as $G_1 = G_2$ and we shall often not distinguish between isomorphic groups.

A *subgroup* H of a group G is a subset H of G, which is itself a group with respect to the multiplication operation defined in G. In other words, for $h, h_1, h_2 \in G$, the product $h_1 h_2$ and the inverse element h^{-1} are defined; $h_1 h_2$ and h^{-1} are required to be elements of the set H, if $h, h_1, h_2 \in H$.

Let us give some examples.

1. The group $U(1)$ is the set of complex numbers z with modulus equal to unity, $|z| = 1$. Multiplication in $U(1)$ is the multiplication of complex numbers (since for $|z_1| = |z_2| = 1$ we have $|z_1 z_2| = 1$, multiplication is indeed an operation in $U(1)$). The unit element is $z = 1$ and the inverse element to $z \in U(1)$ is z^{-1} ($z^{-1} \in U(1)$, since $|z^{-1}| = 1$ for $|z| = 1$).

2. The group Z_n is the set of integers modulo n, i.e. integers k and $(k+n)$ are identified (in other words, the set Z_n consists of n integers $0, 1, \ldots, (n-1)$). Multiplication in Z_n is defined as addition of integers modulo n; in other words, if $0 \le k_1 \le n-1$, $0 \le k_2 \le n-1$, then

$$(k_1 + k_2) \quad (\mathrm{mod}\ n) = \begin{cases} k_1 + k_2 & \text{for } (k_1 + k_2) \le n - 1 \\ k_1 + k_2 - n & \text{for } (k_1 + k_2) > n - 1. \end{cases}$$

Subtraction modulo n is defined analogously. We note that addition modulo n is commutative. The unit element in Z_n is $k = 0$, the inverse to the element k is equal to

$$(-k) \quad (\mathrm{mod}\ n) = \begin{cases} 0 & \text{for } k = 0 \\ n - k & \text{for } 0 < k \le n - 1. \end{cases}$$

Problem 1. *Show that the group Z_n is isomorphic to the group of nth roots of unity, i.e. the group consisting of all complex numbers z such that $z^n = 1$ (group multiplication is multiplication of complex numbers). Thus, Z_n is a subgroup of the group $U(1)$.*

3. The group $GL(n, C)$ is the set of complex $n \times n$ matrices with a non-zero determinant. Multiplication in $GL(n, C)$ is matrix multiplication; the unit element is the unit $n \times n$ matrix, the inverse element to $M \in GL(n, C)$ is the inverse matrix M^{-1} (which always exists because $\det M \ne 0$ by the definition of the group $GL(n, C)$).

Problem 2. *Describe the group $GL(1, C)$.*

The groups $U(1), Z_n$ and $GL(1, C)$ are Abelian groups, the groups $GL(n, C)$ with $n \ge 2$ are non-Abelian.

The groups in the following examples are subgroups of the group $GL(n, C)$. In other words, we are dealing with $n \times n$ matrices and the multiplication operation is matrix multiplication.

4. The group $GL(n, R)$ is the group of real matrices with non-zero determinant.

5. The group $U(n)$ is the group of unitary $n \times n$ matrices, i.e. such that

$$U^\dagger U = 1 \tag{3.1}$$

(we shall write the unit $n \times n$ matrix simply as 1; this is the matrix on the right-hand side of (3.1)). In order to see that $U(n)$ is indeed a group, we shall show that $U_1 U_2$ and U^{-1} are unitary if U_1, U_2, U are unitary. We have

$$(U_1 U_2)^{\dagger} (U_1 U_2) = U_2^{\dagger} U_1^{\dagger} U_1 U_2 = 1$$
$$(U^{-1})^{\dagger} (U^{-1}) = U U^{\dagger} = 1,$$

as required. We note that it follows from (3.1) that

$$|\det U|^2 = \det U \det U^{\dagger} = 1,$$

i.e. $|\det U| = 1$ for all $U \in U(n)$.

6. The group $SU(n)$ is the group of unitary matrices with unit determinant ($SU(n)$ is evidently a subgroup of $U(n)$). The fact that the group operations (matrix multiplication and inversion) are closed in $SU(n)$ (i.e. $SU(n)$ is indeed a group) follows from the equations

$$\det (U_1 U_2) = \det U_1 \det U_2 = 1$$
$$\det U^{-1} = (\det U)^{-1} = 1,$$

when $\det U_1 = \det U_2 = \det U = 1$.

7. The group $O(n)$ is the group of real orthogonal matrices, i.e. such that

$$O^T O = 1. \tag{3.2}$$

$O(n)$ is clearly a subgroup of $GL(n, R)$ and also of $U(n)$. We note that it follows from (3.2) that $\det O = +1$, since

$$\det O^T O = \det O^T \det O = (\det O)^2 = 1.$$

Thus, the group $O(n)$ divides into two disjoint subsets ($\det O = +1$ and $\det O = -1$).

8. The group $SO(n)$ is the subgroup of the group $O(n)$ consisting of the matrices O with $\det O = +1$.

   We note that the subset of $O(n)$ consisting of matrices with $\det O = -1$ is not a subgroup of $O(n)$. Indeed, if $\det O_1 = \det O_2 = -1$, then $\det (O_1 O_2) = +1$, i.e. this subset is not closed under matrix multiplication.

Let us continue with definitions which will be useful in the sequel. The *center* of a group G is the subset of G consisting of all elements $w \in G$, which commute with all elements of the group, i.e. such that for all $g \in G$

$$wg = gw. \tag{3.3}$$

The center of the group $W \subset G$ is a subgroup of G. Indeed, for $w_1, w_2 \in W$, we have

$$(w_1 w_2)g = w_1(w_2 g) = w_1 g w_2 = g(w_1 w_2),$$

so that $w_1 w_2 \in W$. Multiplying (3.3) by w^{-1} on the left and on the right, we obtain

$$gw^{-1} = w^{-1}g,$$

so that the set W is closed under group operations.

Problem 3. *Describe the center of the group $SU(n)$ and show that it is isomorphic to Z_n.*

Problem 4. *Show that the center of the group $GL(n, C)$ consists of matrices of the form $\lambda \cdot 1$, where λ is an arbitrary, non-zero complex number and 1 is the unit $n \times n$ matrix (the non-trivial part of the problem is to show that all matrices which commute with any matrix in $GL(n, C)$ are multiples of unity).*

The *direct product* $G_1 \times G_2$ of the groups G_1 and G_2 is the set of pairs $\{g, h\}$ where $g \in G_1$ and $h \in G_2$, in which the multiplication operation and the inverse element take the form

$$\begin{aligned} \{g, h\}\{g', h'\} &= \{gg', hh'\} \\ \{g, h\}^{-1} &= \{g^{-1}, h^{-1}\}, \end{aligned}$$

the unit element is the pair $\{e_1, e_2\}$ where e_1 and e_2 are the unit elements in G_1 and G_2, respectively. Thus, $G_1 \times G_2$ is a group. We note that G_1 is a subgroup of the group $G_1 \times G_2$; more precisely, G_1 is isomorphic to the subgroup of the group $G_1 \times G_2$, consisting of the elements of the form $\{g, e_2\}$ for $g \in G_1$.

This definition is useful because, if one succeeds in identifying that some group G is a direct product of two other groups G_1 and G_2, then properties of the group G can be determined by studying the properties of the groups G_1 and G_2 individually.

A *group homomorphism* is a mapping f from a group G to a group G', consistent with the multiplication operations, i.e. for all $g, g_1, g_2 \in G$

$$f(g_1 g_2) = f(g_1)f(g_2)$$

(the product $g_1 g_2$ is given in the sense of multiplication in G, while the product $f(g_1)f(g_2)$ is given in the sense of multiplication in G'),

$$f(e) = e'$$

$(e, e'$ are the units of G, G', respectively)

$$f(g^{-1}) = [f(g)]^{-1}$$

(the inverse elements on the left- and right-hand sides of the equation are taken in the sense of the groups G and G', respectively).

Here are some examples of homomorphisms.

1. A homomorphism from $SU(2)$ to $SU(3)$ under which the 2×2 matrix g $(g \in SU(2))$ is mapped to the 3×3 matrix of the form

$$\begin{pmatrix} & & 0 \\ & g & 0 \\ 0 & 0 & 1 \end{pmatrix}, \tag{3.4}$$

   which clearly belongs to the group $SU(3)$.

2. The homomorphism from the group $G_1 \times G_2$ to the group G_1 under which the element $\{g, h\}$ is mapped to $g \in G_1$.

Suppose f is a homomorphism from G to G'. The set of all elements of G' which can be represented in the form $f(g)$ for some $g \in G$ is called the image of the homomorphism, Im f. The set of elements $g \in G$ such that $f(g) = e'$ is called the kernel of the homomorphism, Ker f. In the first example, Im f is the set of all matrices of the form (3.4), and Ker f is the unit 2×2 matrix. In the second example, Im $f = G_1$, while Ker f is the set of elements of the form $\{e, h\}$, where h is arbitrary (i.e. Ker $f = G_2$).

Problem 5. *Show that* Im f *is a subgroup of G' (f is a homomorphism from G to G'). Show that* Ker f *is a subgroup of G.*

Let us now introduce the concept of the (right) *coset space*, G/H of a group G by its subgroup H. Let H be a subgroup of a group G. Let us define equivalence in G: we shall say that g_1 is equivalent to g_2 $(g_1 \sim g_2)$ if $g_1 = g_2 h$ for some $h \in H$. We recall that the following properties are required for an equivalence relation: 1) if $g_1 \sim g_2$, then $g_2 \sim g_1$; 2) if $g_1 \sim g_2$ and $g_2 \sim g_3$, then $g_1 \sim g_3$. In our case, these properties are easy to verify: 1) if $g_1 = g_2 h$, then $g_2 = g_1 h^{-1}$, i.e. $g_2 \sim g_1$, since $h^{-1} \in H$; 2) if $g_1 = g_2 h_{12}$, $g_2 = g_3 h_{23}$, then $g_1 = g_3 (h_{23} h_{12})$, and $g_1 \sim g_3$ since $h_{23} h_{12} \in H$.

This equivalence relation allows us to divide the set G into disjoint sets (cosets): a coset consists of elements of G which are all equivalent to one another. We note that the coset containing the unit element $e \in G$ is the subgroup H itself.

The set of cosets is called the (right) coset space G/H.

Another definition of equivalence is possible: $g_1 \sim g_2$ if $g_1 = h g_2$ for some $h \in H$. This is used to construct the left coset space, which is sometimes denoted by $G \backslash H$.

Problem 6.   *Take the subgroup isomorphic to $SO(2)$ in the group $SO(3)$ to be the group of matrices of the form*

$$\begin{pmatrix} & & 0 \\ & g & 0 \\ 0 & 0 & 1 \end{pmatrix}, \quad g \in SO(2).$$

Show that there is a one-to-one correspondence between the coset space of $SO(3)$ by this subgroup and the two-dimensional sphere

$$SO(3)/SO(2) = S^2.$$

   The coset space G/H is closely related to *homogeneous spaces*. A set A is said to be a homogeneous space with respect to the group G if the group G acts transitively on A, i.e. to each $g \in G$ there corresponds an invertible mapping of the space A to itself, such that

$$a' = F(g)a.$$

Here, the operation F is required to be consistent with the group operations, i.e.

$$\begin{aligned} F(g_1 g_2)a &= F(g_1)F(g_2)a \\ F(e)a &= a \\ F(g^{-1})a &= [F(g)]^{-1}a, \end{aligned} \tag{3.5}$$

where F^{-1} is a mapping from A to A which is the inverse of the mapping F; a is an arbitrary element of A; g, g_1, g_2 are arbitrary elements of the group G. In addition, it is required that for any pair $a, a' \in A$, there exists $g \in G$ such that

$$a' = F(g)a$$

(transitivity of the group action).

   The *stationary subgroup* H for the element $a_0 \in A$ consists of all elements $h \in G$ which leave a_0 unchanged:

$$F(h)a_0 = a_0.$$

The fact that this set is a subgroup can be checked using (3.5); for example, if $h_1, h_2 \in H$, then

$$F(h_1 h_2)a_0 = F(h_1)F(h_2)a_0 = F(h_1)a_0 = a_0,$$

i.e. $h_1 h_2 \in H$.

For a homogeneous space the stationary subgroups for all elements $a \in A$ are the same. Indeed, suppose H_0 and H_1 are stationary subgroups for the elements a_0 and a_1, respectively. Take $g \in G$ such that

$$a_1 = F(g)a_0.$$

Then an isomorphism of the subgroups H_0 and H_1 is given by the mapping

$$h' = ghg^{-1}, \tag{3.6}$$

where h is any element of H_0. First, we check that $h' \in H_1$, i.e. $F(h')a_1 = a_1$. We have

$$
\begin{aligned}
F(h')a_1 &= F(ghg^{-1})F(g)a_0 = F(g)F(h)F(g^{-1}g)a_0 \\
&= F(g)F(h)a_0 = F(g)a_0 = a_1,
\end{aligned}
$$

as required. The correspondence (3.6) is clearly one-to-one: the inverse mapping is given by the formula

$$h = g^{-1}h'g.$$

Finally, the mapping (3.6) is consistent with the group operations, for example, if $h_1, h_2 \in H$, then

$$gh_1h_2g^{-1} = gh_1g^{-1}gh_2g^{-1} = h_1'h_2',$$

where $h_{1,2}' = gh_{1,2}g^{-1}$.

Problem 7. *We define the action of the group $SO(3)$ on the two-dimensional sphere S^2 as follows. Let g be a matrix of $SO(3)$ and $\vec{a}$ a (unit) vector with components a_i, $i = 1, 2, 3$. Every such vector corresponds to a point on the unit two-dimensional sphere in three-dimensional Euclidean space. Define $F(g)\vec{a}$ to be the vector $\vec{b}$ with components $b_i = g_{ij}a_j$. Since $g^T g = 1$, we have $\vec{b}^2 = \vec{a}^2$, i.e. the action of $F(g)$ takes the sphere to the sphere. Show that $SO(3)$ acts transitively on S^2, and that the stationary subgroup of any point of the sphere S^2 is equal to $SO(2)$.*

If the group G acts transitively on the space A (i.e. A is a homogeneous space under G) then there is an isomorphism

$$A = G/H, \tag{3.7}$$

where H is the stationary subgroup of any element of the space A.

Indeed, let a_0 be some element of A, with H its stationary subgroup. Let us define the element $a_k \in A$ which corresponds to the coset $k \in G/H$, as follows

$$a_k = F(g_k)a_0, \tag{3.8}$$

where g_k is a representative of the coset k. The element a_k does not depend on the choice of representative g_k: if $g'_k = g_k h$ is another representative of the coset k, then $F(g'_k)a_0 = F(g_k)F(h)a_0 = F(g_k)a_0$. Thus, the mapping (3.8) is indeed a mapping from G/H to A. Let us check that it is one-to-one. Let a be some element of A. It is always possible to find some $g \in G$ such that $a = F(g)a_0$. It belongs to some coset $k \in G/H$. We show that if $F(g)a_0 = F(g')a_0$, then g and g' belong to the same coset (which proves the invertibility of the mapping (3.8)). From $F(g)a_0 = F(g')a_0$ we have the equation

$$F(g^{-1})F(g')a_0 = a_0,$$

which means that $g^{-1}g' \in H$, i.e. $g^{-1}g' = h$, where $h \in H$. Hence, $g' = gh$ and, consequently, g' and g belong to the same coset.

Illustrations of equation (3.7) are provided by assertions formulated in the following two problems.

Problem 8. *Show that $SO(n)/SO(n-1) = S^n$, where S^n is the n-dimensional sphere. Here, the embedding of $SO(n-1)$ in $SO(n)$ is given by*

$$\begin{pmatrix} SO(n-1) & 0 \\ 0 & 1 \end{pmatrix} \subset SO(n).$$

Problem 9. *Show that $SU(n)/SU(n-1) = S^{2n}$, where the embedding of $SU(n-1)$ in $SU(n)$ is defined analogously to in the previous problem.*

The subgroup H of the group G is said to be a *normal subgroup* of the group G if for all $h \in H$ and all $g \in G$

$$ghg^{-1} \in H.$$

If H is a normal subgroup, then $K = G/H$ is a group. Indeed, we construct the multiplication operation in K as follows. Let $k_1, k_2 \in K$, where k_1 and k_2 are cosets, and choose representatives of these, $g_1 \in k_1$, $g_2 \in k_2$. Then $k_1 k_2$ is the coset which contains the element $g_1 g_2$ of the group G. The unit $e_k \in K$ is the equivalence class which contains the unit element of the group G (observe that, from the definition of the coset space, it follows that $e_k = H$), and k^{-1} is the coset containing g^{-1}, where g is a representative of the coset k.

For these operations indeed to be operations in K, it is required that the result of their actions should not depend on the choice of representatives in the cosets. Let us verify this for the multiplication operation. Suppose $g_1, g'_1 \in k_1$, $g_2, g'_2 \in k_2$ are two sets of representatives, such that

$$g'_1 = g_1 h_1, \quad g'_2 = g_2 h_2,$$

where $h_1, h_2 \in H$. We check that $g_1' g_2' = g_1 g_2 h$ for some $h \in H$. We have

$$g_1' g_2' = g_1 h_1 g_2 h_2 = g_1 g_2 g_2^{-1} h_1 g_2 h_2.$$

But $g_2^{-1} h_1 g_2 \in H$, and so $(g_2^{-1} h_1 g_2) h_2$ also belongs to H, as required.

Problem 10. *Let $G = G_1 \times G_2$. Show that G_2 is a normal subgroup of the group G, and*

$$G_2 = G/G_1.$$

Problem 11. *Show that the subgroup $U(1)$ of the group $U(n)$, consisting of matrices which are multiples of unity, is a normal subgroup of the group $U(n)$.*

Problem 12. *Show that the center of any group is a normal subgroup of that group.*

Problem 13. *Show that*

$$U(n)/U(1) = SU(n)/Z_n,$$

where Z_n is the center of the group $SU(n)$.

3.2 Lic groups and algebras

For simplicity in what follows, we shall consider matrix groups whose elements are matrices (in other words, we shall consider subgroups of the group $GL(n, C)$); although the notions expounded here are of a general nature, they are most easily formulated for matrix groups.

In the space of $n \times n$ matrices the notion of neighborhood (topology) is introduced in a natural way: two matrices are said to be nearby if all their elements are nearby. We also introduce the differentiation of a family of matrices $M(t)$ with respect to a real parameter t: the elements of the matrix $(\frac{dM}{dt})_{ij}$ are the derivatives $\frac{d}{dt} M_{ij}(t)$ of the matrix elements $M_{ij}(t)$. Generally, the space of all complex $n \times n$ matrices can be viewed as a $2n^2$-dimensional (real) Euclidean space R^{2n^2}, whose coordinates are the $2n^2$ matrix elements $\operatorname{Re} M_{ij}$ and $\operatorname{Im} M_{ij}$. Smooth families of matrices are surfaces (manifolds) embedded in this Euclidean space. For example, a smooth family of matrices $M(t)$, depending on a real parameter t, is a curve in R^{2n^2}, and $\frac{dM}{dt}$ corresponds to the tangent vector to this curve.

Smooth (matrix) *groups* are groups which are smooth manifolds[1] in the space R^{2n^2} described above. These groups are called *Lie groups*.

The simplest non-trivial example of a Lie group is the group $U(1)$. It can also be understood as a matrix group by considering complex numbers as 1×1 matrices. The group $U(1)$ is a circle in the complex plane (in the two-dimensional real space of 1×1 matrices). The groups $U(n), SU(n), O(n), SO(n)$ are also Lie groups.

Two manifolds are said to be *homeomorphic* if there exists a smooth one-to-one mapping from one to the other.[2] For example, an ellipsoid is homeomorphic to a sphere, but a torus and a sphere are not homeomorphic.

Problem 14.   *Show that the group $SU(2)$ is homeomorphic to the three-dimensional sphere S^3.*

For each point of a (curved) manifold of dimension k in $2n^2$-dimensional Euclidean space, one can define the tangent space to the manifold at that point: this is a real vector space of dimension k consisting of vectors tangent to the manifold at the given point.

The tangent space for a Lie group at the unit element is the *Lie algebra* of that Lie group (the unit element of the group; the unit matrix is a point of the group manifold). In other words, any curve $g(t)$ in the Lie group G is represented near unity in the form

$$g(t) = 1 + At + O(t^2), \tag{3.9}$$

where unity is the unit matrix, addition is matrix addition and A belongs to the Lie algebra of the group G. In what follows, the Lie algebra of the group G will be denoted by AG.

Equation (3.9) can be viewed as a definition of the algebra AG: its elements are all matrices A, such that (3.9) is a curve in G near unity. Let us check that the algebra AG is a real vector space. If $A \in AG$ corresponds to the curve $g(t)$, then the curve $g'(t) = g(ct)$, where c is a real number, corresponds to the element cA (because, $g'(t) = 1 + (cA)t + O(t^2)$). If $A_1, A_2 \in AG$ correspond to the curves $g_1(t), g_2(t)$ in the group, then the curve

$$g''(t) = g_1(t)g_2(t)$$

corresponds to the sum $(A_1 + A_2)$, since

$$g''(t) = (1 + A_1 t + \cdots)(1 + A_2 t + \cdots) = 1 + (A_1 + A_2)t + O(t^2).$$

---

[1]Here and in what follows, we shall not refine the notion of smoothness. For example, we shall not encounter continuous manifolds which are not infinitely differentiable.

[2]Again, we shall not distinguish between homeomorphism (continuous but not necessarily differentiable mapping) and diffeomorphism.

Thus, the product of an element of AG by a real number and the sum of two elements of AG are also elements of the Lie algebra AG, i.e. A is a real vector space.

One more operation, commutation, is defined in a Lie algebra: the matrix $[A_1, A_2] = A_1 A_2 - A_2 A_1$ belongs to the algebra AG, if $A_1, A_2 \in AG$. Indeed, if

$$g_1(t) = 1 + A_1 t + \cdots, \quad g_2(t) = 1 + A_2(t) + \cdots$$

then the curve

$$g(t) = g_1(\xi) g_2(\xi) g_1^{-1}(\xi) g_2^{-1}(\xi),$$

where $\xi = \sqrt{t}$, corresponds to the matrix $[A_1, A_2]$. To verify this with accuracy up to and including $t \equiv \xi^2$, we write,

$$g(t) = (1 + A_1 \xi + \alpha_1 \xi^2)(1 + A_2 \xi + \alpha_2 \xi^2)(1 - A_1 \xi - \beta_1 \xi^2)(1 - A_2 \xi - \beta_2 \xi^2), \tag{3.10}$$

where $\beta_{1,2} = \alpha_{1,2} - A_{1,2}^2$ (so that the matrix $(1 - A_1 \xi - \beta_1 \xi^2)$ is the inverse to the matrix $(1 + A_1 \xi + \alpha_1 \xi^2)$ with accuracy up to and including ξ^2). Collecting terms in (3.10), we obtain

$$g(t) = 1 + [A_1, A_2]\xi^2 + O(\xi^3),$$

so that to linear order in t,

$$g(t) = 1 + [A_1, A_2]t.$$

Thus, in a Lie algebra, in addition to multiplication by a number and addition, commutation is also defined.

Let us describe the Lie algebras of certain groups.

1. The $U(n)$ algebra (we shall sometimes denote specific groups and their algebras in the same way, provided this does not lead to confusion). Unitary matrices close to unity must have the property

$$(1 + At + O(t^2))(1 + A^\dagger t + O(t^2)) = 1.$$

Therefore

$$A^\dagger = -A,$$

i.e. the Lie algebra of the group $U(n)$ is the algebra of all anti-Hermitian matrices.

Problem 15.   *Check explicitly that addition, multiplication by a number and commutation are defined in the set of anti-Hermitian matrices.*

2. The $SU(n)$ algebra. In addition to unitarity, the matrices of $SU(n)$ close to unity must satisfy the property

$$\det\left(1 + At + O(t^2)\right) = 1.$$

Since, for small t, $\det\left(1 + At\right) = 1 + (\operatorname{Tr} A)t + O(t^2)$, we have the condition

$$\operatorname{Tr} A = 0.$$

The $SU(n)$ algebra is the algebra of all anti-Hermitian matrices with zero trace.

3. The $SO(n)$ algebra. This is the algebra of all real matrices satisfying the condition

$$A^T = -A$$

(in other words, the matrices of the $SO(n)$ algebra are real antisymmetric matrices).

Problem 16.   *Check that the operations of a Lie algebra (addition, multiplication by a real number and commutation) are closed in (a) the set of anti-Hermitian matrices with zero trace; (b) the set of real antisymmetric matrices.*

Since every anti-Hermitian matrix can be represented in the form iA, where A is an Hermitian matrix, the $SU(n)$ algebra in physics is often defined as the algebra of Hermitian matrices with zero trace, and elements of the group $SU(n)$ near unity are written in the form

$$g = 1 + iAt + O(t^2).$$

Problem 17.   *Describe the Lie algebras of the groups $GL(n, C)$ and $GL(n, R)$.*

Two Lie algebras are isomorphic if there exists a one-to-one correspondence between them which preserves addition, multiplication by a real number and commutation.

Problem 18.   *Show that the Lie algebras of $SU(2)$ and $SO(3)$ are isomorphic. Show that the relation between the groups is $SU(2)/Z_2 = SO(3)$, where Z_2 is the center of the group $SU(2)$. Thus, although locally* *(close to unity) the groups $SU(2)$ and $SO(3)$ are the same, on the whole (globally), they are different.*

The dimension of the vector space, which is a Lie algebra, is called the dimension of the algebra. It is equal to the dimension of the group manifold for the corresponding group. Let us find the dimension of the $SU(n)$ algebra. Arbitrary $n \times n$ matrices are characterized by $2n^2$ parameters. In the $SU(n)$ algebra, n^2 linear conditions are imposed upon them:

$$A^\dagger = -A$$

(this is a matrix condition, i.e. $2n^2$ conditions, however, only half of them are independent, since from $A_{ij} = -A_{ji}^*$ we have the complex conjugate condition $A_{ij}^* = -A_{ji}$). In addition, another linear condition is imposed:

$$\text{Tr } A = 0$$

(this is a single condition, since, from $A^\dagger = -A$ it follows that all diagonal elements are imaginary). Thus, the dimension of the $SU(n)$ algebra is equal to $(n^2 - 1)$.

Problem 19. *Show that the dimension of the $SO(n)$ algebra is equal to $n(n-1)/2$.*

In a Lie algebra, as in a vector space, one can choose a basis. The elements of this basis are k matrices T_i ($i = 1, \ldots, k$; where k is the dimension of the algebra), called the generators of the Lie algebra and of the corresponding Lie group. Since the commutator $[T_i, T_j]$ belongs to the algebra, it decomposes in terms of generators, i.e.

$$[T_i, T_j] = C_{ijk}T_k,$$

where C_{ijk} are antisymmetric in the first two indices and real. The C_{ijk} are called the *structure constants* of the algebra, or, which amounts to the same thing, the structure constants of the group. Their values, of course, depend on the choice of basis.

For example, in the space of anti-Hermitian 2×2 matrices, one can choose a basis in the form $T_i = -\frac{i}{2}\tau_i$, where the τ_i are Pauli matrices

$$\tau_1 = \begin{pmatrix} 0 & 1 \\ 1 & 0 \end{pmatrix}, \quad \tau_2 = \begin{pmatrix} 0 & -i \\ i & 0 \end{pmatrix}, \quad \tau_3 = \begin{pmatrix} 1 & 0 \\ 0 & -1 \end{pmatrix}.$$

The structure constants of the $SU(2)$ algebra are obtained from the equations

$$[\tau_i, \tau_j] = 2i\varepsilon_{ijk}\tau_k$$

and are equal to ε_{ijk}. However, the $SU(2)$ algebra in physics is often defined as the algebra of Hermitian 2×2 matrices; the generators (the basis in this algebra) are chosen in the form

$$T_i = \frac{1}{2}\tau_i.$$

Here, the structure constants are purely imaginary and the commutation relation for generators takes the form

$$[T_i, T_j] = i\varepsilon_{ijk}T_k.$$

The generators of the $SU(3)$ algebra (in physics, this is also defined as the algebra of Hermitian matrices with zero trace) are chosen in the form $T_a = \frac{1}{2}\lambda_a$, $a = 1, 2, \ldots, 8$, where the λ_a are the Gell–Mann matrices

$$\lambda_1 = \begin{pmatrix} 0 & 1 & 0 \\ 1 & 0 & 0 \\ 0 & 0 & 0 \end{pmatrix}, \quad \lambda_2 = \begin{pmatrix} 0 & -i & 0 \\ i & 0 & 0 \\ 0 & 0 & 0 \end{pmatrix}, \quad \lambda_3 = \begin{pmatrix} 1 & 0 & 0 \\ 0 & -1 & 0 \\ 0 & 0 & 0 \end{pmatrix},$$

$$\lambda_4 = \begin{pmatrix} 0 & 0 & 1 \\ 0 & 0 & 0 \\ 1 & 0 & 0 \end{pmatrix}, \quad \lambda_5 = \begin{pmatrix} 0 & 0 & -i \\ 0 & 0 & 0 \\ i & 0 & 0 \end{pmatrix},$$

$$\lambda_6 = \begin{pmatrix} 0 & 0 & 0 \\ 0 & 0 & 1 \\ 0 & 1 & 0 \end{pmatrix}, \quad \lambda_7 = \begin{pmatrix} 0 & 0 & 0 \\ 0 & 0 & -i \\ 0 & i & 0 \end{pmatrix},$$

$$\lambda_8 = \frac{1}{\sqrt{3}} \begin{pmatrix} 1 & 0 & 0 \\ 0 & 1 & 0 \\ 0 & 0 & -2 \end{pmatrix}.$$

Problem 20.   *Show that these generators of the group $SU(3)$ are linearly independent.*

Problem 21.   *Calculate the structure constants of the group $SU(3)$ in the Gell–Mann basis (as mentioned earlier, the structure constants of the group and the algebra are the same).*

A *Lie subalgebra* of a Lie algebra is a real vector space in A, which is closed under the operation of commutation (i.e. it is itself a Lie algebra). For example, one subalgebra in the $SU(3)$ algebra is the set of matrices of the form

$$\begin{pmatrix} A & \begin{matrix} 0 \\ 0 \end{matrix} \\ 0 \quad 0 \quad 0 \end{pmatrix},$$

where A is a 2×2 matrix in the $SU(2)$ algebra. This subalgebra is clearly isomorphic to the $SU(2)$ algebra.

Problem 22. *Let H be a Lie subgroup of the Lie group G. Considering H as a Lie group, construct its Lie algebra AH. Show that AH is a subalgebra in AG.*

Let A and B be two Lie algebras of dimensions N_A and N_B, respectively; $T_1^A, \ldots, T_{N_A}^A$ a full set of generators of the algebra A; $T_1^B, \ldots, T_{N_B}^B$ a full set of generators of the algebra B. We shall assume that the elements of the algebra A are $n_A \times n_A$ matrices, and that the elements of the algebra B are $n_B \times n_B$ matrices. We construct the set of $(N_A + N_B)$ matrices of dimension $(n_A + n_B) \times (n_A + n_B)$ such that the first N_A matrices have the form

$$\begin{pmatrix} T_i^A & O_{n_A \times n_B} \\ O_{n_B \times n_A} & O_{n_B \times n_B} \end{pmatrix}, \quad i = 1, \ldots, N_A,$$

where $O_{k \times l}$ is the zero $k \times l$ matrix. We choose the remaining N_B matrices in the form

$$\begin{pmatrix} O_{n_A \times n_A} & O_{n_A \times n_B} \\ O_{n_B \times n_A} & T_q^B \end{pmatrix}, \quad q = 1, \ldots, N_B.$$

The real vector space in which this set of $(N_A + N_B)$ matrices forms a basis is called the *direct sum* of the algebras A and B and is denoted by $(A + B)$. Clearly, the study of the direct sum of two Lie algebras reduces to the study of each algebra individually.

Problem 23. *Let $G = G_1 \times G_2$ be the direct product of the Lie groups G_1 and G_2. Show that the Lie algebra of the group G is isomorphic to the direct sum of the Lie algebras of the groups G_1 and G_2 defined above, i.e.*

$$AG = AG_1 + AG_2.$$

The Lie subalgebra C in the Lie algebra A is said to be an invariant subalgebra (or ideal), if for all $c \in C$ and $a \in A$,

$$[c, a] \in C.$$

Problem 24. *Let the subgroup H be a normal subgroup in the Lie group G. Show that the Lie algebra of the group H is an invariant subalgebra in the Lie algebra G.*

Thus, it is convenient to study local (and only local) properties of Lie groups by considering the corresponding Lie algebras. The main concepts of group theory have analogies in the theory of Lie algebras. At the same time, Lie algebras are relatively simple objects, since they are vector (linear) spaces.

3.3   Representations of Lie groups and Lie algebras

A *representation T of a group G* in a linear space V is a mapping under which each element $g \in G$ is mapped to an invertible linear operator $T(g)$, acting on V; this mapping must be consistent with the group operations, so that the unit element of the group G is mapped to the unit operator and the following equations are satisfied:

$$
\begin{aligned}
T(g_1 g_2) &= T(g_1) T(g_2) \\
T(g^{-1}) &= [T(g)]^{-1}.
\end{aligned}
\tag{3.11}
$$

Correspondingly, *a representation T of the Lie algebra AG* in the space V is a mapping under which each element $A \in AG$ is mapped to a linear operator $T(A)$, where this mapping is consistent with the operations in the algebra AG, i.e.

$$
\begin{aligned}
T(A + B) &= T(A) + T(B) \\
T(\alpha A) &= \alpha T(A) \\
T([A, B]) &= [T(A), T(B)]
\end{aligned}
\tag{3.12}
$$

for all $A, B \in AG$ and any real number α. Here, the commutator of two operators acting on V is, as usual,

$$
[T(A), T(B)] = T(A)T(B) - T(B)T(A).
$$

If $T(G)$ is a representation of the Lie group G in the space V, then it can be used to construct a representation $T(AG)$ of the corresponding Lie algebra AG in the space V, according to the formula

$$
T(1 + \varepsilon A) = 1 + \varepsilon T(A),
\tag{3.13}
$$

where ε is a small parameter. On the left-hand side $T(1 + \varepsilon A)$ is an operator corresponding to the element $(1 + \varepsilon A) \in G$ which is close to the unit element of the group; on the right-hand side $T(A)$ is the operator corresponding to the element of the algebra $A \in AG$ for the representation $T(AG)$. We remark that not every representation of an algebra is generated by a representation of the group (see problem below).

Problem 25. *Check that the mapping of the algebra AG to the set of linear operators acting on V, defined by equation (3.13) is indeed a representation of the algebra AG, i.e. the properties (3.12) are satisfied.*

If V is a real vector space (i.e. only multiplication of vectors by a real number is defined in V), then a representation of a Lie group or algebra in it is said to be a *real representation*.

If $T(g)$ is a unitary operator for all $g \in G$, then the representation of the group is said to be a *unitary representation*. For a unitary representation of a group, the representation of the corresponding Lie algebra defined by formula (3.13) consists of anti-Hermitian operators,

$$[T(A)]^\dagger = -T(A)$$

for all $A \in AG$.

Let us fix a basis e_i in V. If $T(g)$ is the operator corresponding to the element $g \in G$ for the group representation $T(G)$, then its action takes e_i to some vector of V which can again be decomposed with respect to the basis e_i, so that

$$T(g)e_i = T_{ji}(g)e_j. \tag{3.14}$$

Thus, for a fixed basis, every element $g \in G$ is mapped to a matrix $T_{ji}(g)$. For a real representation the matrices $T_{ji}(g)$ are real, for a unitary representation the $T_{ji}(g)$ are unitary matrices. The matrix $T_{ji}(g)$ has dimension $n \times n$, where n is the dimension of the space V (and has nothing in common with the dimension of the group G). Any vector $\psi \in V$ can be represented in the form of a decomposition with respect to the basis e_i,

$$\psi = \psi_i e_i,$$

where the ψ_i are the components of the vector (numbers). Then

$$T(g)\psi = \psi_i(T(g)e_i) = \psi_i T_{ji} e_j.$$

Thus, the components of the vector $T(g)\psi$ are equal to

$$(T(g)\psi)_i = T_{ij}(g)\psi_j. \tag{3.15}$$

This relation explains the somewhat unusual choice of the order of the indices in (3.14).

From equation (3.15) it follows that

$$
\begin{aligned}
T_{ij}(g_1 g_2) &= T_{ik}(g_1)T_{kj}(g_2) & (3.16)\\
T_{ij}(e) &= \delta_{ij} & (3.17)\\
T_{ij}(g^{-1}) &= [T(g)]_{ij}^{-1}, & (3.18)
\end{aligned}
$$

i.e. a product of elements of the group corresponds to a product of matrices, the unit element to the unit matrix, and the inverse element to the inverse matrix. Indeed, for all ψ, we have

$$[T(g_1 g_2)\psi]_i = T_{ij}(g_1 g_2)\psi_j.$$

On the other hand,

$$[T(g_1 g_2)\psi]_i = [T(g_1)T(g_2)\psi]_i = T_{ik}(g_1)[T(g_2)\psi]_k = T_{ik}(g_1)T_{kj}(g_2)\psi_j,$$

which, by virtue of the arbitrariness of ψ, proves the equality (3.16). Properties (3.17) and (3.18) are proved analogously. We note that

equations (3.16)–(3.18) could be used as the basis for the definition of a representation.

Representations of groups (or algebras) $T(G)$ and $T'(G)$ on the same space V are said to be *equivalent* if there exists an invertible operator S, acting on V, such that

$$T'(g) = ST(g)S^{-1}$$

for all $g \in G$.

Let W be a linear subspace in V. It is said to be an *invariant subspace* of the representation $T(G)$ acting on V if for all $\psi \in W$ and $g \in G$,

$$T(g)\psi \in W,$$

i.e. the action of any operator $T(g)$ does not lead out of the subspace W. The trivial invariant subspaces are the space V itself and the space consisting of the zero vector alone. The representation $T(G)$ is said to be an *irreducible* representation of the group G on V if there are no non-trivial invariant subspaces.

We now present examples of representations of Lie groups which are important for what follows.

1. The fundamental representation

Let G be a Lie group consisting of $n \times n$ matrices (for example, $SU(n)$ or $SO(n)$), and V an n-dimensional space of columns

$$\psi = \begin{pmatrix} \psi_1 \\ \vdots \\ \psi_n \end{pmatrix}. \tag{3.19}$$

The fundamental representation $T(g)$ acts on this space V as follows:

$$(T(g)\psi)_i = g_{ij}\psi_j.$$

Another definition is possible: let V be an n-dimensional space, e_i a basis in V; then the action of the operator $T(g)$ on the vector e_i is of the form

$$T(g)e_i = g_{ji}e_j.$$

Problem 26.   *Show that these definitions are equivalent.*

We note that for the groups $SU(n)$ the fundamental representation is complex, while for the groups $SO(n)$ it is real.

Problem 27. *Show that the fundamental representations of the groups $SU(n)$ and $SO(n)$ are irreducible.*

2. Representation conjugate to the fundamental representation

This is a representation of a group of $n \times n$ matrices on an n-dimensional space of columns (3.19), defined by the equation

$$(T(g)\psi)_i = g_{ij}^* \psi_j.$$

Equivalent definition: the conjugate of the fundamental representation is the representation on the space of rows $\phi = (\phi_1, \ldots, \phi_n)$ such that

$$(T(g)\phi)_i = \phi_j g_{ji}^\dagger.$$

Problem 28. *Show that the fundamental representation of the group $SU(2)$ is equivalent to its conjugate.*

The fundamental representation of a Lie algebra AG and the conjugate of the fundamental representation of the algebra are defined analogously.

Problem 29. *As previously mentioned, the $SU(2)$ and $SO(3)$ algebras are isomorphic. Let T be the fundamental representation of the $SU(2)$ algebra. This corresponds to some representation of the $SO(3)$ algebra, to be denoted by $\bar{T}$. Show that no representation of the group $SO(3)$ generates the representation $\bar{T}$ of the $SO(3)$ algebra according to formula (3.13).*

3. The adjoint representation $\mathrm{Ad}\,(G)$ of the Lie group G

Let AG be the Lie algebra of the group G; we shall suppose that both the group G and the algebra AG consist of $n \times n$ matrices. The algebra AG is a real vector space, which is also the space of the adjoint representation. We define the action of the linear operator $\mathrm{Ad}\,(g)$, corresponding to the element $g \in G$, on a matrix $A \in AG$ as follows:

$$\mathrm{Ad}\,(g)A = gAg^{-1}.$$

For this to be a representation, the essential requirement is that gAg^{-1} should be an element of the algebra AG for all $A \in AG$ and $g \in G$. To see this, we construct a curve in the group G of the form

$$h(t) = gg_A(t)g^{-1},$$

where $g_A(t) = 1 + tA + \cdots$ is the curve defining the element $A \in AG$. We have $h(0) = 1$ and

$$h(t) = 1 + tA_h + \cdots,$$

where A_h is some element of the algebra AG. On the other hand,

$$h(t) = 1 + tgAg^{-1} + \cdots,$$

so that $gAg^{-1} = A_h \in AG$, as required.

The properties (3.11) are easily checked; for example,

$$
\begin{aligned}
\text{Ad}\,(g_1 g_2)A &= (g_1 g_2)A(g_1 g_2)^{-1} \\
&= g_1 g_2 A g_2^{-1} g_1^{-1} = g_1 (g_2 A g_2^{-1}) g_1^{-1} \\
&= \text{Ad}\,(g_1)\,\text{Ad}\,(g_2)A
\end{aligned}
$$

(as always, $\text{Ad}\,(g_1)\,\text{Ad}\,(g_2)$ is understood as the consecutive action of first the operator $\text{Ad}\,(g_2)$ and then the operator $\text{Ad}\,(g_1)$).

From formula (3.13) it follows that the *adjoint representation of a Lie algebra* is such that the element $B \in AG$ is mapped to the operator $\text{ad}\,(B)$ acting on elements A of AG (the space of the representation) as follows:

$$\text{ad}\,(B)A = [B, A]. \tag{3.20}$$

Indeed, if $g = 1 + \varepsilon B$, then

$$\text{Ad}\,(g)A = (1 + \varepsilon B)A(1 - \varepsilon B) = A + \varepsilon[B, A],$$

which, together with equation (3.13), which in this case has the form

$$\text{Ad}\,(g)A = A + \varepsilon\,\text{ad}\,(B)A,$$

leads to (3.20).

The matrices of the adjoint representation of a Lie algebra coincide with the structure constants. Indeed, by the definition of a matrix of a representation

$$\text{ad}\,(t_i)t_j = T^{(i)}_{kj}t_k,$$

where t_j are generators (basis elements) in AG, and $T^{(i)}_{kj}$ is the matrix of the linear operator corresponding to the generator t_i. On the other hand,

$$\text{ad}\,(t_i)t_j = [t_i, t_j] = C_{ijk}t_k,$$

where C_{ijk} are the structure constants of the algebra AG. Consequently,

$$T^{(i)}_{kj} = C_{ijk}. \tag{3.21}$$

We again stress that the adjoint representation is always real. This can be seen from (3.21), since the structure constants are real.

The following statement holds. Any compact Lie algebra A is uniquely representable in the form of a direct sum of a certain number of $U(1)$ subalgebras and simple subalgebras.

$$A = U(1) + U(1) + \cdots + U(1) + A_1 + \cdots + A_n, \qquad (3.23)$$

where the A_n are simple algebras. Thus, the study of compact Lie algebras reduces to the study of simple Lie algebras. Equation (3.23) implies that *locally* every compact Lie group is represented in a unique way in the form of a direct product

$$G = U(1) \times U(1) \times \cdots \times U(1) \times G_1 \times \cdots \times G_n,$$

where the G_n are simple groups (simple Lie groups are those which correspond to simple algebras). The global (i.e. valid for groups as a whole) version of this statement is somewhat more complicated; we shall not use it and we shall not formulate it here.[3]

In the case of a simple compact Lie algebra, there exists just one invariant positive-definite scalar product (up to multiplication by a number). If the algebra is semi-simple, the full set of invariants is described as follows. Suppose, for example,

$$A = A_1 + A_2.$$

Then any vector $B \in A$ has the form

$$B = B_1 + B_2 \quad B_1 \in A_1, B_2 \in A_2. \qquad (3.24)$$

Let $(,)_1$ be an invariant scalar product in A_1 and $(,)_2$ an invariant scalar product in A_2. Then all invariant scalar products of vectors of the form (3.24) have the form

$$(B, B') = \alpha_1(B_1, B_1')_1 + \alpha_2(B_2, B_2')_2,$$

where α_1 and α_2 are arbitrary positive numbers. In other words, positive-definite quadratic invariants (relative to the adjoint representation) in a sum of simple algebras are linear combinations of quadratic invariants in each of the simple algebras with arbitrary positive coefficients.

The complete list of simple Lie algebras is known. In addition to the algebras with which we have become acquainted $SU(n)$, $n = 2, 3, \ldots$, and $SO(n)$, $n = 5, 7, 8 \ldots$, ($SO(3)$ and $SO(4)$ reduce to $SU(2)$ and $SO(6)$ to $SU(4)$), there is an infinite set of matrix algebras $Sp(n, C)$, $n = 3, 4, \ldots$, and a finite number (five) of so-called exceptional algebras G_2, F_4, E_6, E_7, E_8.

---

[3]That the analogous assertion to (3.23) for groups as a whole is not completely trivial can be seen from the fact that different Lie groups can correspond to the same Lie algebra. An example is provided by the groups $SU(2)$ and $SO(3)$.

Problem 33.   *Show that the $SO(4)$ algebra is isomorphic to the $(SU(2)+SU(2))$ algebra.*

In the construction of models in particle physics, the groups $SU(n)$ are most often used; the symmetries $SO(n)$ are occasionally considered, while the groups E_6 and E_8 are used in the construction of unified theories of the strong, weak and electromagnetic interactions.

The following statement holds for representations. *Any representation* of a compact Lie group is *equivalent to a unitary* representation, and representations of the Lie algebra are equivalent to anti-Hermitian representations. This property is also important for the theory of gauge fields; in what follows, we shall always assume that group representations are unitary.

As previously mentioned, when the group $SU(n)$ is considered in physics, it is customary to use *Hermitian* (rather than anti-Hermitian) generators (if A is an anti-Hermitian matrix, then $A = iB$, where B is Hermitian). Then, every element of the algebra is represented in the form

$$A = iA^a t_a,$$

where t_i are Hermitian matrices, and the A^a are real coefficients. Elements close to the unit element of the Lie group are written in the form

$$g = 1 + i\varepsilon^a t_a,$$

where ε^a are small real parameters. The relations between the generators explicitly contain the imaginary unit, i,

$$[t_a, t_b] = iC_{abc}t_c,$$

where C_{abc} are fully antisymmetric real structure constants of the algebra. For complex representations of $SU(n)$ and other algebras, Hermitian generators $T(t_a) \equiv T_a$ such that

$$[T_a, T_b] = iC_{abc}T_c$$

are also used.

We shall usually employ this convention in the following study.

Problem 34.   *Show that $SU(n)$, $n = 2, 3, \ldots$, and $SO(n)$, $n = 5, 6, \ldots$, are simple groups.*

Chapter 4

Non-Abelian Gauge Fields

4.1 Non-Abelian global symmetries

In the theory of the complex scalar field (Section 2.4), we encountered global $U(1)$ symmetry: the Lagrangian is invariant under transformations

$$\varphi(x) \to g\varphi(x),$$

where $g = \mathrm{e}^{i\alpha}$ is an arbitrary element of the group $U(1)$, independent of the space–time coordinates. In this section, we consider the generalization of $U(1)$ symmetry (which is Abelian, since $U(1)$ is an Abelian group) to non-Abelian cases.

The simplest model with global non-Abelian symmetry is the model of N complex scalar fields φ_i with Lagrangian

$$\mathcal{L} = \partial_\mu \varphi_i^* \partial_\mu \varphi_i - m^2 \varphi_i^* \varphi_i - \lambda (\varphi_i^* \varphi_i)^2 \qquad (4.1)$$

(here and below, we shall assume summation over the index $i = 1, \ldots, N$). This model clearly has an Abelian $U(1)$ symmetry

$$\varphi_i \to \mathrm{e}^{i\alpha} \varphi_i. \qquad (4.2)$$

In addition, the Lagrangian (4.1) is invariant under global (independent of the space–time point) transformations

$$\varphi_i(x) \to \varphi_i'(x) = \omega_{ij} \varphi_j(x), \qquad (4.3)$$

where ω is an arbitrary matrix of $SU(N)$. The invariance of the Lagrangian (4.1) under the transformations (4.3) is evident from the identity

$$\varphi_i'^* \varphi_i' = \varphi_k^* \omega_{ik}^* \omega_{ij} \varphi_j = \varphi_k^* (\omega^\dagger \omega)_{kj} \varphi_j = \varphi_k^* \varphi_k.$$

We note that the $SU(N)$-invariance of the Lagrangian (4.1) is ensured by the fact that the mass of each of the fields $\varphi_1, \varphi_2, \ldots, \varphi_N$ is the same (and equal to m), while the interaction term is specially selected and has just one coupling constant.

In order to move to further generalizations of the symmetry (4.3), we shall write the Lagrangian (4.1) in a more convenient form. We introduce the column of fields

$$\varphi = \begin{pmatrix} \varphi_1 \\ \vdots \\ \varphi_N \end{pmatrix} \tag{4.4}$$

such that $\varphi^\dagger = (\varphi_1^*, \ldots, \varphi_N^*)$. The Lagrangian (4.1) can be written in the form

$$\mathcal{L} = \partial_\mu \varphi^\dagger \partial_\mu \varphi - m^2 \varphi^\dagger \varphi - \lambda (\varphi^\dagger \varphi)^2, \tag{4.5}$$

where differentiation of the column means, as usual, differentiation of each of its components (the same for rows or matrices). The column of fields $\varphi(x)$ can be understood as a single field with values in the N-dimensional complex space of columns. The transformation (4.3) is a transformation under the action of the fundamental representation of the group $SU(N)$:

$$\varphi(x) \to \varphi'(x) = \omega \varphi(x). \tag{4.6}$$

We note that the invariance of the Lagrangian (4.5) under transformations (4.6) is evident from the unitarity of the matrix ω.

This construction is immediately generalized to more complicated cases. We shall be interested in the situation where the symmetry group is a compact Lie group G. Let $T(G)$ be a unitary representation of the group G (generally speaking, reducible), and let the field $\varphi(x)$ take values in the space of this representation. Let us consider a Lagrangian of the form

$$\mathcal{L} = \partial_\mu \varphi^\dagger \partial_\mu \varphi - V(\varphi^\dagger, \varphi) \tag{4.7}$$

and require that the potential V be invariant under the action of the representation T:

$$V(\varphi^\dagger T^\dagger(\omega), T(\omega)\varphi) = V(\varphi^\dagger, \varphi)$$

for all $\omega \in G$. Then the Lagrangian (4.7) is invariant under the transformations

$$\varphi(x) \to T(\omega)\varphi(x)$$

for which

$$\varphi^\dagger(x) \to \varphi^\dagger(x) T^\dagger(\omega).$$

Indeed, the potential term is invariant and the invariance of the kinetic term follows from the unitarity of the representation T: $T^\dagger(\omega)T(\omega) = 1$ (we assume throughout that ω does not depend on the space–time coordinates, i.e. we consider global transformations).

Let us give some examples.

1. Suppose $\varphi_i(x)$, $\chi_i(x)$ are two sets of N complex scalar fields, $i = 1 \dots, N$. If φ and χ are the corresponding columns, then the Lagrangian

$$\begin{aligned} \mathcal{L} = {} & \partial_\mu \varphi^\dagger \partial_\mu \varphi + \partial_\mu \chi^\dagger \partial_\mu \chi \\ & -m_\varphi^2 \varphi^\dagger \varphi - m_\chi^2 \chi^\dagger \chi \\ & -\lambda_1 (\varphi^\dagger \varphi)^2 - \lambda_2 \left[(\varphi^\dagger \chi)^2 + (\chi^\dagger \varphi)^2 \right] - \lambda_3 (\chi^\dagger \chi)^2 \end{aligned} \tag{4.8}$$

is invariant under transformations of the group $SU(N)$:

$$\begin{aligned} \varphi &\to \omega\varphi \\ \chi &\to \omega\chi, \end{aligned}$$

$\omega \in SU(N)$. We note that the pair of fields φ, χ can be interpreted as a single field taking values in the space of the reducible representation of the group $SU(N)$, formed as the direct sum of the two fundamental representations.

2. Suppose the field $\varphi(x)$ transforms according to the fundamental representation of the group $SU(2)$ (i.e. $\varphi(x)$ is a complex column $\binom{\varphi_1(x)}{\varphi_2(x)}$, while the transformation ω from $SU(2)$ acts as

$$\varphi \to \omega\varphi,$$

where ω does not depend on x). Suppose the field $\xi(x)$ is a real triplet $\xi^a(x)$, $a = 1, 2, 3$, which transforms according to the adjoint representation of the group $SU(2)$. The Lagrangian, which is invariant under the group $SU(2)$, can be constructed as follows:

$$\mathcal{L} = \partial_\mu \varphi^\dagger \partial_\mu \varphi + \partial_\mu \xi^a \partial_\mu \xi^a - \lambda_1 (\varphi^\dagger \varphi)^2 - \lambda_2 (\xi^a \xi^a)^2 - \lambda_3 \varphi^\dagger (\tau^a \xi^a) \varphi, \tag{4.9}$$

where the τ^a are the Pauli matrices (generators of $SU(2)$).

To verify the invariance, it is sufficient to show that $(\xi^a \xi^a)^2$ and $\varphi^\dagger (\tau^a \xi^a) \varphi$ are invariant. We construct the matrix field

$$\xi = \tau^a \xi^a$$

taking values in the Lie algebra of group $SU(2)$ (modulo the imaginary unit). By the definition of the fundamental and adjoint representations, the fields φ and ξ transform under the action of the group $SU(2)$ as follows

$$\varphi \to \varphi' = \omega\varphi \tag{4.10}$$

$$\xi \to \xi' = \omega\xi\omega^{-1}. \tag{4.11}$$

We further note that

$$\xi^a\xi^a = \frac{1}{2}\mathrm{Tr}\,\xi^2. \tag{4.12}$$

The invariance of expression (4.12) under (4.11) is evident, while the invariance of the quantity

$$\varphi^\dagger(\tau^a\xi^a)\varphi = \varphi^\dagger\xi\varphi$$

follows from the chain of equations

$$\varphi'^\dagger\xi'\varphi' = (\varphi^\dagger\omega^\dagger)(\omega\xi\omega^{-1})(\omega\varphi) = \varphi^\dagger(\omega^\dagger\omega)\xi(\omega^{-1}\omega)\varphi = \varphi^\dagger\xi\varphi.$$

We note that the pair of fields (φ, ξ) is a field in the space of the direct sum of the fundamental and adjoint representations of the group $SU(2)$.

3. Let $\varphi_{i\alpha}(x)$ be a set of $m \cdot n$ complex fields, $i = 1, \ldots, n;\ \alpha = 1, \ldots, m$. This set realizes a representation of the group $SU(n) \times SU(m)$ and is the direct product of the fundamental representation of the group $SU(n)$ and the fundamental representation of the group $SU(m)$. This means that the pair (ω, Ω), $\omega \in SU(n)$, $\Omega \in SU(m)$ acts on $\varphi_{i\alpha}$ as follows

$$\varphi_{i\alpha} \to \varphi'_{i\alpha} = \omega_{ij}\Omega_{\alpha\beta}\varphi_{j\beta} \tag{4.13}$$

(in other words, the group $SU(n)$ acts on the first index in $\varphi_{i\alpha}$, and the group $SU(m)$ on the second).

The invariant Lagrangian has the form

$$\mathcal{L} = \partial_\mu\varphi^*_{i\alpha}\partial_\mu\varphi_{i\alpha} - m^2\varphi^*_{i\alpha}\varphi_{i\alpha} - \lambda(\varphi^*_{i\alpha}\varphi_{i\alpha})^2. \tag{4.14}$$

In what follows, we shall somewhat inaccurately omit the indices for the field and write the transformation (4.13) as

$$\varphi \to \varphi^\dagger = \omega\Omega\varphi,$$

and the Lagrangian (4.14) in the form

$$\mathcal{L} = \partial_\mu\varphi^\dagger\partial_\mu\varphi - m^2\varphi^\dagger\varphi - \lambda(\varphi^\dagger\varphi)^2.$$

If the field φ is described and it is known that $\omega \in SU(n)$, $\Omega \in SU(m)$, then this notation does not lead to confusion.

We note that we have, in fact, already encountered a situation, similar to the latter example: the field (4.4) with the Lagrangian (4.5) actually realizes a representation of the group $SU(N) \times U(1)$, where the group $SU(N)$ acts according to (4.6) (fundamental representation) and the group $U(1)$ acts according to (4.2).

4. We construct a non-trivial model, invariant under the group $SU(2) \times U(1)$. Let φ, χ be doublets under the group $SU(2)$ (fundamental representation), ξ a singlet under $SU(2)$ (trivial representation). In other words, the fields φ, χ, ξ transform under $SU(2)$ as follows:

$$\varphi \to \omega\varphi$$
$$\chi \to \omega\chi$$
$$\xi \to \xi.$$

The fields φ, χ are two-component complex columns, ξ is a one-component complex scalar field. Suppose the fields φ, χ and ξ transform under $U(1)$ as

$$\varphi \to e^{iq_\varphi \alpha}\varphi$$
$$\chi \to e^{iq_\chi \alpha}\chi \qquad (4.15)$$
$$\xi \to e^{iq_\xi \alpha}\xi.$$

Here α is a parameter of the transformation, q_φ, q_χ and q_ξ are real numbers. The kinetic term in the Lagrangian has the standard form, and the interaction can be chosen in the form

$$\lambda[(\varphi^\dagger \xi)\chi + \text{c.c.}]. \qquad (4.16)$$

It is invariant under $SU(2) \times U(1)$, if

$$q_\chi + q_\xi = q_\varphi$$

(the numbers q_χ, q_ξ and q_φ can be chosen to be integers, then (4.15) is a single-valued representation of the group $U(1)$).

Problem 1. *Consider the theory of three fields, as in the previous example. Suppose, however, that*

$$q_\varphi + q_\chi + q_\xi = 0.$$

Construct a cubic interaction of the fields φ, χ and ξ invariant under $SU(2) \times U(1)$ (hint: use the fact that for the group $SU(2)$ the fundamental representation is equivalent to its conjugate).

Examples of global symmetries could be extended, by considering various groups $G_1 \times \cdots \times G_N$, where the G_i are simple groups or $U(1)$ factors, different fields, transforming according to the direct product of irreducible representations of the groups $G_1, \ldots, G_N$ and also combinations of such fields.

Until now, we have, for simplicity, discussed scalar fields. However, global internal symmetries can be introduced for fields with any properties under Lorentz transformations: the Lorentz structure and the 'internal' structure of fields do not influence each other (for example, when considering internal symmetries of a vector field one merely makes the substitution $\varphi \to \varphi_\mu$ in the formulae of this section).

Problem 2. *Find the conserved currents corresponding to the global symmetries, in models with Lagrangian (4.1) (equivalently, (4.5)), (4.8), (4.9) and (4.14) and also in the model of the fourth example.*

Symmetries similar to those considered above actually exist in particle physics. For example, the proton and neutron fields are combined in a column $N = \binom{p}{n}$, which belongs to the fundamental representation of the group $SU(2)$ (so-called isotopic symmetry). From the real field of the π^0-meson and the complex field π of charged π-mesons, one can construct three real fields $\pi^1 = \frac{1}{\sqrt{2}}(\pi + \pi^*)$, $\pi^2 = \frac{1}{i\sqrt{2}}(\pi - \pi^*)$, $\pi^3 = \pi^0$, which form a triplet (adjoint representation) under the isotopic group $SU(2)$. Strong interactions are invariant under the isotopic group $SU(2)$ and the pion–nucleon Lagrangian has the structure of (4.9) (with the substitution $\varphi \to N$, $\xi^a \to \pi^a$ and peculiarities due to the fact the nucleons have spin $1/2$ and are fermions; further differences arise because the full Lagrangian contains small terms, which are not invariant under isotopic symmetry).

An interaction Lagrangian of type (4.16) arises in the description of the interaction of a lepton doublet (left-handed electron, neutrino), the right-handed component of an electron and a Higgs field doublet; $SU(2) \times U(1)$ is the group of the electroweak interactions.

Transformations of type (4.13) exist in the theory of light quarks, where the symmetry is $SU(3) \times SU(3)$. The first $SU(3)$ is the color group (in fact, a gauge group), while the second $SU(3)$ is the flavor group with respect to which the light quarks form the triplet

$$\begin{pmatrix} u \\ d \\ s \end{pmatrix}.$$

The first and second $SU(3)$ are denoted by $SU(3)_c$ and $SU(3)_f$, respectively. The group $SU(3)_f$ is not exact; mass terms and also electromagnetic and weak interactions are not invariant under it.

4.2 Non-Abelian gauge invariance and gauge fields: the group $SU(2)$

Our aim is to generalize the construction presented in Section 2.7 for scalar electrodynamics with the gauge group $U(1)$ to the case of a non-Abelian gauge group (Yang and Mills 1954). Let us again consider the theory of two complex scalar fields, forming a column

$$\varphi = \begin{pmatrix} \varphi_1 \\ \varphi_2 \end{pmatrix},$$

whose Lagrangian has the form

$$\mathcal{L} = \partial_\mu \varphi^\dagger \partial_\mu \varphi - m^2 \varphi^\dagger \varphi - \lambda (\varphi^\dagger \varphi)^2. \tag{4.17}$$

This Lagrangian is invariant under global transformations from the group $SU(2)$,

$$\varphi(x) \to \varphi'(x) = \omega \varphi(x), \quad \omega \in SU(2),$$

where ω does not depend on the point of space–time.

We shall attempt to modify the Lagrangian (4.17) in such a way that it becomes invariant under transformations of $SU(2)$, depending in an arbitrary manner on the point of space–time

$$\varphi(x) \to \varphi'(x) = \omega(x)\varphi(x) \tag{4.18}$$

$$\omega(x) \in SU(2). \tag{4.19}$$

(We recall that the analogous requirement in scalar electrodynamics led to the replacement of the conventional derivative in the Lagrangian by the covariant derivative $\partial_\mu \varphi \to (\partial_\mu - ieA_\mu)\varphi$. The potential terms (the last two terms in (4.17)) are invariant under the transformations (4.18), but the kinetic term (containing derivatives) is not invariant. Indeed, under the transformation (4.18), the derivative of the field becomes

$$\partial_\mu \varphi'(x) = \omega(x)\partial_\mu \varphi(x) + \partial_\mu \omega(x) \cdot \varphi(x) \tag{4.20}$$

and the Lagrangian $\mathcal{L}(\varphi')$ now contains terms with $\partial_\mu \omega$. In order to eliminate these terms, we replace, in the Lagrangian (4.17), the conventional derivative by the covariant derivative $\partial_\mu \varphi \to D_\mu \varphi$ and require that under the transformations (4.18) it becomes

$$(D_\mu \varphi)' = \omega D_\mu \varphi. \tag{4.21}$$

From (4.20) it is clear that this can be achieved by introducing a vector field $A_\mu(x)$ and writing

$$D_\mu \varphi = \partial_\mu \varphi + A_\mu \varphi.$$

The structure of the field A_μ (range of its values) is still unknown; finding it is our next problem.

Let us determine, in the first place, how the field A_μ transforms under gauge transformations. For this, we write the left-hand side of equation (4.21) explicitly:

$$D_\mu\varphi' = \partial_\mu\varphi' + A'_\mu\varphi' = \omega\partial_\mu\varphi + \partial_\mu\omega\varphi + A'_\mu\omega\varphi$$

and require that it should equal the right-hand side

$$\omega D_\mu\varphi = \omega\partial_\mu\varphi + \omega A_\mu\varphi.$$

Bearing in mind that φ is an arbitrary column, we thus obtain

$$\partial_\mu\omega + A'_\mu\omega = \omega A_\mu,$$

i.e. the transformation rule for A_μ is of the form

$$A_\mu \to A'_\mu = \omega A_\mu\omega^{-1} + \omega\partial_\mu\omega^{-1} \tag{4.22}$$

(we use the fact that since $\omega\omega^{-1} = 1$, we have $\omega\partial_\mu\omega^{-1} + \partial_\mu\omega \cdot \omega^{-1} = 0$).

We now determine the values taken by the field A_μ. For this, let us consider an infinitesimal gauge transformation, i.e. a transformation (4.22) with

$$\omega = 1 + \varepsilon(x),$$

where $\varepsilon(x)$ takes values in the Lie algebra of the group $SU(2)$ (in other words, $\varepsilon(x)$ is an anti-Hermitian 2×2 matrix with zero trace at each point x). The second term in (4.22) to the lowest order in ε, is

$$\omega\partial_\mu\omega^{-1} = -\partial_\mu\varepsilon(x), \tag{4.23}$$

i.e. it takes values in the Lie algebra. Consequently, the Lie algebra must be contained in the range of values of the field A_μ. It turns out that this is also sufficient: if the field A_μ takes values in the Lie algebra of the group $SU(2)$ then for any $\omega(x)$, both $\omega A_\mu\omega^{-1}$ and $\omega\partial_\mu\omega^{-1}$ belong to the Lie algebra. The fact that $\omega A_\mu\omega^{-1} \in ASU(2)$ is evident, since $\omega A_\mu\omega^{-1}$ is the result of the action of the adjoint representation on the element $A_\mu \in ASU(2)$.

Problem 3. *Show that if $\omega(x)$ belongs to the group $SU(2)$ at each point x, then $\omega\partial_\mu\omega^{-1}$ belongs to the Lie algebra of the group $SU(2)$ at each point x.*

Thus, the gauge field $A_\mu(x)$ (it is called the Yang–Mills field) is the field taking values in the Lie algebra (in this case, of the group $SU(2)$); the transformation rule for the scalar and gauge fields under gauge transformations has the form (as before, we assume that the scalar field transforms according to the fundamental representation of the group $SU(2)$)

$$A_\mu(x) \rightarrow A'_\mu(x) = \omega(x)A_\mu(x)\omega^{-1}(x) + \omega(x)\partial_\mu\omega^{-1}(x) \qquad (4.24)$$

$$\varphi(x) \rightarrow \varphi'(x) = \omega(x)\varphi(x), \qquad (4.25)$$

and the Lagrangian of the scalar field, which is invariant under gauge transformations, is equal to

$$\mathcal{L} = (D_\mu\varphi)^\dagger D_\mu\varphi - m^2\varphi^\dagger\varphi - \lambda(\varphi^\dagger\varphi)^2,$$

where

$$D_\mu\varphi = \partial_\mu\varphi + A_\mu\varphi \equiv (\partial_\mu + A_\mu)\varphi \qquad (4.26)$$

is the covariant derivative of the scalar field, which transforms according to (4.21).

We note one of the main differences of the non-Abelian gauge field from the vector potential of electrodynamics. Under global (independent of x) transformations the electrodynamic vector potentials do not change, but non-Abelian potentials transform non-trivially,

$$A_\mu(x) \rightarrow A'_\mu(x) = \omega A_\mu(x)\omega^{-1}, \qquad (4.27)$$

i.e. according to the adjoint representation of the group.

Let us now construct the Lagrangian for the field A_μ itself, which is analogous to $-\frac{1}{4}F^2_{\mu\nu}$ in electrodynamics. For this we first find the strength tensor for the non-Abelian field. By analogy with electrodynamics, we expect that the strength tensor will contain a term

$$\partial_\mu A_\nu - \partial_\nu A_\mu. \qquad (4.28)$$

Hence, from (4.27), it is clear that the strength tensor must transform non-trivially under gauge transformations: expression (4.28) transforms according to the adjoint representation of the group in the case of global transformations. We require the strength tensor to transform according to the adjoint representation for all gauge transformations,

$$F_{\mu\nu}(x) \rightarrow F'_{\mu\nu}(x) = \omega(x)F_{\mu\nu}\omega^{-1}(x). \qquad (4.29)$$

Then the gauge-invariant Lagrangian will be constructed from the invariant $\text{Tr}\,(F_{\mu\nu}F_{\mu\nu})$.

Expression (4.28) itself does not have the property (4.29). Indeed, differentiating (4.24), we obtain

$$\begin{aligned} \partial_\mu A'_\nu - \partial_\nu A'_\mu &= \omega(\partial_\mu A_\nu - \partial_\nu A_\mu)\omega^{-1} \qquad (4.30)\\ &\quad +\partial_\mu\omega A_\nu\omega^{-1} + \omega A_\nu\partial_\mu\omega^{-1}\\ &\quad -\partial_\nu\omega A_\mu\omega^{-1} - \omega A_\mu\partial_\nu\omega^{-1}\\ &\quad +\partial_\mu\omega\partial_\nu\omega^{-1} - \partial_\nu\omega\partial_\mu\omega^{-1}. \end{aligned}$$

Terms with second derivatives of the function ω cancel out, but terms with first derivatives remain. Their elimination would require the addition to (4.28) of terms not containing derivatives of the field A_μ. Analogously to (4.28), these terms must be elements of the Lie algebra, they must have indices μ, ν and be antisymmetric in these indices. The only candidate is the commutator

$$[A_\mu, A_\nu] = A_\mu A_\nu - A_\nu A_\mu.$$

From (4.24), it follows that

$$
\begin{aligned}
[A'_\mu, A'_\nu] &= \omega[A_\mu, A_\nu]\omega^{-1} \\
&+ \omega\partial_\mu\omega^{-1}\omega A_\nu\omega^{-1} + \omega A_\mu\omega^{-1}\omega\partial_\nu\omega^{-1} \\
&- \omega\partial_\nu\omega^{-1}\omega A_\mu\omega^{-1} - \omega A_\nu\omega^{-1}\omega\partial_\mu\omega^{-1} \\
&+ \omega\partial_\mu\omega^{-1}\omega\partial_\nu\omega^{-1} - \omega\partial_\nu\omega^{-1}\omega\partial_\mu\omega^{-1}.
\end{aligned}
\tag{4.31}
$$

Comparing (4.30) and (4.31), we see that the covariant quantity (in the sense of (4.29)) is the tensor

$$F_{\mu\nu} = \partial_\mu A_\nu - \partial_\nu A_\mu + [A_\mu, A_\nu].\tag{4.32}$$

Indeed, the undesirable terms (all terms, except the first) in (4.30) and (4.31) cancel out when one takes into account the identities

$$
\begin{aligned}
\omega\partial_\mu\omega^{-1}\omega &= -\partial_\mu\omega\omega^{-1}\omega = -\partial_\mu\omega \\
\omega\partial_\mu\omega^{-1}\omega\partial_\nu\omega^{-1} &= -\omega\omega^{-1}\partial_\mu\omega\partial_\nu\omega^{-1} = -\partial_\mu\omega\partial_\nu\omega^{-1},
\end{aligned}
$$

which follow from $\omega\omega^{-1} = 1$, $\omega^{-1}\omega = 1$ and the derivatives of these equations with respect to x^μ.

Thus, the strength tensor is of the form (4.32) and transforms according to the rule (4.29).

Problem 4. *Show that in electrodynamics $[D_\mu, D_\nu] = -ieF_{\mu\nu}$, where $D_\mu = \partial_\mu - ieA_\mu$ is understood in the sense of an operator acting on the scalar field, $D_\mu D_\nu$ is the consecutive action of first D_ν and then D_μ; as usual, $[D_\mu, D_\nu] = D_\mu D_\nu - D_\nu D_\mu$.*

Problem 5. *Show that for the gauge theory with the gauge group $SU(2)$, $[D_\mu, D_\nu] = F_{\mu\nu}$, where the covariant derivative is defined by formula (4.26). Using this equation, show once again that the strength tensor transforms according to (4.29).*

We choose the gauge-invariant Lagrangian of the gauge field in the form

$$\mathcal{L}_A = \frac{1}{2g^2}\operatorname{Tr} F_{\mu\nu}F_{\mu\nu},\tag{4.33}$$

where g^2 is some positive constant. We shall discuss the choice of sign in (4.33) a little later.

We note that the important distinction between the non-Abelian gauge theory and electrodynamics is the presence in the Lagrangian $\mathcal{L}_A$ of terms of third and fourth order in the field A_μ. (They have a structure of the type $\mathrm{Tr}\,(\partial_\mu A_\nu \cdot A_\mu A_\nu)$ and $\mathrm{Tr}\,(A_\mu A_\nu A_\mu A_\nu)$). This means that the gauge field equations are nonlinear, even in the absence of other fields. In this connection, the field A_μ is said to be self-interacting.

A gauge field $A_\mu(x)$, taking values in the $SU(2)$ algebra, can be expressed in terms of three real fields (according to the number of generators of the $SU(2)$ algebra),

$$A_\mu(x) = -ig\frac{\tau^a}{2}A_\mu^a(x), \tag{4.34}$$

where $a = 1, 2, 3$; $A_\mu^a(x)$ are real fields; $\tau^a/2$ are Hermitian generators of the $SU(2)$ algebra; the factor g, which is the same as in (4.33), is introduced for convenience. Analogously, the strength tensor can be written as:

$$F_{\mu\nu}(x) = -ig\frac{\tau^a}{2}F_{\mu\nu}^a(x). \tag{4.35}$$

From the definition (4.32) we obtain

$$
\begin{aligned}
F_{\mu\nu}(x) &= -ig\frac{\tau^a}{2}(\partial_\mu A_\nu^a - \partial_\nu A_\mu^a) + (ig)^2 A_\mu^a A_\nu^b\left[\frac{\tau^a}{2}, \frac{\tau^b}{2}\right] \\
&= -ig\frac{\tau^a}{2}(\partial_\mu A_\nu^a - \partial_\nu A_\mu^a) - g^2 A_\mu^a A_\nu^b i\varepsilon^{abc}\frac{\tau^c}{2} \\
&\quad - ig\frac{\tau^a}{2}(\partial_\mu A_\nu^a - \partial_\nu A_\mu^a + g\varepsilon^{abc}A_\mu^b A_\nu^c).
\end{aligned}
$$

Consequently, the real components of the strength tensor $F_{\mu\nu}^a$ are expressed in terms of the real fields A_μ^a as follows:

$$F_{\mu\nu}^a = \partial_\mu A_\nu^a - \partial_\nu A_\mu^a + g\varepsilon^{abc}A_\mu^b A_\nu^c. \tag{4.36}$$

We note that the factors ε^{abc} arise here as a result of commutation of the generators $\tau^a/2$, i.e. they appear as structure constants of the group $SU(2)$.

Problem 6. *Let*

$$\omega(x) = 1 + i\frac{\tau^a}{2}\varepsilon^a(x)$$

be infinitesimal gauge transformations with real parameters $\varepsilon^a(x)$. Find the transformations of the components $A_\mu^a(x)$ and express them in terms of $\varepsilon^a(x)$. Do the same for $F_{\mu\nu}^a(x)$.

The Lagrangian of the gauge field (4.33) can be expressed in terms of real components $F^a_{\mu\nu}$:

$$\mathcal{L}_A = \frac{1}{2g^2} F^a_{\mu\nu} F^b_{\mu\nu} (-ig)^2 \operatorname{Tr} \frac{\tau^a}{2} \frac{\tau^b}{2} = -\frac{1}{4} F^a_{\mu\nu} F^a_{\mu\nu}.$$

In what follows, we shall use both the matrix fields A_μ and the real components A^a_μ, and employ the terms 'gauge fields' for both. The same applies to the matrix strength tensor $F_{\mu\nu}$ and its real components $F^a_{\mu\nu}$. We note that the covariant derivative for a scalar field doublet (fundamental representation of $SU(2)$) has the form

$$D_\mu \varphi = \left(\partial_\mu - ig \frac{\tau^a}{2} A^a_\mu \right) \varphi. \tag{4.37}$$

Let us now discuss the choice of sign in the Lagrangian (4.33) and the occurrence of the factor g in (4.34) and (4.35). Let us consider small (linear) perturbations of the field about the state $A^a_\mu = 0$. For this, in the Lagrangian of the gauge field

$$\mathcal{L}_A = -\frac{1}{4} F^a_{\mu\nu} F^a_{\mu\nu}, \tag{4.38}$$

we neglect terms of third and fourth order in the field A^a_μ, which are small in comparison with quadratic terms, if the field A^a_μ is small at all points of space–time. Then the Lagrangian for small perturbations will have the form

$$\mathcal{L}^{(2)}_A = -\frac{1}{4} (\partial_\mu A^a_\nu - \partial_\nu A^a_\mu)(\partial_\mu A^a_\nu - \partial_\nu A^a_\mu). \tag{4.39}$$

This clearly decomposes into the sum of the Lagrangians for the fields $A^1_\mu, A^2_\mu, A^3_\mu$, where each of these Lagrangians coincides with the Lagrangian of electrodynamics. This was made possible thanks to the factor g in (4.34) and (4.35). Equation (4.39) explains also the choice of sign in (4.33) (or, what amounts to the same thing, in (4.38)): for this sign the energy of fields with small amplitude will be positive.

It follows immediately from (4.39) that small physical excitations of the field $A^a_\mu(x)$ comprise three types ($a = 1, 2, 3$) of transverse massless (i.e. moving with the speed of light) waves, each of which is completely analogous to electromagnetic waves in vacuum. The constant g appears in the Lagrangian of the scalar field

$$\mathcal{L}_\varphi = (D_\mu \varphi)^\dagger (D_\mu \varphi) - V(\varphi^\dagger, \varphi)$$

and in the Lagrangian of the gauge field (4.38) only in terms of third and fourth order (recall equation (4.37)), i.e. only in interaction terms. Thus g is called the *gauge coupling constant*.

Problem 7. *Find the dimension of the field A_μ^a and the constant g in the system of units $\hbar = c = 1$.*

Problem 8. *Both in the theory of electromagnetic fields and non-Abelian gauge theory, there is another gauge-invariant and Lorentz-invariant quantity $\varepsilon_{\mu\nu\lambda\rho} \mathrm{Tr}\,(F_{\mu\nu} F_{\lambda\rho})$ (for the electromagnetic field, $\varepsilon_{\mu\nu\lambda\rho} F_{\mu\nu} F_{\lambda\rho}$) which is quadratic in $F_{\mu\nu}$. One could try to use this as the additional term in the Lagrangian of the gauge field. Show that, in the Abelian and non-Abelian cases, this quantity is the full four-divergence $\partial_\mu K_\mu$. Find an expression for K_μ. Show that adding the full divergence of any vector, depending on the field, to the Lagrangian, does not change the field equations. Thus (4.33) is the only non-trivial classical Lagrangian, quadratic in $F_{\mu\nu}$. (In quantum theory, adding a term constant $\times \varepsilon_{\mu\nu\lambda\rho} \mathrm{Tr}\,(F_{\mu\nu} F_{\lambda\rho})$ to the Lagrangian leads to non-trivial consequences).*

4.3   Generalization to other groups

The generalization of the concept of the gauge field A_μ to other simple Lie groups G is evident. The field A_μ takes values in the Lie algebra of the group G, i.e.

$$A_\mu(x) = g t^a A_\mu^a(x),$$

where t^a are generators of the group (anti-Hermitian generators for the groups $SU(n)$, antisymmetric real generators for $SO(n)$, etc.). The tensor $F_{\mu\nu}$ also takes values in the Lie algebra and is given by formula (4.32). Gauge transformations of the field A_μ and the strength $F_{\mu\nu}$, as before, have the form

$$A_\mu \to A_\mu' = \omega A_\mu \omega^{-1} + \omega \partial_\mu \omega^{-1}$$
$$F_{\mu\nu} \to F_{\mu\nu}' = \omega F_{\mu\nu} \omega^{-1},$$

where $\omega(x) \in G$ at each point x (the arguments of Section 4.2, leading to these formulae, did not use the fact that the group $SU(2)$ was chosen for the group G). As before, the only quadratic invariant is $\mathrm{Tr}\, F_{\mu\nu} F_{\mu\nu}$, so that the Lagrangian of the gauge field has the form

$$\mathcal{L}_A = \frac{1}{2g^2} \mathrm{Tr}\, F_{\mu\nu} F_{\mu\nu}.$$

We note that, if the Lie group were not compact, then its algebra would not contain a positive-definite quadratic invariant, which, ultimately, would lead to the energy of small perturbations of the gauge field being unbounded from below: a quadratic Lagrangian would contain both terms

$$-\frac{1}{4}(\partial_\mu A_\nu^i - \partial_\nu A_\mu^i)^2,$$

and terms

$$+\frac{1}{4}(\partial_\mu A_\nu^j - \partial_\nu A_\mu^j)^2.$$

The terms of the second type would give rise to a negative field energy.

The expression for the tensor $F_{\mu\nu}^a$ in terms of the components A_μ^a contains the structure constants of the group:

$$F_{\mu\nu}^a = \partial_\mu A_\nu^a - \partial_\nu A_\mu^a + gC_{abc}A_\mu^b A_\nu^c.$$

The derivation of this equation repeats the derivation of formula (4.36).

The number of field components A_μ^a is equal to the dimension of the group G (for example, for the groups $SU(n)$ the index a takes the values $1, 2, \ldots, n^2 - 1$). In a linear approximation, each of these components describes a massless vector field, completely analogous to the electromagnetic field in vacuum.

In the case when the gauge algebra is compact, but not simple, it is convenient to work with each component $U(1)$ and each simple component separately. To each of these components correspond its own gauge field and its own coupling constant. For example, a gauge-invariant analogue of example (3) of Section 4.1 is as follows. As before, the field $\varphi_{i\alpha}(x)$ transforms according to the fundamental representation of the group $SU(n)$ ($i = 1, \ldots, n$) and according to the fundamental representation of the group $SU(m)$ ($\alpha = 1, \ldots, m$). We introduce the gauge field of the group $SU(n)$

$$A_\mu(x) = -igt^a A_\mu^a(x),$$

where $(t^a)_{ij}$ are Hermitian generators of the group $SU(n)$, acting on the first index of the field $\varphi_{i\alpha}$, $a = 1, \ldots, n^2 - 1$. The constant g is the gauge coupling constant for the group $SU(n)$. Analogously, we introduce the gauge fields for the group $SU(m)$,

$$B_\mu(x) = -i\tilde{g}\tilde{t}^p B_\mu^p(x),$$

where $(\tilde{t}^p)_{\alpha\beta}$ are generators of the group $SU(m)$ acting on the second index of the field $\varphi_{i\alpha}$, $p = 1, \ldots, m^2 - 1$, and $\tilde{g}$ is the coupling constant for the group $SU(m)$. We generalize the derivative of the field $\varphi_{i\alpha}$,

$$\partial_\mu \varphi_{i\alpha} \equiv \delta_i^j \delta_\alpha^\beta \partial_\mu \varphi_{j\beta}$$

to the covariant derivative

$$(D_\mu\varphi)_{i\alpha} = (\delta_i^j\delta_\alpha^\beta\partial_\mu - ig\delta_\alpha^\beta t_{ij}^a A_\mu^a - i\tilde{g}\delta_i^j \tilde{t}_{\alpha\beta}^p B_\mu^p)\varphi_{j\beta}.$$

In other words,

$$(D_\mu\varphi)_{i\alpha} = \partial_\mu\varphi_{i\alpha} + (A_\mu)_{ij}\varphi_{j\alpha} + (B_\mu)_{\alpha\beta}\varphi_{i\beta}$$

or, symbolically,

$$D_\mu\varphi = (\partial_\mu + A_\mu + B_\mu)\varphi,$$

where the matrix A_μ acts on the first index of the field φ and "does not act" on the second (the converse holds for B_μ). The Lagrangian of the model is of the form

$$\begin{aligned}\mathcal{L} &= (D_\mu\varphi)_{i\alpha}^*(D_\mu\varphi)_{i\alpha} - m^2\varphi_{i\alpha}^*\varphi_{i\alpha} - \lambda(\varphi_{i\alpha}^*\varphi_{i\alpha})^2 \\ &\quad -\frac{1}{4}F_{\mu\nu}^a F_{\mu\nu}^a - \frac{1}{4}\tilde{F}_{\mu\nu}^p\tilde{F}_{\mu\nu}^p,\end{aligned}$$

where $F_{\mu\nu}^a$ is constructed from the field A_μ^a,

$$F_{\mu\nu}^a = \partial_\mu A_\nu^a - \partial_\nu A_\mu^a + gC_{abc}A_\mu^b A_\nu^c,$$

where C_{abc} are the structure constants of the group $SU(n)$; analogously,

$$\tilde{F}_{\mu\nu}^p = \partial_\mu B_\nu^p - \partial_\nu B_\mu^p + \tilde{g}\tilde{C}_{pqr}B_\mu^q B_\nu^r,$$

where $\tilde{C}_{pqr}$ are the structure constants of the group $SU(m)$.

Thus, in models, where the gauge group is not simple, the number of gauge coupling constants is the same as the number of $U(1)$ components and simple components in the gauge group. We note, however, that it is by no means mandatory to generalize *all* global symmetries to gauge symmetries: depending on the physical situation, the model may simultaneously have gauge invariance under the gauge group G, and global invariance with respect to another group G'. For example, if we did not introduce the vector potential B_μ into the previous model, it would have $SU(n)$ gauge symmetry and $SU(m)$ global symmetry. Such a situation occurs in strong interactions of light quarks, where there is a gauge group $SU(3)_c$ and a global group $SU(3)_f$. Another example of global invariance in physics is the symmetry leading to conservation of the baryon number.

Problem 9. *Construct a gauge generalization of the model of example (4) of Section 4.1.*

Further generalization is obtained by employing arbitrary group representations, which transform scalar fields (in general, "matter fields" as opposed to gauge fields). We shall always assume that representations of Lie groups and Lie algebras are unitary and anti-Hermitian, respectively. Let $T(\omega)$ be a representation of a gauge group, and $T(A)$ the corresponding representation of the Lie algebra of the group. Without loss of generality, we shall assume that φ comprises columns, so that $T(\omega)$ and $T(A)$ are unitary and anti-Hermitian matrices, respectively. Under gauge transformations, φ becomes

$$\varphi'(x) = T(\omega(x))\varphi(x) \tag{4.40}$$

and, as before,

$$A_\mu \to A'_\mu = \omega A_\mu \omega^{-1} + \omega \partial_\mu \omega^{-1}. \tag{4.41}$$

We define the covariant derivative of the field φ as follows (assuming that the gauge group is simple and that the representation is irreducible):

$$D_\mu \varphi = [\partial_\mu + T(A_\mu)]\varphi, \tag{4.42}$$

or

$$D_\mu \varphi = (\partial_\mu - igT^a A_\mu^a)\varphi,$$

where T^a are generators in the representation T:

$$T^a = T(t^a)$$

(with the group $SU(n)$ in mind, we assume here that the generators of the algebra t^a are Hermitian, so that T^a are also Hermitian matrices). The covariant derivative defined in this way indeed transforms under the gauge transformations (4.40), (4.41) in a covariant manner:

$$D_\mu \varphi \to (D_\mu \varphi)' = T(\omega)D_\mu \varphi. \tag{4.43}$$

To verify this equation, it is sufficient to see that

$$T(\omega A \omega^{-1}) = T(\omega)T(A)T(\omega^{-1}) \tag{4.44}$$

and

$$T(\omega \partial_\mu \omega^{-1}) = T(\omega)\partial_\mu T(\omega^{-1}), \tag{4.45}$$

where $T(A)$ and $T(\omega \partial_\mu \omega^{-1})$ are representations of elements of the algebra, and $T(\omega)$ and $T(\omega^{-1})$ are representations of elements of the group. For the rest, the reasoning is completely analogous to that given in Section 4.2.

Problem 10. *Show that equations (4.44) and (4.45) are indeed satisfied.*

The gauge-invariant Lagrangian of the scalar field is constructed completely analogously to Section 4.2:

$$\mathcal{L}_\varphi = (D_\mu\varphi)^\dagger D_\mu\varphi - m^2\varphi^\dagger\varphi - \lambda(\varphi^\dagger\varphi)^2$$

(generally speaking, it is possible to construct more than one fourth-order invariant for the field φ in an irreducible representation of the group G; occasionally it is also possible to construct cubic invariants, so the self-interaction $(\varphi^\dagger\varphi)^2$ is only an example).

As an example, we consider the field φ in the adjoint representation, assuming the gauge group to be the group $SU(n)$. Then $\varphi(x)$ are matrices in the Lie algebra (which we take to be Hermitian), such that

$$\varphi(x) = t^a\varphi^a(x),$$

where t^a are Hermitian generators of $SU(n)$ (for $n = 2$ the generators are equal to $t^a = \tau^a/2$, where τ^a are the Pauli matrices), the $\varphi^a(x)$ are real fields, $a = 1, 2, \ldots, n^2 - 1$. The covariant derivative (4.42) in this case is equal to

$$D_\mu\varphi = \partial_\mu\varphi + \mathrm{ad}\,(A_\mu)\varphi = \partial_\mu\varphi + [A_\mu, \varphi]. \tag{4.46}$$

It can be written in the form ($D_\mu\varphi$ is again an element of the $SU(N)$ algebra)

$$D_\mu\varphi = t^a(D_\mu\varphi)^a,$$

where $(D_\mu\varphi)^a$ are real coefficients. We express these coefficients in terms of the real fields A_μ^a and φ^a. From (4.46) we have

$$t^a(D_\mu\varphi)^a = t^a\partial_\mu\varphi^a + (-igA_\mu^b\varphi^c)[t^b, t^c].$$

For Hermitian generators one has $[t^b, t^c] = iC^{bca}t^a$, so that

$$(D_\mu\varphi)^a = \partial_\mu\varphi^a + gC^{abc}A_\mu^b\varphi^c. \tag{4.47}$$

The invariant quantities are

$$\mathrm{Tr}\,(D_\mu\varphi)(D_\mu\varphi) = \frac{1}{2}(D_\mu\varphi)^a(D_\mu\varphi)^a$$

and

$$\mathrm{Tr}\,\varphi^2 = \frac{1}{2}\varphi^a\varphi^a,$$

thus the gauge-invariant Lagrangian of the scalar field can be chosen in the form

$$\mathcal{L}_\varphi = \mathrm{Tr}\,(D_\mu\varphi)^2 - m^2\mathrm{Tr}\,\varphi^2 - \lambda(\mathrm{Tr}\,\varphi^2)^2$$

or, in components,

$$\mathcal{L}_\varphi = \frac{1}{2}(D_\mu\varphi)^a(D_\mu\varphi)^a - \frac{m^2}{2}\varphi^a\varphi^a - \frac{\lambda}{4}(\varphi^a\varphi^a)^2. \qquad (4.48)$$

We note that the last expression contains only real fields φ^a and A_μ^a.

Problem 11. *Construct all invariants of order up to and including four, for the scalar field in the adjoint representation of the group $SU(2)$.*

Problem 12. *The same for the group $SU(3)$.*
    Fields, which transform under reducible representations of a gauge group can conveniently be regarded as sets of independent fields, each of which transforms under an irreducible representation.
    We note one more distinction between non-Abelian theories and electrodynamics. In electrodynamics, the field charge may take any value and the ratio of the charges of two different fields may be arbitrary (integer, rational, irrational). In other words, nothing prevents one from choosing fields φ and χ, to transform under gauge transformations of electrodynamics as

$$\varphi \to e^{i\alpha}\varphi$$
$$\chi \to e^{iq\alpha}\chi.$$

The arbitrary constant will then appear in the covariant derivative

$$D_\mu\chi = (\partial_\mu - ieqA_\mu)\chi$$

i.e. ultimately, in the interaction of the field χ with the electromagnetic field. In non-Abelian theories, there is only one free constant g, determining the interaction of the matter fields with the gauge field for each simple component of the gauge group. If g is fixed, then the interaction of the matter field with the gauge field is uniquely determined by the representation under which the matter fields are transformed. In this connection, non-Abelian charges are said to be "quantized", unlike in electrodynamics, where charges are not necessarily quantized, i.e. they are not necessarily integer numbers (it is another matter, that the electric charges of known particles are multiples of the electron charge; this is an experimental fact, established with high accuracy, but which it is evidently not possible to explain in the framework of electrodynamics).
    The constructions considered in this section essentially exhaust generalizations of the gauge model with the group $SU(2)$ and the scalar field doublet considered in Section 4.2; they can be used to construct a gauge theory for any compact Lie group and any representation of scalar fields.

4.4   Field equations

We now obtain the field equations for gauge fields and matter fields in non-Abelian gauge theories. We shall consider the case of a simple gauge group and a scalar field, transforming according to an irreducible representation. The action for such a system has the form

$$S = S_A + S_\varphi, \tag{4.49}$$

where

$$S_A = \int d^4x \left(-\frac{1}{4} F^a_{\mu\nu} F^a_{\mu\nu} \right) \tag{4.50}$$

$$F^a_{\mu\nu} = \partial_\mu A^a_\nu - \partial_\nu A^a_\mu + gC^{abc} A^b_\mu A^c_\nu$$

and

$$S_\varphi = \int d^4x \left[(D_\mu\varphi)^\dagger (D_\mu\varphi) - m^2 \varphi^\dagger \varphi - \lambda (\varphi^\dagger \varphi)^2 \right] \tag{4.51}$$

$$D_\mu\varphi = \partial_\mu\varphi - igT^a A^a_\mu \varphi. \tag{4.52}$$

We have chosen the self-interaction of the scalar field in the simplest form $(\varphi^\dagger \varphi)^2$. We assume that T^a are Hermitian generators in the representation T.

Let us first consider the variation of the action S_A with respect to the real fields A^a_μ; this leads to equations for the gauge field in the absence of matter fields. We obtain

$$\delta S_A = \int d^4x \left(\frac{1}{2} F^a_{\mu\nu} \delta F^a_{\mu\nu} \right), \tag{4.53}$$

where

$$\delta F^a_{\mu\nu} = \partial_\mu \delta A^a_\nu - \partial_\nu \delta A^a_\mu + gC^{abc} A^b_\mu \delta A^c_\nu + gC^{abc} \delta A^b_\mu A^c_\nu.$$

We note that the second and fourth terms in the last expression differ from the first and the third by the substitution $\mu \leftrightarrow \nu$ and by their sign. Using this fact and the antisymmetry of the tensor $F^a_{\mu\nu}$ in the indices μ, ν, for (4.53) we write

$$\delta S_A = \int d^4x \left[-F^a_{\mu\nu} (\partial_\mu \delta A^a_\nu + gC^{abc} A^b_\mu \delta A^c_\nu) \right].$$

In the first term, we integrate by parts; in the second, we rename the indices a, b, c and use the antisymmetry of C^{abc}. We obtain

$$\delta S_A = \int d^4x [\partial_\mu F^a_{\mu\nu} + gC^{abc} A^b_\mu F^c_{\mu\nu}] \delta A^a_\nu. \tag{4.54}$$

Hence, we obtain equations for the gauge fields without matter:

$$\partial_\mu F^a_{\mu\nu} + gC^{abc}A^b_\mu F^c_{\mu\nu} = 0. \tag{4.55}$$

We recall that the strength tensor $F_{\mu\nu}$ transforms according to the adjoint representation of the gauge group. Thus, the covariant derivative for it is equal to (see (4.47), the Lorentz indices of $F_{\mu\nu}$ are unimportant)

$$(D_\mu F_{\lambda\rho})^a = \partial_\mu F^a_{\lambda\rho} + gC^{abc}A^b_\mu F^c_{\lambda\rho}. \tag{4.56}$$

Consequently, the field equation (4.55) can be represented in the form

$$(D_\mu F_{\mu\nu})^a = 0. \tag{4.57}$$

We note that the left-hand side of this equation transforms covariantly (according to the adjoint representation) under gauge transformations.

Problem 13. *Show that the tensor $F_{\mu\nu}$ satisfies the Bianchi identity*

$$\varepsilon_{\mu\nu\lambda\rho}(D_\nu F_{\lambda\rho})^a = 0. \tag{4.58}$$

Equation (4.57) and the Bianchi identity (4.58) are non-Abelian analogues of the Maxwell equations in electrodynamics. However, unlike the Maxwell equations, the non-Abelian equations (4.57) and (4.50) contain, in addition to the strength tensor, the vector potential A^a_μ; this is clear from (4,56). We also remark that the covariant derivative (4.56) can be written in matrix form

$$D_\mu F_{\lambda\rho} = \partial_\mu F_{\lambda\rho} + [A_\mu, F_{\lambda\rho}],$$

where, as usual, $F_{\lambda\rho} = -igt^a F^a_{\lambda\rho}$.

Let us now consider the variation of the action S_φ with respect to the field A^a_μ. It follows from (4.52) that

$$(D_\mu\varphi)^\dagger = \partial_\mu\varphi^\dagger + igA^a_\mu\varphi^\dagger T^a \tag{4.59}$$

so that

$$\delta(D_\mu\varphi)^\dagger = ig\varphi^\dagger T^a \delta A^a_\mu \tag{4.60}$$

(we assume that φ is a column and, correspondingly, $\varphi^\dagger$ is a row; the Hermitian matrices T^a act from the right on $\varphi^\dagger$, so that (4.60) is an equation between rows). Furthermore, from (4.52) we have

$$\delta(D_\mu\varphi) = -igT^a\varphi\delta A^a_\mu,$$

thus

$$\delta S_\varphi = \int d^4x[(D_\nu\varphi)^\dagger(-igT^a\varphi) + ig\varphi^\dagger T^a D_\nu\varphi]\delta A^a_\nu.$$

We introduce the current

$$j_\nu^a = -i[\varphi^\dagger T^a D_\nu \varphi - (D_\nu \varphi)^\dagger T^a \varphi]. \tag{4.61}$$

Then

$$\delta S_\varphi = \int d^4 x (-g j_\nu^a) \delta A_\nu^a.$$

Taking into account (4.54) we hence obtain that the requirement that the variation of the total action $(S_A + S_\varphi)$ be equal to zero leads to a gauge field equation involving matter fields

$$(D_\mu F_{\mu\nu})^a = g j_\nu^a, \tag{4.62}$$

which is analogous to Maxwell's equation for electrodynamics with matter fields.

We now find equations, following from the variation of the action with respect to the scalar fields. The scalar fields are contained only in S_φ, thus it suffices to vary only this part of the action. As in electrodynamics, we shall consider $\varphi^\dagger$ and φ as independent variables and find the variation of S_φ with respect to $\varphi^\dagger$ (in other words, if

$$\varphi = \begin{pmatrix} \varphi_1 \\ \vdots \\ \varphi_N \end{pmatrix},$$

then $\varphi^\dagger = (\varphi_1^*, \ldots, \varphi_N^*)$ and we take variations with respect to all φ_i^*, $i = 1, \ldots, N$, assuming they are independent of φ_i). Taking into account (4.59) we obtain

$$\delta S_\varphi = \int d^4 x [(\partial_\mu \delta\varphi^\dagger + ig A_\mu^a \delta\varphi^\dagger T^a) D_\mu \varphi - m^2 \delta\varphi^\dagger \varphi - 2\lambda(\varphi^\dagger \varphi)\delta\varphi^\dagger \varphi].$$

Integrating by parts in the first term and requiring that $\delta S_\varphi = 0$, we find the equation

$$(\partial_\mu - ig A_\mu^a T^a) D_\mu \varphi + m^2 \varphi + 2\lambda(\varphi^\dagger \varphi)\varphi = 0. \tag{4.63}$$

We remark that the quantity $D_\mu \varphi$ transforms under gauge transformations according to the same representation $T(\omega)$ as the field φ itself (see (4.43); as usual, the Lorentz index is unimportant). Thus, the covariant derivative of $D_\mu \varphi$ is written in the form

$$D_\nu D_\mu \varphi \equiv D_\nu (D_\mu \varphi) = (\partial_\nu - ig A_\nu^a T^a) D_\mu \varphi.$$

Consequently, equation (4.63) takes on the form

$$D_\mu D_\mu \varphi + m^2 \varphi + 2\lambda(\varphi^\dagger \varphi)\varphi = 0. \tag{4.64}$$

Variation of the action with respect to φ leads to the Hermitian conjugate equation.

The system of equations (4.62) and (4.64) is the complete system of field equations in the model with action (4.49). Its generalization to more complicated cases (semi-simple gauge group, several matter fields) can be quite simply obtained for each specific model.

Problem 14.   *Show that the current j_μ^a, defined according to (4.61), transforms under gauge transformations according to the adjoint representation of the gauge group. Thus, equation (4.62) contains covariant quantities on the left- and right-hand sides (in other words, the left- and right-hand sides transform in the same way under gauge transformations).*

Problem 15.   *Verify the identity*

$$(D_\mu D_\nu F_{\mu\nu})^a = 0.$$

Show that if equations (4.64) are satisfied, then the equation

$$(D_\mu j_\mu)^a = 0$$

for the current j_μ^a is satisfied, where the covariant derivative is understood in the sense of the adjoint representation of the gauge group.   Thus, equations (4.62) and (4.64) are consistent with one another.

Problem 16.   *Obtain equation (4.64) in the case of the gauge group $SU(n)$ and a scalar field in the adjoint representation, directly from the Lagrangian (4.48). Write this equation and the current j_ν^a, in (4.62), in terms of real fields φ^a and A_μ^a.*

Problem 17.   *The gauge theory with the action (4.49) is invariant, in particular, under the global transformations*

$$\begin{aligned} A_\mu &\rightarrow A_\mu' = \omega A_\mu \omega^{-1} \\ \varphi &\rightarrow \varphi' = T(\omega)\varphi, \end{aligned}$$

where ω does not depend on x. Find the Noether current corresponding to these transformations. Does it coincide with the current j_μ^a in the field equation (4.62)? Is the current covariant under gauge transformations? Is *the Noether current equal to zero in the absence of matter fields? Write equation (4.62) in terms of the Noether current and the tensor $F_{\mu\nu}^a$.*

Let us find the energy–momentum tensor in the model with action (4.49). For this we use the method presented at the end of Section 2.8. Namely, we introduce into the action the metric tensor $g^{\mu\nu}$; for example, instead of (4.50), we write

$$S_A = \int d^4x \sqrt{-g} \left(-\frac{1}{4} g^{\mu\nu} g^{\lambda\rho} F_{\mu\lambda}^a F_{\nu\rho}^a \right). \qquad (4.65)$$

The symmetric energy–momentum tensor is related to the variation of the action with respect to $g^{\mu\nu}$,

$$\delta S = \frac{1}{2} \int d^4x \tilde{T}_{\mu\nu} \delta g^{\mu\nu}, \qquad (4.66)$$

where we consider small deviations of $g^{\mu\nu}$ from the Minkowski tensor $\eta^{\mu\nu}$,

$$g^{\mu\nu} = \eta^{\mu\nu} + \delta g^{\mu\nu}.$$

Bearing in mind that $g^{\mu\nu}$ is the inverse matrix to $g_{\mu\nu}$, whence

$$g = \det(g_{\mu\nu}) = \frac{1}{\det(g^{\mu\nu})}, \qquad (4.67)$$

we write

$$\delta g = -\eta_{\lambda\rho} \delta g^{\lambda\rho}$$

(this formula is analogous to the relation $\det(1 + \varepsilon) = 1 + \mathrm{Tr}\,\varepsilon$, if ε is a small matrix). Using (4.66) and (4.67), we obtain the energy–momentum tensor for the gauge field without matter, whose action has the form (4.65):

$$\tilde{T}_{\mu\nu}^{(A)} = -F_{\mu\lambda}^a F_\nu^{a\lambda} + \frac{1}{4} \eta_{\mu\nu} (F_{\lambda\rho}^a F^{a\lambda\rho}). \qquad (4.68)$$

In particular, the energy density of the gauge field is equal to

$$\tilde{T}_{00}^{(A)} = F_{0i}^a F_{0i}^a + \frac{1}{4} (-F_{0i}^a F_{0i}^a - F_{i0}^a F_{i0}^a + F_{ij}^a F_{ij}^a),$$

where the sum over spatial indices is taken with the Euclidean metric δ_{ij}. Finally, the energy of the gauge field has the form

$$E^{(A)} = \int d^3x \tilde{T}_{00} = \int d^3x \left(\frac{1}{2} F_{0i}^a F_{0i}^a + \frac{1}{4} F_{ij}^a F_{ij}^a \right). \qquad (4.69)$$

We shall sometimes use the notation

$$\begin{aligned} F_{0i}^a &= E_i^a \\ F_{ij}^a &= -\varepsilon_{ijk} H_k^a, \end{aligned}$$

where $\mathbf{E}^a$ and $\mathbf{H}^a$ are non-Abelian analogues of the electric and magnetic fields. In these terms, the energy of the gauge field is equal to

$$E^{(A)} = \int d^3x \left(\frac{1}{2}\mathbf{E}^a\mathbf{E}^a + \frac{1}{2}\mathbf{H}^a\mathbf{H}^a \right).$$

Analogously one can obtain an expression for the symmetric energy–momentum tensor of the scalar field, interacting with the gauge field, the action for which has the form (4.51):

$$\tilde{T}_{\mu\nu}^{(\varphi)} = 2(D_\mu\varphi)^\dagger(D_\nu\varphi) - \eta_{\mu\nu}\mathcal{L}_\varphi, \tag{4.70}$$

where

$$\mathcal{L}_\varphi = (D_\lambda\varphi)^\dagger(D_\lambda\varphi) - m^2\varphi^\dagger\varphi - \lambda(\varphi^\dagger\varphi)^2$$

is the Lagrangian of the scalar field. From (4.70), we obtain an expression for the energy of the scalar field

$$E^{(\varphi)} = \int d^3x[(D_0\varphi)^\dagger(D_0\varphi) + (D_i\varphi^\dagger)(D_i\varphi) + m^2\varphi^\dagger\varphi + \lambda(\varphi^\dagger\varphi)^2]. \tag{4.71}$$

The full energy–momentum tensor in the model (4.49) is equal to the sum of the contributions (4.68) and (4.70),

$$\tilde{T}_{\mu\nu} = \tilde{T}_{\mu\nu}^{(A)} + \tilde{T}_{\mu\nu}^{(\varphi)}. \tag{4.72}$$

Correspondingly, the energy of the scalar and gauge fields is equal to

$$E = E^{(A)} + E^{(\varphi)},$$

where $E^{(A)}$ and $E^{(\varphi)}$ are given by formulae (4.69) and (4.71).

We note the evident fact that the symmetric energy–momentum tensor is gauge invariant, while the energy is positive definite (we assume that for the scalar field $m^2 \geq 0$). The positivity of the energy of the gauge field is ensured, in particular, by the compactness of the Lie algebra (by the existence of a positive-definite invariant of type $\mathbf{E}^a\mathbf{E}^a$).

Problem 18. *Find the Noether energy–momentum tensor $T_{\mu\nu}$ for models of scalar and gauge fields with action (4.49). Choose the tensor $\Omega_\nu^{\mu\lambda}$, antisymmetric in the indices μ, λ, such that*

$$\tilde{T}_\nu^\mu = T_\nu^\mu + \partial_\lambda\Omega_\nu^{\mu\lambda},$$

provided the field equations are satisfied, where $\tilde{T}_{\mu\nu}$ is given by the formula (4.72).

4.5   Cauchy problem and gauge conditions

In field theory models without gauge symmetries, the Cauchy problem for the field equations is formulated in a quite evident way. For example, in the model of a single scalar field with Lagrangian

$$\mathcal{L} = \frac{1}{2}(\partial_\mu \varphi)^2 - V(\varphi)$$

the field equations take the form

$$\partial_\mu \partial_\mu \varphi + \frac{\partial V}{\partial \varphi} = 0,$$

or

$$\partial_t^2 \varphi = -\frac{\partial V}{\partial \varphi} + \nabla^2 \varphi, \tag{4.73}$$

where $\partial_t \equiv \partial/\partial t$. Since equation (4.73) allows us to express the second derivative of the field with respect to time in terms of φ and $\partial_t \varphi$, but $\partial_t \varphi$ is not determined by this equation, in order to find a solution near the surface $t = 0$, we need to assign $\varphi(\mathbf{x})$ and $\partial_t \varphi(\mathbf{x})$ on this surface; in other words, $\varphi(\mathbf{x}, t = 0)$ and $\partial_t \varphi(\mathbf{x}, t = 0)$ are Cauchy data.

The fact that formulation of the Cauchy problem in gauge theory is less trivial can already be seen from the existence of different solutions of the field equations with the same values of the fields and of their derivatives with respect to time at the initial surface (let us say $t = 0$). Indeed, if A_μ and φ is a solution of the field equations, then the gauge-transformed configuration

$$\begin{aligned} A'_\mu &\rightarrow \quad \omega A_\mu \omega^{-1} + \omega \partial_\mu \omega^{-1} \\ \varphi' &\rightarrow \quad T(\omega)\varphi \end{aligned} \tag{4.74}$$

is also a solution of the field equations, where $\omega(\mathbf{x}, t)$ can be chosen such that $\omega(\mathbf{x}, t = 0) = 1$, and the derivative of ω with respect to t is equal to zero at $t = 0$ (here, $T(\omega)$ is the representation of the gauge group, under which matter fields are transformed). We shall discuss the Cauchy problem for the system of equations (4.62) and (4.64)

$$\begin{aligned} (D_\mu F_{\mu\nu})^a &= g j_\nu^a \tag{4.75} \\ D_\mu D_\mu \varphi + \frac{\partial V}{\partial \varphi} &= 0 \tag{4.76} \end{aligned}$$

in somewhat more detail.

Let us first consider equation (4.75) for $\nu = 0$ (it is often called Gauss's equation or Gauss's constraint),

$$(D_i F_{i0})^a = -g j_0^a. \tag{4.77}$$

Since F_{i0}^a and j_0^a contain only first derivatives of the fields with respect to time, Gauss's constraint does not contain second derivatives with respect to time. Moreover, it does not contain first derivatives of A_0^a with respect to time. Thus, on the Cauchy surface, it is not possible to assign $A_i^a, \partial_t A_i^a, \varphi, \partial_t \varphi$ and A_0^a arbitrarily: N conditions (4.77) are imposed upon them, where N is the dimension of the gauge group.

Equations (4.75) with $\nu = i$ (spatial components) have the form

$$(D_0 F_{0i})^a - (D_j F_{ji})^a = g j_i^a. \tag{4.78}$$

They contain second derivatives of A_i^a with respect to time and first derivatives of A_0^a with respect to time. One important property of Gauss's constraint (4.77) and equations (4.78) and (4.76) is their consistency; if Gauss's constraint is satisfied at the initial moment in time, then it is satisfied at subsequent time instants if the fields satisfy equations (4.78) and (4.76).

To check this assertion, we find

$$\partial_0 \Gamma \equiv \partial_0 (D_i F_{i0} + g j_0),$$

where we use the matrix form for fields and current. We show that $\partial_0 \Gamma = 0$ at time t, if at that time Gauss's constraint and equations (4.78) and (4.76) are satisfied. Using Gauss's constraint at time t, we write

$$\partial_0 \Gamma = D_0 D_i F_{i0} + g D_0 j_0.$$

Furthermore, $[D_0, D_i] = F_{0i}$ (in this case, this equation has to be thought of as the action of an operator in the adjoint representation), thus

$$D_0 D_i F_{i0} = D_i D_0 F_{i0} + [F_{0i}, F_{i0}] = D_i D_0 F_{i0}.$$

We use equation (4.78) and write

$$\partial_0 \Gamma = -D_i (D_j F_{ji} + g j_i) + g D_0 j_0.$$

Furthermore, $D_i D_j F_{ji} = \frac{1}{2}([D_i, D_j]) F_{ji} = \frac{1}{2}[F_{ij}, F_{ji}] = 0$, thus

$$\partial_0 \Gamma = g(-D_i j_i + D_0 j_0) = g D_\mu j_\mu.$$

Finally, from the results of Problem 15 of Section 4.4, it follows that when equation (4.76) is satisfied, we have $D_\mu j_\mu = 0$, so that

$$\partial_0 \Gamma = 0,$$

as required.

This property can be viewed in two ways. First, it is possible to satisfy Gauss's constraint at the initial time and then "forget" about it and use equations with second derivatives with respect to time, (4.76) and (4.78). On the other hand, it is possible to use Gauss's constraint for all times, in which case N of the $3N$ equations of (4.78) will not be independent. These two approaches are realized in different gauges.

1. *The gauge $A_0 = 0$.* To fix the gauge freedom (4.74), we set

$$A_0^a = 0$$

for all $\mathbf{x}$ and t. For the initial data we choose $\mathbf{A}^a, \partial_t \mathbf{A}^a, \varphi, \partial_t \varphi$, which, however, satisfy Gauss's constraint at the initial moment in time. The second-order equations (4.76) and (4.78) are then used to evaluate $\mathbf{A}^a$ and φ at $t > 0$. Gauss's constraint will be satisfied automatically for $t > 0$. This approach is convenient for real computations; its disadvantage is the need to take explicit account of Gauss's constraint at $t = 0$.

2. *Hamiltonian gauges.* These are conditions imposed on the vector potentials $\mathbf{A}^a$, for example:

$$A_3^a = 0$$

(axial gauge) or

$$\partial_i A_i^a = 0$$

(Coulomb gauge). By fixing the gauge in this way, the fields A_i^a can (at least in principle) be expressed in terms of independent components a_α^a, $\alpha = 1, 2$. For the initial data, we assign $a_\alpha^a(\mathbf{x}, t = 0)$ and $\partial_t a_\alpha^a(\mathbf{x}, t = 0)$, together with, if matter is present, $\varphi(\mathbf{x}, t = 0)$ and $\partial_t \varphi(\mathbf{x}, t = 0)$. Then A_0^a at the initial moment in time is determined from Gauss's constraint. At subsequent times $2N$ out of $3N$ equations (4.78) determine $a_\alpha^a(\mathbf{x}, t)$, while A_0^a will be found from Gauss's constraint, and the remaining N equations of (4.78) will be identically satisfied.

This approach is applied to the construction of the canonical (Hamiltonian) formalism in quantum gauge theories.

Problem 19. *Write down equations with only first derivatives with respect to time for Yang–Mills fields without matter in the axial gauge $A_3^a = 0$, assuming that independent variables are A_α^a and $E_\alpha^a \equiv F_{0\alpha}^a$, $\alpha = 1, 2$ and expressing A_0^a in terms of these variables using Gauss's constraint. Show that these equations can be represented in the Hamiltonian form*

$$\frac{\partial A_\alpha^a}{\partial t} = \frac{\delta H}{\delta E_\alpha^a}, \qquad \frac{\partial E_\alpha^a}{\partial t} = -\frac{\delta H}{\delta A_\alpha^a},$$

where

$$H = \int \left(\frac{1}{2} E_\alpha^a E_\alpha^a + \frac{1}{4} F_{\alpha\beta}^a F_{\alpha\beta}^a + \frac{1}{2} E_3^a E_3^a + \frac{1}{2} F_{3\alpha}^a F_{3\alpha}^2 \right) d^3x.$$

In the latter expression, it is necessary to set $A_3^a = 0$ in E_3^a and $F_{3\alpha}^a$, and express A_0^a in terms of A_α^a and the E_α^a using Gauss's constraint.

For an almost unique choice of solution of the gauge theory equations one can also use covariant gauges like the Lorentz gauge:

$$\partial_\mu A_\mu^a = 0.$$

As in Abelian gauge theory, a solution is determined modulo a function satisfying some (nonlinear) equation, but otherwise arbitrary. Hence, it is clear that gauges of the Lorentz type are not suitable for formulation of the Cauchy problem. However, they are very convenient for computations in quantum theory.

Chapter 5

Spontaneous Breaking of Global Symmetry

In the situations considered up to now, symmetries (global and gauge) have led to certain properties of small perturbations of the fields. Namely, global symmetries in scalar theories have implied equality of the masses of all small linear waves for fields belonging to the same representation of the symmetry group, and also the same interaction properties of these fields (see Section 4.1). Gauge symmetries have led to the masslessness of vector gauge fields: indeed, expressions of the type

$$m^2 A_\mu^a A_\mu^a$$

are not invariant under gauge transformations (and generally, it is impossible to construct an invariant expression quadratic in A_μ^a which does not contain derivatives), thus, adding these to the Lagrangian would clearly violate the gauge invariance.

Here and in the next chapter, we consider a dynamical mechanism, which leads to a dramatic modification of these properties in theories, where the Lagrangians are invariant under global (Chapter 5) and gauge (Chapter 6) transformations.

As previously mentioned, small perturbations of fields correspond to particles. In nature, there exist both massless vector bosons (photons, and also gluons, carriers of strong interactions between quarks) and massive vector bosons (charged $W^\pm$ and neutral Z^0, mediating weak interactions). The mechanism considered in Chapter 6 (and its generalization) enables us to describe the $W^\pm$ and the Z^0 in the framework of gauge theories. The breaking of global symmetry, studied in detail in Chapter 5, also occurs in nature; an important example is the spontaneous breaking of chiral symmetry in strong interactions. There are numerous examples of spontaneous symmetry breaking in condensed matter; one such is superconductivity.

5.1 Spontaneous breaking of discrete symmetry

Let us consider the theory of a single scalar field with Lagrangian

$$\mathcal{L} = \frac{1}{2}(\partial_\mu \varphi)^2 - \frac{m^2}{2}\varphi^2 - \frac{\lambda}{4}\varphi^4. \tag{5.1}$$

This Lagrangian is invariant under the transformation

$$\varphi(x) \rightarrow -\varphi(x), \tag{5.2}$$

which are a discrete symmetry of the model. The energy functional for this model has the form

$$E = \int d^3x \left[\frac{1}{2}(\partial_0 \varphi)^2 + \frac{1}{2}(\partial_i \varphi)^2 + \frac{m^2}{2}\varphi^2 + \frac{\lambda}{4}\varphi^4 \right]. \tag{5.3}$$

For this energy to be bounded from below we require that

$$\lambda > 0$$

(the case $\lambda = 0$ is trivial and will not be considered), but there is no constraint on the parameter m^2.

Let us find the *ground state* in this model, the field configuration $\varphi(x)$ with minimal energy. From (5.3) it is clear that minimum energy is achieved for fields which do not depend on time

$$\partial_t \varphi(\mathbf{x}, t) = 0 \tag{5.4}$$

and are homogeneous in space

$$\partial_i \varphi(\mathbf{x}, t) = 0 \tag{5.5}$$

(the first two terms in (5.3) are non-negative and are equal to zero only if equations (5.4) and (5.5) are satisfied). In other words, a field configuration in the ground state actually does not depend on $\mathbf{x}$ and t,

$$\varphi = \text{constant}.$$

We determine the constant from the requirement of minimality of the potential

$$V(\varphi) = \frac{m^2}{2}\varphi^2 + \frac{\lambda}{4}\varphi^4.$$

The cases $m^2 \geq 0$ and $m^2 < 0$ need to be considered separately.

1. $m^2 \geq 0$. The minimum of the potential is at

$$\varphi = 0.$$

This state is invariant under the transformation (5.2); we say that the ground state does not break the symmetry of the model. Perturbations about the ground state are described by the field φ itself, and the Lagrangian for these perturbations coincides with the original Lagrangian (5.1); the Lagrangian for perturbations about the ground state (field excitations) is invariant under the discrete symmetry (5.1).

2. $m^2 < 0$. In this case, the potential has the form illustrated in Figure 5.1. The field $\varphi = 0$, which is symmetric under the transformation $\varphi \to -\varphi$, corresponds to the maximum of the potential, and there are two ground states:

$$\begin{aligned} \varphi &= \pm\varphi_0 \\ \varphi_0 &= \frac{\mu}{\sqrt{\lambda}}, \end{aligned}$$

where we have changed the notation,

$$\mu^2 = -m^2 > 0, \quad \mu > 0.$$

Indeed,

$$\frac{\partial V}{\partial \varphi}(\varphi = \pm\varphi_0) = \pm\varphi_0(-\mu^2 + \lambda\varphi_0^2) = 0,$$

so that the minima of the potential are at $\pm\varphi$.

If we "take away" all available energy from the field, then it will be in one of the ground states, let us say at $\varphi = +\varphi_0$. In order to transfer the field from one ground state to the other, one has to add an energy, proportional to the volume of the space (we recall that $V(\varphi)$ is the energy density of a homogeneous field; the potential energy of the homogeneous field φ is equal to $\Omega V(\varphi)$, where Ω is the volume of the space). Thus, for a system with a large spatial volume ($\Omega \to \infty$) we have to choose one of the ground states and consider perturbations about it. Let us choose as the ground state

$$\varphi = +\varphi_0 \equiv \frac{\mu}{\sqrt{\lambda}}$$

(we could equally well choose $\varphi = -\varphi_0$, however, a specific choice is necessary). This state *is not invariant* under the transformation $\varphi \to -\varphi$;

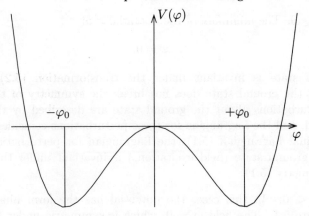

Figure 5.1.

we say that the symmetry is spontaneously broken. The energy of the ground state is equal to

$$E_0 = \Omega V(\varphi_0) = -\Omega \frac{\mu^4}{4\lambda}.$$

In what follows, we shall calculate the energy with respect to the energy of the ground state (the zero-point energy can be chosen arbitrarily, if gravitational interactions are not considered). For this, it is convenient to consider, instead of the Lagrangian (5.1), a Lagrangian differing from that by a constant:

$$\mathcal{L} = \frac{1}{2}(\partial_\mu \varphi)^2 + \frac{\mu^2}{2}\varphi^2 - \frac{\lambda}{4}\varphi^4 + \frac{\mu^4}{4\lambda} \tag{5.6}$$

or, which amounts to the same thing,

$$\mathcal{L} = \frac{1}{2}(\partial_\mu \varphi)^2 - \frac{\lambda}{4}(\varphi^2 - \varphi_0^2)^2.$$

Here, the energy density of a homogeneous field (potential) is equal to

$$V(\varphi) = \frac{\mu}{2}\varphi^2 - \frac{\lambda}{4}\varphi^4 + \frac{\mu^4}{4\lambda} = \frac{\lambda}{4}(\varphi^2 - \varphi_0^2)^2,$$

so that $V(\varphi_0) = 0$.

It is impossible to observe a homogeneous field itself in the ground state directly; any observation has to do with changes of physical quantities in space and in time. However, the fact that the ground state is non-trivial leads to a number of consequences for perturbations about it.

Let us consider perturbations $\chi(x)$ about $\varphi = +\varphi_0$, i.e. we write

$$\varphi(x) = \varphi_0 + \chi(x). \tag{5.7}$$

The Lagrangian for $\chi(x)$ is obtained by substituting (5.7) in the original Lagrangian (5.6),

$$\mathcal{L}_\chi(\chi) = \mathcal{L}(\varphi_0 + \chi).$$

We have

$$\partial_\mu(\varphi_0 + \chi) = \partial_\mu \chi,$$

$$V_\chi(\chi) = V(\varphi_0 + \chi) = \frac{\lambda}{4}((\varphi_0 + \chi)^2 - \varphi_0^2)^2.$$

Hence

$$
\begin{aligned}
V_\chi(\chi) &= \lambda \varphi_0^2 \chi^2 + (\lambda \varphi_0)\chi^3 + \frac{\lambda}{4}\chi^4 \\
&= \mu^2 \chi^2 + \sqrt{\lambda}\mu\chi^3 + \frac{\lambda}{4}\chi^4.
\end{aligned}
$$

Consequently, the Lagrangian for perturbations has the form

$$\mathcal{L}_\chi = \frac{1}{2}(\partial_\mu \chi)^2 - \mu^2 \chi^2 - \sqrt{\lambda}\mu\chi^3 - \frac{\lambda}{4}\chi^4. \tag{5.8}$$

This Lagrangian *is not invariant* under discrete transformations $\chi \to -\chi$, which might have been anticipated, since the ground state is not invariant. The remnant of the symmetry $\varphi \to -\varphi$ in the Lagrangian (5.8) is a relation between the mass of the field χ and cubic and quartic interaction constants: an arbitrary polynomial Lagrangian of up to fourth order in field looks like

$$\mathcal{L}_\chi = \frac{1}{2}(\partial_\mu \chi)^2 - \frac{m_\chi}{2}\chi^2 - \alpha\chi^3 - \frac{\beta}{4}\chi^4, \tag{5.9}$$

where m_χ, α, β are, generally speaking, arbitrary. For the theory with spontaneous symmetry breaking we have, on the other hand,

$$
\begin{aligned}
m_\chi^2 &= 2\mu^2 \\
\alpha &= \sqrt{\lambda}\mu \\
\beta &= \lambda,
\end{aligned}
$$

and so there exists one relation between α, β and m_χ.

Problem 1. *Show that the Lagrangian for perturbations about the ground state $\varphi = -\varphi_0$ is equivalent to (5.8) (i.e. it reduces to (5.8) by change of variables).*

Thus, the phenomenon of spontaneous breaking of global symmetry amounts to the fact that the ground state is not invariant under the symmetry of the Lagrangian. Neither does the Lagrangian for perturbations have the original symmetry; this symmetry manifests itself in the relations between coupling constants and the masses of perturbations.

In quantum mechanics with two symmetric wells (potential of the type illustrated in Figure 5.1) the ground state is a *symmetric* linear superposition of wave functions, concentrated near each well. This property of the ground state arises as a consequence of tunneling between the wells. In field theory, this tunneling must take place simultaneously in the whole space, thus its amplitude vanishes in the limit $\Omega \to \infty$. Indeed, let us write down the action for spatially homogeneous fields $\varphi(t)$:

$$S = \Omega \int dt \left[\frac{\dot{\varphi}}{2} - V(\varphi) \right].$$

This coincides with the action for a particle with mass

$$\mathcal{M} = \Omega$$

in the potential

$$\mathcal{U}(\varphi) = \Omega V(\varphi).$$

The amplitude of the tunneling between $+\varphi_0$ and $-\varphi_0$ is described by the quantity

$$A \propto \exp\left(-\int_{-\varphi_0}^{+\varphi_0} \sqrt{2\mathcal{M}\mathcal{U}} d\varphi \right),$$

therefore

$$A \propto \exp\left(-\Omega \int_{-\varphi_0}^{+\varphi_0} \sqrt{2V(\varphi)} d\varphi \right),$$

i.e. the amplitude of the tunneling tends exponentially to zero as $\Omega \to \infty$. Thus, in quantum theory, it is legitimate to consider $\varphi = +\varphi_0$ as the ground state, if the spatial volume is sufficiently large.

5.2 Spontaneous breaking of global $U(1)$ symmetry. Nambu–Goldstone bosons

Let us consider the simplest case of continuous symmetry, $U(1)$ symmetry. Let

$$\varphi = \frac{1}{\sqrt{2}}(\varphi_1 + i\varphi_2)$$

be a complex scalar field, with Lagrangian of the form

$$\mathcal{L} = \partial_\mu \varphi^* \partial_\mu \varphi - m^2 \varphi^* \varphi - \lambda(\varphi^*\varphi)^2 - c, \qquad (5.10)$$

where the constant c is introduced for convenience in what follows. The Lagrangian (5.10) can also be written in terms of real fields $(\varphi_1, \varphi_2) = \frac{1}{\sqrt{2}}(\mathrm{Re}\,\varphi, \mathrm{Im}\,\varphi)$,

$$\mathcal{L} = \frac{1}{2}\partial_\mu\varphi_i\partial_\mu\varphi_i - \frac{m^2}{2}\varphi_i\varphi_i - \frac{\lambda}{4}(\varphi_i\varphi_i)^2 - c,$$

where the summation over $i = 1, 2$ is understood.

The Lagrangian (5.10) is invariant under global $U(1)$ transformations

$$\varphi(x) \to \varphi'(x) = e^{i\alpha}\varphi(x) \qquad (5.11)$$

or, in terms of the fields $\varphi_{1,2}$,

$$\varphi_1 \to \cos\alpha\varphi_1 - \sin\alpha\varphi_2 \qquad (5.12)$$
$$\varphi_2 \to \sin\alpha\varphi_1 + \cos\alpha\varphi_2.$$

Let us consider the field energy

$$E = \int d^3x (\partial_0\varphi^*\partial_0\varphi + \partial_i\varphi^*\partial_i\varphi + V(\varphi^*, \varphi)),$$

where

$$V(\varphi^*, \varphi) = m^2\varphi^*\varphi + \lambda(\varphi^*\varphi)^2 + c. \qquad (5.13)$$

The ground state is again homogeneous in space–time, $\varphi = $ constant, and is a minimum of the potential (5.13).

For $m^2 \geq 0$, the ground state is $\varphi = 0$, and the excitations represent two real fields φ_1 and φ_2 of equal mass with a special choice of the interaction $(\varphi_1^2 + \varphi_2^2)^2$. The $U(1)$ symmetry is not broken.

For $m^2 = -\mu^2 < 0$, the potential $V(\varphi)$ is a solid of revolution, with "a Mexican hat shape", as shown in Figure 5.2. It depends only on one variable

$$|\varphi| = \sqrt{\frac{\varphi_1^2 + \varphi_2^2}{2}}$$

$$V(\varphi_1, \varphi_2)$$

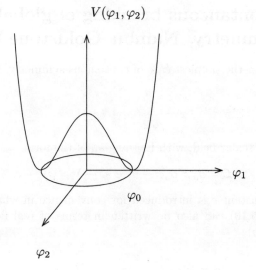

Figure 5.2.

and has a continuous set of minima

$$\varphi = e^{i\alpha} \frac{\varphi_0}{\sqrt{2}},$$

where φ_0 is determined from the condition

$$\left(\frac{\partial V}{\partial |\varphi|}\right)\left(\frac{\varphi_0}{\sqrt{2}}\right) = 0$$

and is equal to

$$\varphi_0 = \frac{\mu}{\sqrt{\lambda}}.$$

We again need to choose one of the minima as the ground state and consider excitations about that.

Although transitions between different minima can be completed without increasing the potential energy by moving along the circle of minima, they still require an infinite energy in the limit of infinite volume of the space, $\Omega \to \infty$. Indeed, for a homogeneous field $\varphi(t)$, the kinetic term in the energy is proportional to the volume

$$E_{\text{kin}} = \Omega |\dot{\varphi}|^2.$$

Therefore a change in the field in the whole space requires an infinite energy. Thus, again, it is sufficient to consider just one of the minima.

Let us consider the ground state

$$\varphi = \varphi_0/\sqrt{2}, \tag{5.14}$$

i.e. $\varphi_1 = \varphi_0$, $\varphi_2 = 0$, and perturbations about that, described by the fields

$$\begin{aligned} \varphi_1(x) &= \varphi_0 + \chi(x) \\ \varphi_2(x) &= \theta(x). \end{aligned} \tag{5.15}$$

We restrict ourselves to small perturbations, so we retain terms in the Lagrangian which are quadratic in the perturbations χ and θ. We have $\partial_\mu \varphi_1 = \partial_\mu \chi$, $\partial_\mu \varphi_2 = \partial_\mu \theta$, and

$$V = -\frac{\mu^2}{2}[(\varphi_0 + \chi)^2 + \theta^2] + \frac{\lambda}{4}[(\varphi_0 + \chi)^2 + \theta^2]^2 + \frac{\mu^4}{4\lambda},$$

where the constant c in (5.10) is chosen such that the energy of the ground state is equal to zero.

To quadratic order in the fields χ and θ we obtain

$$V = \mu^2 \chi^2.$$

There are no terms of type θ^2 or $\chi\theta$ in the potential. This can be seen from Figure 5.2: quadratic terms in χ and θ in the potential are the curvatures of the potential along the directions φ_1 and φ_2; the curvature of the potential along the direction φ_2 is equal to zero at the point $(\varphi_1 = \varphi_0, \varphi_2 = 0)$, by virtue of the $U(1)$ symmetry (5.12).

Thus, the quadratic Lagrangian is equal to

$$\mathcal{L}_{\chi,\theta}^{(2)} = \frac{1}{2}(\partial_\mu \chi)^2 + \frac{1}{2}(\partial_\mu \theta)^2 - \mu^2 \chi^2.$$

The field χ has mass $m_\chi = \sqrt{2}\mu$, while the field θ remains massless. The occurrence of the massless mode is directly related to the presence of the $U(1)$-symmetry in the Lagrangian and to the non-symmetric nature of the ground state (Nambu 1960, Vaks and Larkin 1961, Goldstone 1961). This massless field is called the Nambu–Goldstone field and the corresponding particle is the Nambu–Goldstone boson.

The connection between the Nambu–Goldstone mode and symmetries of the Lagrangian can also be illustrated as follows. Let us consider small perturbations of the field about a non-symmetric ground state, such that φ is not equal to zero anywhere in space–time (this is only possible if the symmetry is spontaneously broken!). Then, we can introduce variables $\rho(x)$ and $\alpha(x)$,

$$\varphi(x) = e^{i\alpha(x)}\rho(x). \tag{5.16}$$

Since the Lagrangian is symmetric under the transformations (5.11) the potential does not contain the field $\alpha(x)$ (this is also clear from (5.13)), i.e. the field $\alpha(x)$ only appears in the Lagrangian via the derivative $\partial_\mu \alpha(x)$. The absence of a term of type $\alpha^2(x)$ in the Lagrangian means that $\alpha(x)$ is a massless field.

Thus, in the example of the model with $U(1)$ symmetry, we have seen that spontaneous breaking of continuous global symmetry (the Lagrangian is symmetric, the ground state is not) leads to the occurrence of massless perturbations, called Nambu–Goldstone modes. This statement is of a general nature, as we shall see in Sections 5.3 and 5.4.

The particles closest to Nambu–Goldstone bosons are the $\pi^\pm$ and π^0-mesons. The corresponding symmetry is the chiral invariance of strong interactions. The fact that the masses of π-mesons are non-zero has to do with small terms in the Lagrangian, which explicitly break the chiral symmetry.

Problem 2.　*Find the full Lagrangian for the fields χ and θ, defined by (5.15). How many independent constants does it contain?*

Problem 3.　*Show that choice of the ground state in the form*

$$\varphi_1 = \varphi_0 \cos \alpha$$
$$\varphi_2 = \varphi_0 \sin \alpha$$

leads to the full Lagrangian for the perturbations, equivalent to that found in the previous example.

Problem 4.　*Write the full Lagrangian for perturbations about the ground state in terms of the fields $\alpha(x)$ and $\Delta\rho(x) = \rho(x) - \rho_0$. Find the relation between $\alpha(x)$ and the Nambu–Goldstone mode $\theta(x)$ to the leading order in the fields.*

5.3　Partial symmetry breaking: the $SO(3)$ model

Global symmetry of the Lagrangian may be broken incompletely. An example of partial symmetry breaking arises in the model with three real fields φ^a, $a = 1, 2, 3$ and Lagrangian

$$\mathcal{L} = \frac{1}{2}\partial_\mu \varphi^a \partial_\mu \varphi^a - V(\varphi), \tag{5.17}$$

where

$$V(\varphi) = -\frac{\mu^2}{2}\varphi^a\varphi^a + \frac{\lambda}{4}(\varphi^a\varphi^a)^2 + \frac{\mu^4}{4\lambda}. \tag{5.18}$$

In (5.17) and (5.18) summation over a is understood; we suppose that $\mu^2 > 0$.

The Lagrangian (5.17) has a global $SO(3)$ symmetry (rotation in three-dimensional internal space). The ground state is a homogeneous field in space–time, minimizing the potential $V(\varphi)$. The potential $V(\varphi)$ has minima at

$$\varphi^a\varphi^a = \varphi_0^2,$$

where

$$\varphi_0 = \frac{\mu}{\sqrt{\lambda}}.$$

Thus, the set of all possible ground states is a two-dimensional sphere of radius φ_0. As the ground state (we shall call it the classical vacuum of the model) one can choose any point on the sphere; let us choose

$$\begin{aligned} \varphi^1 &= \varphi^2 = 0 \\ \varphi^3 &= \varphi_0. \end{aligned}$$

Unlike in the previous example, the vacuum vector $\vec{\varphi}^{(0)} = (0, 0, \varphi_0)$ does not break the symmetry completely; there is a non-trivial subgroup of the group $SO(3)$, under which the vacuum vector is invariant:

$$\omega\vec{\varphi}^{(0)} = \vec{\varphi}^{(0)}. \tag{5.19}$$

This subgroup is the group $SO(2)$ of rotations in the space of the fields about the third axis

$$\begin{aligned} \varphi^1 &\to \cos\alpha\varphi^1 - \sin\alpha\varphi^2 \\ \varphi^2 &\to \sin\alpha\varphi^1 + \cos\alpha\varphi^2 \\ \varphi^3 &\to \varphi^3. \end{aligned}$$

It is clear that the Lagrangian for perturbations about the chosen classical vacuum will be invariant under this group $SO(2)$. Let us find the quadratic part of this Lagrangian and, in particular, determine which perturbations are massless Nambu–Goldstone modes.

Let us introduce the fields of perturbations $\chi(x), \theta^1(x), \theta^2(x)$, such that

$$\begin{aligned} \varphi^1(x) &= \theta^1(x) \\ \varphi^2(x) &= \theta^2(x) \\ \varphi^3(x) &= \varphi_0 + \chi(x). \end{aligned} \tag{5.20}$$

The potential term in the Lagrangian for the perturbations has the form

$$
\begin{aligned}
V &= -\frac{\mu^2}{2}\left[(\theta^1)^2 + (\theta^2)^2\right] - \frac{\mu^2}{2}(\varphi_0 + \chi)^2 \\
&\quad + \frac{\lambda}{4}\left[(\theta^1)^2 + (\theta^2)^2 + (\varphi_0 + \chi)^2\right]^2 + \frac{\mu^4}{4\lambda},
\end{aligned}
\tag{5.21}
$$

and the kinetic term is equal to

$$
\mathcal{L}_{\text{kin}} = \frac{1}{2}(\partial_\mu \theta^1)^2 + \frac{1}{2}(\partial_\mu \theta^2)^2 + \frac{1}{2}(\partial_\mu \chi)^2.
\tag{5.22}
$$

From (5.21) and (5.22) we see that the Lagrangian of the perturbations $\mathcal{L} = \mathcal{L}_{\text{kin}} - V$ is invariant under $SO(2)$ rotations of the fields θ^1 and θ^2, since it contains only combinations of type $(\theta^1)^2 + (\theta^2)^2$. Of course, it does not have full $SO(3)$ symmetry.

Evaluating (5.21), we see that the quadratic part of the potential contains only χ^2, so that the quadratic Lagrangian for the perturbations is of the form

$$
\mathcal{L}^{(2)} = \frac{1}{2}(\partial_\mu \theta^1)^2 + \frac{1}{2}(\partial_\mu \theta^2)^2 + \frac{1}{2}(\partial_\mu \chi)^2 - \mu^2 \chi^2.
$$

Consequently, θ^1 and θ^2 are massless Nambu–Goldstone fields. The fact that there must indeed be two real Nambu–Goldstone modes is clear from the following arguments. Of the three generators of the group $SO(3)$, one generator annihilates the vacuum $\vec{\varphi}^{(0)} = (0, 0, \varphi_0)$:

$$
t_h \vec{\varphi}^{(0)} = 0.
\tag{5.23}
$$

This is the generator of the unbroken subgroup $SO(2)$: equation (5.23) is equivalent to (5.19) for ω close to unity, $\omega = 1 + \varepsilon t_h$, where ε is a small parameter. The two other generators (and any linear combinations of them) do not annihilate the vacuum, otherwise, the unbroken subgroup should be larger than $SO(2)$. We form two vectors

$$
\begin{aligned}
\vec{n}_1 &= t_1 \vec{\varphi}^{(0)} \\
\vec{n}_2 &= t_2 \vec{\varphi}^{(0)}
\end{aligned}
\tag{5.24}
$$

where t_1, t_2 are broken generators (i.e. they do not annihilate the vacuum $\vec{\varphi}^{(0)}$). The vectors $\vec{n}_1$ and $\vec{n}_2$ are linearly independent (otherwise a linear combination of t_1 and t_2 would annihilate the vacuum). Moreover, if $\tilde{\theta}^1$ and $\tilde{\theta}^2$ are small then the vector

$$
\vec{\varphi} = \vec{\varphi}^{(0)} + \tilde{\theta}^1 \vec{n}_1 + \tilde{\theta}^2 \vec{n}_2
$$

is one of classical vacua, near $\vec{\varphi}^{(0)}$: indeed

$$
\vec{\varphi} = (1 + \tilde{\theta}^1 t_1 + \tilde{\theta}^2 t_2)\vec{\varphi}^{(0)} = \omega \vec{\varphi}^{(0)},
$$

where ω is close to the unit element of $SO(3)$. Thus to the first non-trivial order in $\tilde{\theta}^1, \tilde{\theta}^2$ we have

$$V(\vec{\varphi}) = V(\vec{\varphi}^{(0)}),$$

i.e.

$$V(\vec{\varphi}^{(0)} + \tilde{\theta}^1 \vec{n}_1 + \tilde{\theta}^2 \vec{n}_2) = 0$$

(we have taken into account that $V(\vec{\varphi}^{(0)}) = 0$). Since $\vec{\varphi}^{(0)}$ is the point of the minimum of the potential V, the first non-trivial order is quadratic. Thus, for deviations from the vacuum $\vec{\varphi}^{(0)}$ of the form

$$\vec{\varphi}(x) = \vec{\varphi}^{(0)} + \tilde{\theta}^1(x)\vec{n}_1 + \tilde{\theta}^2(x)\vec{n}_2 \tag{5.25}$$

the quadratic part of the potential does not contain $\tilde{\theta}^1$ and $\tilde{\theta}^2$, i.e. the fields $\tilde{\theta}^1$ and $\tilde{\theta}^2$ are massless.

For the group $SO(3)$ the generators have the form

$$(t_a)_{bc} = \varepsilon_{abc},$$

and the chosen classical vacuum is equal to

$$\varphi^{(0)a} = \delta^{a3}\varphi_0.$$

The unbroken generator is t_3, since

$$(t_3)_{bc}\varphi^{(0)c} = \varepsilon_{3bc}\delta^{c3}\varphi_0 = 0.$$

The broken generators are t_1 and t_2 and the vectors in (5.24) are equal to

$$
\begin{aligned}
n_1^a &= (t_1)_{ab}\varphi^{(0)b} = \varepsilon_{1ab}\delta^{b3}\varphi_0 = \delta^{a2}\varphi_0 \\
n_2^a &= \varepsilon_{2ab}\delta^{b3}\varphi_0 = -\delta^{a1}\varphi_0.
\end{aligned}
$$

It follows from (5.25) that the field with massless perturbations has the form

$$\vec{\varphi}(x) = (-\tilde{\theta}^2(x)\varphi_0, \tilde{\theta}^1(x)\varphi_0, \varphi_0).$$

Comparing this expression with (5.20) at $\chi = 0$, we have a relation between the fields $\tilde{\theta}^{1,2}$ constructed above and the fields $\theta^{1,2}$ used in the explicit calculations

$$
\begin{aligned}
\theta^1 &= -\tilde{\theta}^2\varphi_0 \\
\theta^2 &= \tilde{\theta}^1\varphi_0.
\end{aligned}
$$

Thus, modulo the notation and the normalization, the fields constructed using (5.24) coincide with the Nambu–Goldstone modes which we found explicitly.

The construction described here generalizes to the case of an arbitrary compact global group and an arbitrary unitary representation of scalar fields. This generalization is called the Goldstone theorem and is studied in the next section.

Generally speaking, we might not be so lucky with the choice of fields of perturbations $\theta^1(x), \theta^2(x)$ and $\chi(x)$ (see formula (5.20)) that the fields θ^1 and θ^2 are at once Nambu–Goldstone modes. It would be possible to introduce three linearly independent vectors $\vec{n}_1, \vec{n}_2$ and $\vec{n}_3$ and write for the field with perturbations, instead of (5.20),

$$\vec{\varphi}(x) = \vec{\varphi}^{(0)} + \vec{n}_i \xi^i(x),$$

where $\vec{\varphi}^{(0)} = (0, 0, \varphi_0)$, and the ξ^i are the fields of perturbations. Then, the kinetic term in the Lagrangian of the perturbations would have the form

$$\mathcal{L}_{\mathrm{kin}} = \frac{1}{2} \partial_\mu \xi^i \partial_\mu \xi^j \cdot (\vec{n}_i \cdot \vec{n}_j),$$

and the quadratic contribution to the potential would have the structure

$$V^{(2)}(\vec{\varphi}) = \frac{1}{2} M_{ij} \xi_i \xi_j, \tag{5.26}$$

where M_{ij} is some real symmetric matrix, independent of the coordinates (mass matrix for the fields ξ^i). To bring the quadratic Lagrangian to canonical form (5.22), we need to do the following.

1. Choose the vectors $\vec{n}_i$ to be orthogonal $(\vec{n}_i \cdot \vec{n}_j) = \delta_{ij}$. Then the kinetic term in the quadratic Lagrangian will have the canonical form

$$\mathcal{L}_{\mathrm{kin}} = \frac{1}{2} \partial_\mu \xi^i \partial_\mu \xi^i,$$

   but the potential terms, as before, will have the structure (5.26).

2. Next perform an orthogonal transformation of the fields ξ^i, i.e. introduce fields

$$\xi'^i = O^i_j \xi^j,$$

   where O^i_j is an orthogonal matrix. In terms of the fields ξ'^i, the kinetic part of the Lagrangian, as before, is of canonical form. The matrix

O^i_j should be chosen such that the quadratic term in the potential takes the form

$$V^{(2)} = \frac{1}{2} \sum_i m_i^2 \xi'^i \xi'^i,$$

i.e. so that the matrix $(O^T M O)$ is diagonal. Such a matrix O always exists; it follows from (5.21) that in the model in question, two eigenvalues m_i^2 of the matrix M are equal to zero.

Problem 5. *Consider the theory of three complex scalar fields $f_i(x)$, $i = 1, 2, 3$, with Lagrangian*

$$\mathcal{L} = \partial_\mu f_i^* \partial_\mu f_i + \mu^2 f_i^* f_i - \lambda (f_i^* f_i)^2.$$

a) Find the global symmetry group for this Lagrangian. b) Find the set of ground states of the model. Choosing one of them, find the unbroken subgroup. c) By considering perturbations about the ground state find the Nambu–Goldstone modes and the masses of the remaining perturbations. d) To which representations of the unbroken subgroup do the Nambu–Goldstone modes and massive modes belong?

5.4   General case. Goldstone's theorem

As the general case, we shall consider the theory with scalar fields which, for definiteness, we shall assume to be real (a complex field is equivalent to a pair of real fields). Suppose G is the global symmetry group of the Lagrangian; we shall restrict ourselves to the physically interesting case of a compact group G. We shall denote the set of scalar fields by $\varphi(x)$; for each x, the field $\varphi(x)$ takes values in the space of the unitary (generally speaking, reducible) representation $T(\omega)$ of the group G. We choose the Lagrangian in the form

$$\mathcal{L} = \frac{1}{2} (\partial_\mu \varphi, \partial_\mu \varphi) - V(\varphi), \qquad (5.27)$$

where (φ_1, φ_2) is the scalar product in the space of the fields. The unitarity of the representation $T(\omega)$ implies that the kinetic term in the Lagrangian is invariant under the action of the group G; for invariance of the potential term, we require

$$V(T(\omega)\varphi) = V(\varphi)$$

for all $\omega \in G$.

Suppose the minimum of the potential $V(\varphi)$ is non-trivial. We choose as the ground state the homogeneous field

$$\varphi(x) = \varphi_0.$$

The homogeneity of the field configuration of the ground state arises, as in the examples considered earlier, from the requirement of minimality (equality to zero) of the gradient terms in the energy; the value of φ_0 realizes a minimum of the potential, which we write symbolically as

$$\frac{\partial V}{\partial \varphi}(\varphi = \varphi_0) = 0.$$

Suppose further that H is the subgroup of the group G which is the stationary subgroup of the classical vacuum φ_0, i.e.

$$T(h)\varphi_0 = \varphi_0 \tag{5.28}$$

for all $h \in H$. The fact that the set of all elements $h \in G$ satisfying (5.28) actually forms a subgroup of G follows from the basic properties of group representations: indeed, for all $h, h_1, h_2 \in H$ (see also Section 3.1)

$$T(h_1 h_2)\varphi_0 = T(h_1)T(h_2)\varphi_0 = T(h_1)\varphi_0 = \varphi_0,$$

and from

$$T(h^{-1})T(h)\varphi_0 = \varphi_0$$

it follows that

$$T(h^{-1})\varphi_0 = \varphi_0,$$

i.e. $h_1 h_2$ and h^{-1} belong to H. We shall call H the unbroken subgroup for the model (5.27).

Let t_h be generators of the subgroup H. Since $(1 + \varepsilon^h t_h)$ are close to the unit element of H, the following holds for them

$$T(1 + \varepsilon^h t_h)\varphi_0 = \varphi_0. \tag{5.29}$$

On the other hand, by the definition of the representation of the algebra $T(1 + \varepsilon^h t_h) = 1 + \varepsilon^h T(t_h)$, and introducing the notation $T_h = T(t_h)$ for the representation of the generators, from (5.29) we have

$$T_h \varphi_0 = 0.$$

We divide the generators of the group G into two families $\{t_h\}$ and $\{t'_\alpha\}$, where t_h are the generators of the group H and t'_α supplement the family

$\{t_h\}$ to form a full orthonormalized set. If R_G and R_H are the dimensions of the groups G and H, respectively, then the family $\{t_h\}$ consists of R_H generators, and the family $\{t'_\alpha\}$ of $(R_G - R_H)$ generators. We note that any element A of the Lie algebra, annihilating the classical vacuum, i.e.

$$T(A)\varphi_0 = 0,$$

is a linear combination of the generators t_h:

$$A = a_h t_h$$

(any element of G near the unit element which has the form $(1 + \varepsilon A)$ and leaves vacuum invariant, by definition belongs to H). Thus, none of the generators t'_α or their linear combinations annihilate the vacuum,

$$T(c^\alpha t'_\alpha)\varphi_0 \neq 0 \qquad (5.30)$$

for non-zero c^α. Generators of type t'_α will be called broken generators.

Let us now consider perturbations of the field $\varphi(x)$ about the ground state φ_0, i.e. we write

$$\varphi(x) = \varphi_0 + \chi(x)$$

where the $\chi(x)$ are new dynamical variables. The Lagrangian for the fields $\chi(x)$ has the form

$$\mathcal{L}_\chi(\chi) = \frac{1}{2}(\partial_\mu\chi, \partial_\mu\chi) - V(\varphi_0 + \chi).$$

Let us show, in the first place, that $\mathcal{L}_\chi$ is invariant under the global group H. Let h be any element of the group H. We need to show that

$$\mathcal{L}_\chi(T(h)\chi) = \mathcal{L}_\chi(\chi). \qquad (5.31)$$

Since

$$\mathcal{L}_\chi(\chi) = \mathcal{L}(\varphi_0 + \chi) \qquad (5.32)$$

we can write

$$\mathcal{L}_\chi(T(h)\chi) = \mathcal{L}(\varphi_0 + T(h)\chi). \qquad (5.33)$$

Moreover, by virtue of the fact that $T(h)\varphi_0 = \varphi_0$ and the linearity of the operator $T(h)$ we have

$$\mathcal{L}(\varphi_0 + T(h)\chi) = \mathcal{L}(T(h)(\varphi_0 + \chi)). \qquad (5.34)$$

Furthermore, $\mathcal{L}(\varphi)$ is invariant under the whole group G and, in particular, under its subgroup H. Consequently,

$$\mathcal{L}(T(h)(\varphi_0 + \chi)) = \mathcal{L}(\varphi_0 + \chi). \qquad (5.35)$$

The chain of equations (5.33), (5.34) and (5.35) proves the desired relation (5.31).

Among the perturbations $\chi(x)$ one can identify those which have a structure of the form

$$\chi_\alpha(x) = \theta_\alpha(x) T'_\alpha \varphi_0 \tag{5.36}$$

(no summation with respect to α), where $\alpha = 1, \ldots, R_G - R_H$; and the $\theta_\alpha(x)$ are $(R_G - R_H)$ real scalar fields. As always, $T'_\alpha = T(t'_\alpha)$ are broken generators in the representation T. By virtue of (5.30), these perturbations are linearly independent, i.e. the $\theta_\alpha(x)$ are independent fields. *Goldstone's theorem* says that the fields $\theta_\alpha(x)$ are massless. To convince ourselves of its validity, we write an arbitrary perturbation in the form

$$\chi(x) = \theta_\alpha(x) T'_\alpha \varphi_0 + \eta(x), \tag{5.37}$$

where the field $\eta(x)$ comprises modes differing from θ_α (in (5.37) summation over α is understood). We have to show that the expansion of the potential

$$V(\varphi_0 + \theta_\alpha T'_\alpha \varphi_0 + \eta) \tag{5.38}$$

for small θ_α and η does not contain terms in $\theta_\alpha \theta_\beta$ or $\theta_\alpha \eta$, i.e. the quadratic part of the potential contains only $\eta(x)$ (since the classical vacuum is the minimum of $V(\varphi)$, (5.38) does not contain any terms linear in θ_α and η).

Let us consider the expression

$$V[T(g)(\varphi_0 + \eta)],$$

where g is an element of the group G of the form

$$g = 1 + \theta_\alpha t'_\alpha + B\theta^2$$

(terms of the order of θ^2 are written in symbolic form). By virtue of the invariance of the potential under the group G we have

$$V[T(g)(\varphi_0 + \eta)] = V(\varphi_0 + \eta). \tag{5.39}$$

On the other hand,

$$V[T(g)(\varphi_0 + \eta)] = V(\varphi_0 + \theta_\alpha T'_\alpha \varphi_0 + \eta + \tilde{B}\theta^2 + \theta_\alpha T'_\alpha \eta), \tag{5.40}$$

where we have omitted terms of cubic and higher order in the perturbations θ_α and η, terms of the order of θ^2 are again written in symbolic form. We expand the right-hand side of (5.40) into a series around the point $(\varphi_0 + \theta_\alpha T'_\alpha \varphi_0 + \eta)$:

$$
\begin{aligned}
V[T(g)(\varphi_0 + \eta)] = {} & V(\varphi_0 + \theta_\alpha T'_\alpha \varphi_0 + \eta) \\
& + \left[\frac{\partial V}{\partial \varphi}(\varphi_0 + \theta_\alpha T'_\alpha \varphi_0 + \eta) \right] (B\theta^2 + \theta_\alpha T'_\alpha \eta),
\end{aligned}
\tag{5.41}
$$

where we have again neglected terms of cubic and higher order and have written the second term in a very sketchy form. The second term, in reality, is equal to 0 in quadratic order in the fields: the potential has a minimum at the point φ_0, thus its derivative in (5.41) is at least linear in the perturbations, and is multiplied by a quadratic expression. Using (5.39), we obtain to quadratic order:

$$V(\varphi_0 + \theta_\alpha T'_\alpha \varphi_0 + \eta) = V(\varphi_0 + \eta),$$

so that the quadratic part of the potential does not contain the fields θ_α, as required.

Problem 6. *Show that the Nambu–Goldstone fields $\theta_\alpha(x)$ and the massive scalar fields $u_\alpha(x)$, which together form all possible excitations over the vacuum, can be chosen such that the quadratic part of the Lagrangian has the canonical form*

$$\mathcal{L} = \frac{1}{2} \sum_\alpha \partial_\mu \theta_\alpha \partial_\mu \theta_\alpha + \sum_\alpha \left(\frac{1}{2} \partial_\mu u_a \partial_\mu u_\alpha + \frac{m_a^2}{2} u_a u_a \right).$$

Thus, Goldstone's theorem asserts that in a theory with spontaneously broken global symmetry, there are at least[1] as many massless scalar (or pseudoscalar) fields, as there are broken generators. It also provides a technique for identifying explicitly massless fields in the set of all perturbations about a non-trivial vacuum φ_0 (formula (5.36)).

To end this chapter, we note that we have mainly considered potentials $V(\varphi)$ having the form of a polynomial of degree four in the fields. In classical theory, one might consider arbitrary functions $V(\varphi)$, compatible with invariance under the symmetry group of the model. The results of this chapter remain valid for any $V(\varphi)$ having non-trivial minima, where symmetry is broken spontaneously. As mentioned earlier, in quantum theory, there are constraints on the structure of the scalar potential $V(\varphi)$, depending on the space–time dimensionality.

---

[1] There may be more massless fields than the number of broken generators. Such a situation often arises in supersymmetric theories.

Chapter 6

Higgs Mechanism

6.1   Example of an Abelian model

In this chapter we consider the situation with a non-trivial ground state of a scalar field in models with gauge invariance (Anderson 1963, Englert and Brout 1964, Higgs 1964, Guralnik *et al.* 1964). We shall sometimes use the term "spontaneous symmetry breaking" in this case also, although, unlike in the models with global symmetry considered in Chapter 5, breaking of gauge invariance does not occur in reality. As the simplest example, we consider a model with $U(1)$ gauge symmetry. We choose the Lagrangian in the form

$$\mathcal{L} = -\frac{1}{4}F_{\mu\nu}F_{\mu\nu} + (D_\mu\varphi)^* D_\mu\varphi - [-\mu^2\varphi^*\varphi + \lambda(\varphi^*\varphi)^2], \qquad (6.1)$$

where φ is a complex scalar field, $F_{\mu\nu} = \partial_\mu A_\nu - \partial_\nu A_\mu$, $D_\mu\varphi = (\partial_\mu - ieA_\mu)\varphi$. From the beginning we choose the square of the mass to have negative sign in the scalar field potential

$$V(\varphi^*, \varphi) = -\mu^2\varphi^*\varphi + \lambda(\varphi^*\varphi)^2,$$

since that is our interest in this chapter. We recall that the Lagrangian (6.1) is invariant under the gauge transformations

$$A_\mu(x) \quad \rightarrow \quad A'_\mu(x) = A_\mu(x) + \frac{1}{e}\partial_\mu\alpha(x)$$

$$\varphi(x) \quad \rightarrow \quad \varphi'(x) = e^{i\alpha(x)}\varphi(x),$$

where $\alpha(x)$ is an arbitrary real function.

To find the ground state, we write down the energy functional for the fields A_μ and φ:

$$E(A_\mu, \varphi) = \int d^3x \left[\frac{1}{2}(F_{0i})^2 + \frac{1}{4}(F_{ij})^2 \right. \tag{6.2}$$

$$\left. + (D_0\varphi)^* D_0\varphi + (D_i\varphi)^* D_i\varphi + V(\varphi^*\varphi) \right].$$

The ground state is that configuration of the fields A_μ, φ which minimizes the energy. It is immediately clear that there is a functional freedom in the choice of the ground state: the energy $E(A_\mu, \varphi)$ is gauge invariant, thus if $(A_\mu^{(v)}, \varphi^{(v)})$ is a ground state, then $(A_\mu^{(v)} + \frac{1}{e}\partial_\mu\alpha, e^{i\alpha}\varphi^{(v)})$ is also a ground state for any function $\alpha(x)$. As in Chapter 5, we need to choose a ground state from this family of classical vacua and study excitations about it.

The first two terms in the integrand (6.2) are minimal (equal to zero), when the electric and magnetic fields are equal to zero, i.e. $A_\mu(x)$ is a pure gauge,

$$A_\mu = \frac{1}{e}\partial_\mu\alpha(x). \tag{6.3}$$

The third and fourth terms are minimal (equal to zero) when

$$D_\mu\varphi = (\partial_\mu - i\partial_\mu\alpha)\varphi = 0,$$

i.e.

$$\varphi(x) = e^{i\alpha(x)} \frac{1}{\sqrt{2}}\varphi_0, \tag{6.4}$$

where φ_0 does not depend on x (the factor $1/\sqrt{2}$ was introduced for convenience). The constant φ_0 is determined from minimization of the potential $V(\varphi^*, \varphi)$ and is equal to

$$\varphi_0 = \frac{\mu}{\sqrt{\lambda}}. \tag{6.5}$$

Thus, all possible ground states are determined by formulae (6.3), (6.4) and (6.5); as mentioned earlier, we need to choose one of these (it does not matter which one). Let us choose $\alpha = 0$, so that the vacuum configuration has the form

$$A_\mu^{(v)} = 0, \quad \varphi^{(v)} = \frac{1}{\sqrt{2}}\varphi_0. \tag{6.6}$$

Let us now consider excitations about the ground state. Excitations of the field A_μ are described by the vector potential itself and excitations of the scalar field by two real fields $\chi(x)$ and $\theta(x)$, such that

$$\varphi(x) = \frac{1}{\sqrt{2}}(\varphi_0 + \chi(x) + i\theta(x)). \tag{6.7}$$

We recall that in the analogous model with global $U(1)$ symmetry (Section 5.2) the field θ was a massless Nambu–Goldstone field and χ was massive.

Let us find the spectrum of small (linear) waves about the ground state (6.6). For this, we calculate the Lagrangian in terms of the fields A_μ, χ and θ to a quadratic order in these fields. We use the fact that to quadratic order

$$V(\varphi) = \mu^2 \chi^2$$

modulo an irrelevant additive constant (we performed this computation in Section 5.2), and also

$$D_\mu \varphi = \frac{1}{\sqrt{2}} (\partial_\mu \chi + i\partial_\mu \theta - ie\varphi_0 A_\mu)$$

modulo quadratic terms in the fields A_μ, χ and θ. Thus, the quadratic Lagrangian has the form

$$\mathcal{L}^{(2)} = -\frac{1}{4} F_{\mu\nu}^2 + \frac{1}{2} |\partial_\mu \chi + i\partial_\mu \theta - ie\varphi_0 A_\mu|^2 - \mu^2 \chi^2$$

(recall that $F_{\mu\nu}^2$ is quadratic in A_μ). Evaluating the square of the modulus, we obtain

$$\mathcal{L}^{(2)} = -\frac{1}{4} F_{\mu\nu}^2 + \frac{1}{2} (\partial_\mu \chi)^2 - \mu^2 \chi^2 + \frac{e^2 \varphi_0^2}{2} \left(A_\mu - \frac{1}{e\varphi_0} \partial_\mu \theta \right)^2. \tag{6.8}$$

We have encountered a somewhat unusual situation: the last term in (6.8) contains, in addition to the expressions A_μ^2 and $(\partial_\mu \theta)^2$, a cross term $A_\mu \partial_\mu \theta$. In order to bring the quadratic Lagrangian to canonical form (sum of the Lagrangians of the individual fields) we change the field variables: instead of the fields A_μ and θ, we introduce the fields

$$B_\mu = A_\mu - \frac{1}{e\varphi_0} \partial_\mu \theta$$

and θ. Then, the quadratic Lagrangian will have the form

$$\mathcal{L}^{(2)} = -\frac{1}{4} B_{\mu\nu}^2 + \frac{e^2 \varphi_0^2}{2} B_\mu B_\mu + \frac{1}{2} (\partial_\mu \chi)^2 - \mu^2 \chi^2, \tag{6.9}$$

where $B_{\mu\nu} = \partial_\mu B_\nu - \partial_\nu B_\mu$.

The Lagrangian (6.9) is the sum of the Lagrangian of the massive vector field B_μ with mass

$$m_V = e\varphi_0 = \frac{e}{\sqrt{\lambda}} \mu$$

and the Lagrangian of the massive scalar field χ with mass

$$m_\chi = \sqrt{2}\mu.$$

The field $\theta(x)$ does not appear in the Lagrangian at all: it need not satisfy any field equations, i.e. an arbitrary function of the coordinates $\theta(x)$ is an extremum of the action with respect to the field θ.

The most interesting things about the Lagrangian (6.9) are the appearance of *a mass for the vector field* and the disappearance of the field $\theta(x)$. The field $\theta(x)$ would be a Nambu–Goldstone field if the symmetry were global rather than gauge. Loosely speaking, the vector field has "eaten up" a Nambu–Goldstone field and has acquired a mass. Herein lies the essence of the Higgs mechanism. We stress that the massive vector field appeared in the theory with a gauge-invariant Lagrangian.

In addition to the vector field B_μ, the spectrum of the excitations includes the scalar field χ. We shall see that this always occurs in models where vector bosons acquire mass via the Higgs mechanism; this scalar field is called the Higgs field and the corresponding particle is the Higgs boson (the term "Higgs field" is also applied to the whole of the scalar field $\varphi(x)$, whose vacuum value is non-trivial).

Problem 1. *Choose the ground state in the form*

$$A_\mu^{(v)}(x) = \frac{1}{e}\partial_\mu\alpha(x)$$

$$\varphi^{(v)}(x) = \frac{1}{\sqrt{2}}e^{i\alpha(x)}\varphi_0,$$

where $\alpha(x)$ is a fixed function. By considering fields of the form

$$A_\mu = A_\mu^{(v)} + a_\mu$$

$$\varphi = \varphi^{(v)} + \xi,$$

where a_μ and ξ are small perturbations, show that the spectrum of linear waves is described by a Lagrangian equivalent to (6.9).

Let us make some remarks in connection with the above. First, we determine the properties of the fields B_μ and χ under gauge transformations. The gauge function $\alpha(x)$ is assumed to be small, such that if A_μ is small, then

$$A'_\mu = A_\mu + \frac{1}{e}\partial_\mu\alpha(x)$$

is also small, and may be treated in a linear approximation. We note that for small gauge transformations

$$\varphi \to \varphi' = \varphi + i\alpha\varphi. \tag{6.10}$$

To find the form of transformations for the fields χ and θ, we write

$$\varphi = \frac{1}{\sqrt{2}}(\varphi_0 + \chi + i\theta)$$

$$\varphi' = \frac{1}{\sqrt{2}}(\varphi_0 + \chi' + i\theta'),$$

and from (6.10) we obtain to linear order in χ, θ and α that

$$\chi' + i\theta' = (\chi + i\theta) + i\alpha\varphi_0$$

or

$$\begin{aligned} \chi' &= \chi \\ \theta' &= \theta + \alpha\varphi_0. \end{aligned} \qquad (6.11)$$

Thus, the field χ is gauge invariant, but the field $\theta(x)$ is shifted by $\alpha(x)\varphi_0$. The field B_μ is also gauge invariant: for gauge transformations we have

$$\begin{aligned} B'_\mu &= A'_\mu - \frac{1}{e\varphi_0}\partial_\mu\theta' = A_\mu + \frac{1}{e}\partial_\mu\alpha - \frac{1}{e\varphi_0}\partial_\mu\theta - \frac{1}{e\varphi_0}\varphi_0\partial_\mu\alpha \\ &= A_\mu - \frac{1}{e\varphi_0}\partial_\mu\theta = B_\mu. \end{aligned}$$

Thus,[1] the Lagrangian (6.9) contains only gauge-invariant variables. The fact that the field $\theta(x)$ is arbitrary is explained by formula (6.11); for a given $\theta(x)$, the function $\alpha(x)$ can be chosen to make the gauge-transformed field $\theta'(x)$ an arbitrary preassigned function.

Let us now compare the number of degrees of freedom in the theory with a trivial vacuum (i.e. $m^2 > 0$) and in the theory with the Higgs mechanism. In the theory with a trivial vacuum, there is a massless vector field (two physical degrees of freedom plus an arbitrary gauge function) and a complex massive scalar field (two degrees of freedom), making a total of four physical degrees of freedom (and one gauge function). In the theory with the Higgs mechanism, the massive vector field has three degrees of freedom (we recall that the field B_μ satisfies the condition $\partial_\mu B_\mu = 0$, and each of its components is subject to the Klein–Gordon equation), while the field χ has one degree of freedom, making a total of four physical degrees of freedom (plus one gauge function, or, what amounts to the same thing, an arbitrary function $\theta(x)$). Thus, the number of degrees of freedom is the same in these cases; however, for the Higgs mechanism, the vector field takes away one degree of freedom from the scalar field.

---

[1]This has only been shown here for small gauge functions $\alpha(x)$. The general case is contained in Problem 2.

At first glance, it seems that the unphysical degree of freedom of the model appears in the Lagrangian in terms of order greater than quadratic: if the field (6.7) is substituted into the Lagrangian (6.1), then the field $\theta(x)$ actually appears in the higher-order terms. However, the "superfluous" field can be removed completely by the change of the field variables.

Problem 2. *Find the full Lagrangian in terms of the fields A_μ, χ and θ. Find the change of the variables which reduces this Lagrangian to a Lagrangian of interacting massive vector field and scalar field. Find the transformation rule for these massive fields under general (not necessarily small) gauge transformations.*

That the unphysical field vanishes from the full Lagrangian for field excitations can be easily seen as follows. Since the vacuum value of the scalar field is non-zero, for fields sufficiently close to the classical vacuum, one can introduce the representation

$$\varphi(x) = \frac{1}{\sqrt{2}}\rho(x)e^{i\beta(x)}, \tag{6.12}$$

where $\rho(x)$ and $\beta(x)$ are new real fields (such a representation is possible for fields $\varphi(x)$ which are non-zero everywhere in space–time, whereas the fluctuations of phase should not reach π or $-\pi$). We note that for gauge transformations $\varphi'(x) = e^{i\alpha(x)}\varphi(x)$ we have

$$\rho' e^{i\beta'} = \rho e^{i\beta + i\alpha},$$

i.e. $\rho(x)$ is a gauge-invariant field, and $\beta(x)$ is shifted

$$\beta'(x) = \beta(x) + \alpha(x). \tag{6.13}$$

For the covariant derivative we have

$$D_\mu\varphi = \frac{1}{\sqrt{2}}[\partial_\mu\rho + i\rho(\partial_\mu\beta - eA_\mu)]e^{i\beta},$$

thus the Lagrangian in terms of A_μ, ρ and β has the form

$$\mathcal{L} = -\frac{1}{4}F_{\mu\nu}F_{\mu\nu} + \frac{1}{2}(\partial_\mu\rho)^2 - V(\rho) + \frac{1}{2}e^2\rho^2\left(A_\mu - \frac{1}{e}\partial_\mu\beta\right)^2, \tag{6.14}$$

where

$$V(\rho) = -\frac{\mu^2}{2}\rho^2 + \frac{\lambda}{4}\rho^4.$$

Let us introduce the field

$$B_\mu = A_\mu - \frac{1}{e}\partial_\mu\beta, \tag{6.15}$$

then the full Lagrangian will not contain the field β:

$$\mathcal{L} = -\frac{1}{4} B_{\mu\nu} B_{\mu\nu} + \frac{e^2}{2} \rho^2 B_\mu^2 + \frac{1}{2}(\partial_\mu \rho)^2 - V(\rho). \qquad (6.16)$$

The field B_μ is gauge invariant (this follows from (6.13) and (6.15)), so that the Lagrangian (6.16) contains only gauge-invariant fields. The ground state for such a Lagrangian is the configuration

$$B_\mu^{(v)} = 0$$
$$\rho^{(v)} = \frac{\mu}{\sqrt{\lambda}},$$

and the quadratic Lagrangian for small perturbations B_μ and χ, where

$$\rho = \rho^{(v)} + \chi,$$

has precisely the form (6.9). Thus, in terms of the variables ρ, β and B_μ, the full Lagrangian does not contain the unphysical field β and is the Lagrangian of gauge-invariant interacting massive vector and scalar fields. We note that in the case of small fields, β is related to θ via $\theta = \varphi_0 \beta$. We again stress that transition to fields $\rho(x)$ and $\theta(x)$ according to formula (6.12) is only possible if the vacuum value of the scalar field is non-zero, i.e. only when the Higgs mechanism is realized, and only when small perturbations of fields about the classical vacuum are considered (this last requirement is trivially satisfied for linear waves).

To end this section, we note that until now we have not fixed the gauge. The results of this section, however, could have been obtained more easily by fixing the gauge

$$\beta(x) = 0.$$

Indeed, it is always possible to satisfy this condition by choosing the gauge function $\alpha(x)$ in (6.13). This condition can be written in the form

$$\text{Im}\,\varphi = 0,$$

where $\text{Re}\,\varphi = \frac{1}{\sqrt{2}}\rho$, i.e. in the field $\varphi(x)$ only the physical degree of freedom remains. Such a gauge is called unitary. In a unitary gauge, B_μ coincides with A_μ and the Lagrangian (6.16) is obtained immediately if in the original Lagrangian (6.1) one sets

$$\text{Im}\,\varphi = 0$$
$$\text{Re}\,\varphi = \frac{1}{\sqrt{2}}\rho.$$

Such a choice of gauge is convenient for analysis of the spectrum of physical perturbations in non-Abelian theories.

6.2   Non-Abelian case: model with complete breaking of $SU(2)$ symmetry

As the simplest model with non-Abelian gauge invariance, we choose a model with gauge group $SU(2)$ and a scalar-field doublet (fundamental representation)

$$\varphi = \begin{pmatrix} \varphi_1 \\ \varphi_2 \end{pmatrix}, \tag{6.17}$$

where φ_1, φ_2 are complex scalar fields. The Lagrangian has the form ($a = 1, 2, 3$)

$$\mathcal{L} = -\frac{1}{4} F^a_{\mu\nu} F^a_{\mu\nu} + (D_\mu \varphi)^\dagger D_\mu \varphi - [-\mu^2 \varphi^\dagger \varphi + \lambda(\varphi^\dagger \varphi)^2]. \tag{6.18}$$

The ground state is the minimum of the energy functional

$$E(A^a_\mu, \varphi) = \int d^3x \left[\frac{1}{2} F^a_{0i} F^a_{0i} + \frac{1}{4} F^a_{ij} F^a_{ij} + (D_0\varphi)^\dagger D_0\varphi + (D_i\varphi)^\dagger D_i\varphi + V(\varphi^\dagger, \varphi) \right],$$

where

$$V(\varphi^\dagger, \varphi) = -\mu^2 \varphi^\dagger \varphi + \lambda(\varphi^\dagger \varphi)^2 \tag{6.19}$$

is the potential of the scalar fields.

In the ground state

$$F^a_{\mu\nu} = 0,$$

thus A^a_μ is a pure gauge; in matrix form

$$A_\mu(x) = \omega(x) \partial_\mu \omega^{-1}(x). \tag{6.20}$$

The field $\varphi(x)$ is covariantly constant,

$$D_\mu \varphi(x) = 0,$$

i.e.

$$\varphi(x) = \omega(x) \varphi^{(v)}, \tag{6.21}$$

where $\varphi^{(v)}$ is a constant column. It is determined from minimization of the potential $V(\varphi^\dagger, \varphi)$. The requirement

$$\frac{\partial V}{\partial \varphi^\dagger} = \frac{\partial V}{\partial \varphi} = 0$$

gives

$$\varphi^\dagger \varphi = \frac{\mu^2}{2\lambda}. \tag{6.22}$$

From the family of gauge-equivalent vacua, we choose one; for simplicity, we take it in the form

$$A_\mu^a = 0 \tag{6.23}$$

$$\varphi^{(v)} = \begin{pmatrix} 0 \\ \frac{1}{\sqrt{2}}\varphi_0 \end{pmatrix},$$

where

$$\varphi_0 = \frac{\mu}{\sqrt{\lambda}}.$$

Problem 3. *Show that any configuration, satisfying (6.20), (6.21) and (6.22), can be obtained from the classical vacuum (6.23) by a gauge transformation (generally speaking, not decreasing at infinity).*

To find the spectrum of small perturbations about the ground state (6.23), we use the technique described at the end of Section 6.1, namely, we fix a unitary gauge. An arbitrary field close to $\varphi^{(v)}$ can be represented in the form

$$\varphi(x) = \omega(x)\begin{pmatrix} 0 \\ \frac{1}{\sqrt{2}}(\varphi_0 + \chi(x)) \end{pmatrix}, \tag{6.24}$$

where $\chi(x)$ is a real function of the coordinates and $\omega(x)$ is a function close to unity with values in $SU(2)$. In order to check the legitimacy of the representation (6.24), we note that the left-hand side is equal to

$$\varphi(x) = \begin{pmatrix} \xi_1(x) + i\xi_2(x) \\ \frac{1}{\sqrt{2}}\varphi_0 + \xi_3(x) + i\xi_4(x) \end{pmatrix}, \tag{6.25}$$

where $\xi_1(x), \ldots, \xi_4(x)$, are four small real functions. For $\omega(x)$ close to unity we have

$$\omega(x) = 1 + i\tau^a u^a(x),$$

where the $u^a(x)$ are small. The right-hand side of (6.24) equals, in a linear approximation in u^a and χ,

$$\begin{pmatrix} 0 \\ \frac{1}{\sqrt{2}}\varphi_0 \end{pmatrix} + \begin{pmatrix} 0 \\ \frac{1}{\sqrt{2}}\chi \end{pmatrix} + i\tau^a u^a \begin{pmatrix} 0 \\ \frac{1}{\sqrt{2}}\varphi_0 \end{pmatrix} = \begin{pmatrix} \frac{i\varphi_0}{\sqrt{2}}u^1(x) + \frac{\varphi_0}{\sqrt{2}}u^2(x) \\ \frac{1}{\sqrt{2}}\varphi_0 + \frac{1}{2}\chi(x) - \frac{i\varphi_0}{\sqrt{2}}u^3(x) \end{pmatrix}.$$

Comparing this expression with (6.25), we see that u^a and χ can be chosen so that equation (6.24) is satisfied to linear order in the deviations of the field from the vacuum value.

Representation (6.24) is valid in some finite neighborhood of the classical vacuum, however, it does not hold near $\varphi = 0$.

From the representation (6.24) it is clear that any configuration of the field $\varphi(x)$ close to the classical vacuum is gauge equivalent to the configuration

$$\varphi(x) = \begin{pmatrix} 0 \\ \frac{1}{\sqrt{2}}(\varphi_0 + \chi(x)) \end{pmatrix}, \tag{6.26}$$

where $\chi(x)$ is a real field. Thus, the gauge can be chosen where the field has the form (6.26) (unitary gauge).

We use the unitary gauge to find the structure of linear perturbations in the model. For this, we identify the linear parts in

$$F_{\mu\nu}^a = \partial_\mu A_\nu^a - \partial_\nu A_\mu^a + g\varepsilon^{abc} A_\mu^b A_\nu^c$$

$$D_\mu \varphi = \partial_\mu \varphi - ig\frac{\tau^a}{2} A_\mu^a \varphi, \tag{6.27}$$

assuming that the field A_μ^a is small and that the field $\varphi(x)$ has the form (6.26) with small $\chi(x)$. To linear order, we have

$$F_{\mu\nu}^a \simeq \mathcal{F}_{\mu\nu}^a, \tag{6.28}$$

where

$$\mathcal{F}_{\mu\nu}^a = \partial_\mu A_\nu^a - \partial_\nu A_\mu^a.$$

Furthermore, substituting (6.26) in (6.27) we obtain to linear order

$$D_\mu \varphi = \begin{pmatrix} 0 \\ \frac{1}{\sqrt{2}}\partial_\mu\chi \end{pmatrix} - ig\frac{\tau^a}{2} A_\mu^a \begin{pmatrix} 0 \\ \frac{1}{\sqrt{2}}\varphi_0 \end{pmatrix} = \begin{pmatrix} -i\frac{g}{2\sqrt{2}}\varphi_0 A_\mu^1 - \frac{g}{2\sqrt{2}}\varphi_0 A_\mu^2 \\ \frac{1}{\sqrt{2}}\partial_\mu\chi + i\frac{g}{2\sqrt{2}}\varphi_0 A_\mu^3 \end{pmatrix}. \tag{6.29}$$

Finally, to quadratic order in χ, the potential (6.19) equals, modulo an irrelevant additive constant,

$$V = \mu^2 \chi^2. \tag{6.30}$$

Using equations (6.28), (6.29) and (6.30), we write the quadratic part of the Lagrangian (6.18) in the unitary gauge

$$\mathcal{L} = -\frac{1}{4}\mathcal{F}_{\mu\nu}^a \mathcal{F}_{\mu\nu}^a + \frac{g^2\varphi_0^2}{8} A_\mu^a A_\mu^a + \frac{1}{2}(\partial_\mu\chi)^2 - \mu^2\chi^2. \tag{6.31}$$

Thus, as a result of the Higgs mechanism the model has three massive vector fields A_μ^a, $a = 1, 2, 3$ with the same mass

$$m_V = \frac{g\varphi_0}{2},$$

and one massive scalar field χ (Higgs boson field) with mass

$$m_\chi = \sqrt{2}\mu = \sqrt{2\lambda}\varphi_0.$$

We note that if the symmetry were global rather than gauge, the model would have three Nambu–Goldstone bosons (components ξ_1, ξ_2, ξ_4 in formula (6.25)), according to the number of broken generators (the group $SU(2)$ is completely broken, the number of broken generators is three). These Goldstone bosons are "eaten up" in gauge theory by the three vector bosons, which become massive. We note that when the square of the mass is positive ($\mu^2 < 0$ in the Lagrangian (6.18)) the model has three massless vector fields (six degrees of freedom) and four massive real scalar fields (the real and imaginary components of the fields φ_1 and φ_2 in (6.17), in total four degrees of freedom). As a result of the Higgs mechanism, for $\mu^2 > 0$, the three vector bosons have nine degrees of freedom and the Higgs boson χ has one. Thus, the model has 10 degrees of freedom, in both the unbroken and the broken phase.

Problem 4. *Find the spectrum of linear physical excitations in the model of this section, without fixing the gauge.*

Problem 5. *Find the full Lagrangian of the fields A_μ^a and χ in the model of this section in the unitary gauge.*

Problem 6. *In the model of this section, the field φ can be written in the form*

$$\varphi = \begin{pmatrix} \eta_1 + i\eta_2 \\ u + i\eta_3 \end{pmatrix},$$

where $u, \eta_1, \eta_2, \eta_3$ are real scalar fields. Show that the scalar potential is invariant, not only under the original group $SU(2)$, but also under the global $SO(3)$ symmetry, where u is an $SO(3)$ singlet, and η_a is an $SO(3)$ triplet (vector). Choose a transformation rule for the vector fields under this group $SO(3)$, for which the full Lagrangian (6.18) is invariant under this global group $SO(3)$. The vacuum value of the field (6.23) does not break this global $SO(3)$ symmetry. Show that the equality of the masses of the three vector fields discussed in connection with formula (6.31) is a consequence of the unbroken global $SO(3)$ symmetry. Show that the full Lagrangian obtained in the previous problem is invariant under this $SO(3)$.

Problem 7. *Let us introduce into the theory with gauge group $SU(2)$, in addition to the doublet φ, three extra real scalar fields $f^a(x)$, $a = 1, 2, 3$,* *forming a triplet under the gauge group $SU(2)$. Choose a gauge-invariant scalar potential such that one of the vacua of the model is*

$$\varphi = \begin{pmatrix} 0 \\ \varphi_0/\sqrt{2} \end{pmatrix}$$
$$f^1 = f^2 = 0$$
$$f^3 = v,$$

where φ_0 and v are some constants. Find the spectrum of small physical excitations about this vacuum. Are the masses of the three vector bosons equal? Does the model with a triplet have unbroken global symmetry (not necessarily $SO(3)$), analogous to that considered in the previous problem?

Problem 8. *Construct a model with fully broken $SU(2)$ gauge symmetry, in which the masses of all three vector bosons are different.*

6.3 Example of partial breaking of gauge symmetry: bosonic sector of standard electroweak theory

Like global symmetry, gauge symmetry may be only partially broken.

We recall again, that the term "spontaneous symmetry breaking" is very loose as far as gauge symmetry is concerned; for example, the Lagrangian of an Abelian Higgs model, written in terms of deviations of the fields from the vacuum values, is still gauge invariant. In the case of a quadratic approximation, this is evident from formula (6.9): the Lagrangian $\mathcal{L}^{(2)}$ is invariant under the transformations (6.11), since it is written in terms of the gauge-invariant variables B_μ and χ. In the general case, this is clear from the expression for the full Lagrangian (6.14) (or (6.16)), which is invariant under the transformations (6.13) and under the corresponding transformation of the field $A_\mu(x)$. Here, and in what follows, the term "broken symmetry" will be applied to that which would actually be broken if the original symmetry were global rather than gauge.

The general situation is the following. Let G be a gauge group, $(A_\mu^{(v)}, \varphi^{(v)})$ a ground state. Without loss of generality, we may suppose that

$$A_\mu^{(v)} = 0,$$

and that $\varphi^{(v)}$ does not depend on the coordinates; here, the superscript (v) denotes the field vacuum values. Let us for a moment switch off gauge fields, then the group G will be a global symmetry group. The vacuum value $\varphi^{(v)}$ breaks G down to some subgroup H; then, the generators of the algebra of G can be divided into unbroken generators $\{t_h\}$ (which are the generators of the algebra of H) and broken generators $\{t'_\alpha\}$, as we saw in Section 5.4. If the group G were a global symmetry group, then the model would contain massless Nambu–Goldstone fields, in number equal to the number of broken generators.

If the group G is a gauge group, then the gauge fields corresponding to the subgroup H remain massless, and the H gauge symmetry is realized as discussed in Chapter 4. In other words, the model retains an explicit H gauge symmetry. The gauge fields A^α_μ corresponding to the broken generators t'_α become massive. Furthermore, the Nambu–Goldstone bosons corresponding to the generators t'_α disappear from the spectrum.

Problem 9. *Prove the above statements in the general case of a compact gauge group G.*

We shall illustrate this situation on the example of the standard electroweak theory of Glashow–Weinberg–Salam (more precisely, its bosonic sector since we do not include fermion fields in our analysis). The gauge group of this theory is the group $SU(2) \times U(1)$. As mentioned in Chapter 4, each of the two components, $SU(2)$ and $U(1)$, has its own gauge coupling constant; we shall denote these by g and g', respectively. Let the field A^a_μ ($a = 1, 2, 3$) be the gauge field of the group $SU(2)$, and B_μ be the gauge field of the group $U(1)$. The model includes one scalar-field doublet (with respect to $SU(2)$) φ, with $U(1)$ charge $\frac{1}{2}$. The $U(1)$ charge is often called the weak hypercharge Y; thus, the Higgs doublet has $Y = \frac{1}{2}$. These properties (together with the requirement that the scalar-field potential should be a fourth-order polynomial in φ) determine the Lagrangian of the theory uniquely:

$$\mathcal{L} = -\frac{1}{4} F^a_{\mu\nu} F^a_{\mu\nu} - \frac{1}{4} B_{\mu\nu} B_{\mu\nu} + (D_\mu \varphi)^\dagger D_\mu \varphi - \lambda \left(\varphi^\dagger \varphi - \frac{v^2}{2} \right)^2,$$

where

$$\begin{aligned} F^a_{\mu\nu} &= \partial_\mu A^a_\nu - \partial_\mu A^a_\mu + g\varepsilon^{abc} A^b_\mu A^c_\nu \\ B_{\mu\nu} &= \partial_\mu B_\nu - \partial_\nu B_\mu, \end{aligned}$$

and the covariant derivative of the field φ is equal to

$$D_\mu \varphi = \partial_\mu \varphi - i\frac{g}{2} \tau^a A^a_\mu \varphi - i\frac{g'}{2} B_\mu \varphi, \tag{6.32}$$

where the matrices τ^a act on the column

$$\varphi = \begin{pmatrix} \varphi_1 \\ \varphi_2 \end{pmatrix}.$$

As the ground state, we choose

$$A_\mu^a = B_\mu = 0$$

$$\varphi = \begin{pmatrix} 0 \\ v/\sqrt{2} \end{pmatrix} \equiv \varphi^{(v)}. \tag{6.33}$$

The generators of the group $SU(2) \times U(1)$ are matrices $T^a = \frac{\tau^a}{2}$ and $Y = \frac{1}{2}$ (we take the generators to be Hermitian). Let us switch off the gauge fields and find the broken and unbroken generators. The unbroken generators Q are Hermitian matrices such that

$$Q\varphi^{(v)} = 0. \tag{6.34}$$

For the specific choice (6.33), the condition (6.34) means that T has the form

$$Q = \begin{pmatrix} a & 0 \\ b & 0 \end{pmatrix},$$

and from the Hermiticity we have

$$Q = \begin{pmatrix} 1 & 0 \\ 0 & 0 \end{pmatrix}$$

or

$$Q = T^3 + Y. \tag{6.35}$$

Thus, there is an unbroken subgroup with a single generator Q: this is the subgroup $U(1)_{\text{e.m.}}$ of $SU(2) \times U(1)$. We expect this subgroup to correspond to a massless gauge field, which will be identified with the electromagnetic field. We stress that $U(1)_{\text{e.m.}}$ does not coincide with the factor $U(1)$ in $SU(2) \times U(1)$; in other words, the electromagnetic potential, which we shall denote by A_μ (without superscript) is a linear combination of the fields A_μ^a and B_μ (from (6.35), it is clear that A_μ is actually a linear combination of A_μ^3 and B_μ).

Let us consider small (linear) perturbations of the fields about the vacuum (6.33). As in Section 6.2, we use the unitary gauge, in which

$$\varphi = \begin{pmatrix} 0 \\ \frac{v}{\sqrt{2}} + \frac{\chi}{\sqrt{2}} \end{pmatrix}, \tag{6.36}$$

where $\chi(x)$ is a real scalar field. In the case under study, the unitary gauge fixes the gauge freedom only partially: gauge transformations from $U(1)_{\text{e.m.}}$ do not change the field (6.36), so that the full Lagrangian in the unitary gauge remains invariant under the gauge group $U(1)_{\text{e.m.}}$.

To find the quadratic Lagrangian, we need to calculate the covariant derivative of the field φ. In the unitary gauge,

$$
\begin{aligned}
D_\mu \varphi = \partial_\mu \varphi + \Bigg[&-\frac{ig}{2} A_\mu^1 \begin{pmatrix} 0 & 1 \\ 1 & 0 \end{pmatrix} - \frac{ig}{2} A_\mu^2 \begin{pmatrix} 0 & -i \\ i & 0 \end{pmatrix} \\
&-\frac{ig}{2} A_\mu^3 \begin{pmatrix} 1 & 0 \\ 0 & -1 \end{pmatrix} - \frac{ig'}{2} B_\mu \begin{pmatrix} 1 & 0 \\ 0 & 1 \end{pmatrix} \Bigg] \varphi.
\end{aligned}
$$

We obtain

$$
D_\mu \varphi = \begin{pmatrix} -\frac{ig}{2\sqrt{2}} (A_\mu^1 - i A_\mu^2)(v + \chi) \\ -\frac{i}{2\sqrt{2}} (g' B_\mu - g A_\mu^3)(v + \chi) + \frac{1}{\sqrt{2}} \partial_\mu \chi \end{pmatrix}. \tag{6.37}
$$

We introduce the complex fields

$$
W_\mu^\pm = \frac{1}{\sqrt{2}} (A_\mu^1 \mp i A_\mu^2)
$$

such that $(W_\mu^-)^* = W_\mu^+$. We also introduce two real fields

$$
Z_\mu = \frac{1}{\sqrt{g^2 + g'^2}} (g A_\mu^3 - g' B_\mu) \tag{6.38}
$$

$$
A_\mu = \frac{1}{\sqrt{g^2 + g'^2}} (g B_\mu + g' A_\mu^3). \tag{6.39}
$$

The fields Z_μ and A_μ are chosen so that the covariant derivative (6.37) contains only the field Z_μ and the following property is satisfied:

$$
Z_\mu^2 + (A_\mu)^2 = (A_\mu^3)^2 + B_\mu^2. \tag{6.40}
$$

Thus, the covariant derivative (6.37) is equal to

$$
D_\mu \varphi = \begin{pmatrix} -i \frac{gv}{2} W_\mu^+ \\ \frac{1}{\sqrt{2}} \partial_\mu \chi + \frac{i\sqrt{g^2 + g'^2}}{2\sqrt{2}} v Z_\mu \end{pmatrix} + \begin{pmatrix} -i \frac{g}{2} W_\mu^+ \chi \\ \frac{i\sqrt{g^2 + g'^2}}{2\sqrt{2}} Z_\mu \chi \end{pmatrix}. \tag{6.41}
$$

Here, the first column is linear in the perturbations (the fields $W_\mu^\pm, \chi, Z_\mu$), while the second is quadratic. Thus, the contribution of the covariant derivative to the quadratic part of the Lagrangian is equal to

$$
[(D_\mu \varphi)^\dagger D_\mu \varphi]^{(2)} = \frac{1}{2} (\partial_\mu \chi)^2 + \frac{g^2 v^2}{4} W_\mu^+ W_\mu^- + \frac{1}{2} \left(\frac{(g^2 + g'^2) v^2}{4} \right) Z_\mu^2. \tag{6.42}
$$

We now consider the kinetic term of the vector fields in the quadratic Lagrangian. We obtain, to quadratic order

$$
-\frac{1}{4} \mathcal{F}_{\mu\nu}^a \mathcal{F}_{\mu\nu}^a - \frac{1}{4} B_{\mu\nu}^2 = -\frac{1}{2} \mathcal{W}_{\mu\nu}^+ \mathcal{W}_{\mu\nu}^- - \frac{1}{4} (\mathcal{F}_{\mu\nu}^3)^2 - \frac{1}{4} (B_{\mu\nu})^2. \tag{6.43}
$$

Here

$$\begin{aligned}
\mathcal{F}_{\mu\nu}^a &= \partial_\mu A_\nu^a - \partial_\nu A_\mu^a \\
\mathcal{W}_{\mu\nu}^\pm &= \partial_\mu W_\nu^\pm - \partial_\nu W_\mu^\pm.
\end{aligned}$$

Furthermore, using the property (6.40), we write, instead of (6.43),

$$-\frac{1}{2}\mathcal{W}_{\mu\nu}^+\mathcal{W}_{\mu\nu}^- - \frac{1}{4}\mathcal{Z}_{\mu\nu}\mathcal{Z}_{\mu\nu} - \frac{1}{4}F_{\mu\nu}F_{\mu\nu}, \tag{6.44}$$

where

$$\begin{aligned}
\mathcal{Z}_{\mu\nu} &= \partial_\mu Z_\nu - \partial_\nu Z_\mu \tag{6.45} \\
F_{\mu\nu} &= \partial_\mu A_\nu - \partial_\nu A_\mu.
\end{aligned}$$

Thus, the quadratic Lagrangian contains standard kinetic terms for the complex vector field $W_\mu^\pm$ and the real vector fields Z_μ and A_μ.

Finally, modulo an irrelevant additive constant, the quadratic part of the potential equals

$$\lambda v^2 \chi^2. \tag{6.46}$$

We introduce the notation

$$\begin{aligned}
m_W &= \frac{gv}{2} \\
m_Z &= \frac{\sqrt{g^2 + g'^2}\, v}{2} \\
m_\chi &= \sqrt{2\lambda}\, v.
\end{aligned}$$

Collecting together terms in (6.42), (6.44) and (6.46), we obtain the quadratic Lagrangian in the form

$$\begin{aligned}
\mathcal{L}^{(2)} = \; & -\frac{1}{2}\mathcal{W}_{\mu\nu}^+\mathcal{W}_{\mu\nu}^- + m_W^2 W_\mu^+ W_\mu^- \\
& -\frac{1}{4}F_{\mu\nu}F_{\mu\nu} \\
& -\frac{1}{4}\mathcal{Z}_{\mu\nu}\mathcal{Z}_{\mu\nu} + \frac{m_Z^2}{2}Z_\mu Z_\mu \\
& +\frac{1}{2}(\partial_\mu\chi)^2 - \frac{m_\chi^2}{2}\chi^2.
\end{aligned}$$

This Lagrangian describes the massive complex vector field W_μ^+ (here $W_\mu^- = (W_\mu^+)^*$) with mass m_W (W-boson field), a massless vector field (photon field A_μ), a massive real vector field with mass m_Z (Z-boson field) and a massive real scalar field χ (Higgs boson field). It is convenient to introduce the weak mixing angle θ_W, such that

$$\cos\theta_W = \frac{g}{\sqrt{g^2 + g'^2}}$$

$$\sin\theta_W = \frac{g'}{\sqrt{g^2 + g'^2}}.$$

The reason for this name comes from formulae (6.38) and (6.39), which take the form

$$Z_\mu = \cos\theta_W A_\mu^3 - \sin\theta_W B_\mu \qquad (6.47)$$
$$A_\mu = \cos\theta_W B_\mu + \sin\theta_W A_\mu^3,$$

i.e. the fields Z_μ and A_μ are "mixtures" of the fields A_μ^3 and B_μ. We note that the masses of the W- and Z-bosons are related by the equation

$$m_Z = \frac{m_W}{\cos\theta_W}. \qquad (6.48)$$

W- and Z-bosons have been detected experimentally; their masses are $m_W = 80$ GeV and $m_Z = 91$ GeV. The angle θ_W is measured independently by studying the interaction of photons and W- and Z-bosons with other particles (quarks and leptons); this is possible because θ_W is determined by the coupling constants g and g'. The experimental value of $\sin\theta_W$ is $\sin^2\theta_W = 0.23$. Thus, equation (6.48) is satisfied in nature with good accuracy.

Problem 10. *Let us introduce into the model additional real scalar fields, forming a triplet with respect to the group $SU(2)$. Select the scalar potential and the value for the hypercharge of the scalar-field triplet such that, as before, $SU(2) \times U(1)$ is broken down to $U(1)_{\text{e.m.}}$. Find the masses of the W- and Z-bosons. Is equation (6.48) satisfied?*

Of course, the full Lagrangian for the fields $W_\mu^\pm, Z_\mu, A_\mu$ and χ contains nonlinear terms describing the field interactions. Let us determine, in particular, the interaction of all the fields with the electromagnetic field. We expect the full Lagrangian to be invariant under the gauge group $U(1)_{\text{e.m.}}$, since the vacuum is invariant under $U(1)_{\text{e.m.}}$. This can be seen by direct calculation.

Since the covariant derivative (6.37) contains only the fields $W_\mu^\pm$, Z_μ and χ (see formula (6.41)), and the potential $V(\varphi)$ never contains vector fields, the interaction with the electromagnetic field A_μ is contained solely in the term

$$-\frac{1}{4}F_{\mu\nu}^a F_{\mu\nu}^a$$

in the original Lagrangian. We write this term in terms of the fields $W_\mu^\pm$, Z_μ and A_μ. We have

$$
\begin{aligned}
F_{\mu\nu}^1 &= \partial_\mu A_\nu^1 - \partial_\nu A_\mu^1 + g\varepsilon^{123} A_\mu^2 A_\nu^3 + g\varepsilon^{132} A_\mu^3 A_\nu^2 \qquad (6.49) \\
&= \partial_\mu A_\nu^1 - g A_\mu^3 A_\nu^2 - (\mu \leftrightarrow \nu),
\end{aligned}
$$

where $(\mu \leftrightarrow \nu)$ denotes terms with interchange of indices. Analogously,

$$
F_{\mu\nu}^2 = \partial_\mu A_\nu^2 + g A_\mu^3 A_\nu^1 - (\mu \leftrightarrow \nu). \qquad (6.50)
$$

We form the combination

$$
W_{\mu\nu}^\pm = \frac{F_{\mu\nu}^1 \mp i F_{\mu\nu}^2}{\sqrt{2}}. \qquad (6.51)
$$

From (6.49) and (6.50) we obtain

$$
W_{\mu\nu}^\pm = \partial_\mu W_\nu^\pm \mp ig A_\mu^3 W_\nu^\pm - (\mu \leftrightarrow \nu).
$$

It follows from (6.47) that the field A_μ^3 is expressed in terms of the Z-boson and photon fields as follows

$$
A_\mu^3 = \cos\theta_W Z_\mu + \sin\theta_W A_\mu. \qquad (6.52)
$$

Thus

$$
W_{\mu\nu}^\pm = \partial_\mu W_\nu^\pm \mp ig\sin\theta_W A_\mu W_\nu^\pm \mp ig\cos\theta_W Z_\mu W_\nu^\pm - (\mu \leftrightarrow \nu).
$$

Let us introduce the further notations

$$
e = g\sin\theta_W = \frac{gg'}{\sqrt{g^2 + g'^2}} \qquad (6.53)
$$

and

$$
\mathcal{D}_\mu W_\nu^\pm = (\partial_\mu \mp ieA_\mu)W_\nu^\pm.
$$

Then the field strength $W_{\mu\nu}^\pm$ will have the form

$$
W_{\mu\nu}^\pm = \mathcal{D}_\mu W_\nu^\pm - \mathcal{D}_\nu W_\mu^\pm \mp ig\cos\theta_W(Z_\mu W_\nu^\pm - Z_\nu W_\mu^\pm). \qquad (6.54)
$$

It is clear that this expression transforms covariantly under $U(1)_{\text{e.m.}}$ transformations of the form

$$
W_\mu^\pm \quad \rightarrow \quad W_\mu'^\pm = e^{\pm i\alpha} W_\mu^\pm \qquad (6.55)
$$

$$
A_\mu \quad \rightarrow \quad A_\mu' = A_\mu + \frac{1}{e}\partial_\mu\alpha \qquad (6.56)
$$

$$
Z_\mu \quad \rightarrow \quad Z_\mu' = Z_\mu, \qquad (6.57)
$$

where $\alpha(x)$ is an arbitrary real function: for these transformations $\mathcal{D}_\mu W_\mu^\pm$ is the covariant derivative of the fields $W_\nu^\pm$, so that $W_{\mu\nu}^\pm$ transforms as

$$W_{\mu\nu}^\pm \to W_{\mu\nu}'^\pm = e^{\pm i\alpha} W_{\mu\nu}^\pm.$$

The constant e is identified with the unit of electric charge and the fields W_μ^+ and W_μ^- have charges $(+1)$ and (-1), respectively.

Let us now consider the third component of the field strength $F_{\mu\nu}^a$. We have

$$
\begin{aligned}
F_{\mu\nu}^3 &= \partial_\mu A_\nu^3 - \partial_\nu A_\mu^3 + g\varepsilon^{312} A_\mu^1 A_\nu^2 + g\varepsilon^{321} A_\mu^2 A_\nu^1 \\
&= \partial_\mu A_\nu^3 - \partial_\nu A_\mu^3 + g(A_\mu^1 A_\nu^2 - A_\mu^2 A_\nu^1).
\end{aligned}
$$

Again using expression (6.52), and expressing $A_\mu^{1,2}$ in terms of the fields $W_\mu^\pm$, we obtain

$$F_{\mu\nu}^3 = F_{\mu\nu} \sin\theta_W + \mathcal{Z}_{\mu\nu} \cos\theta_W + ig(W_\mu^- W_\nu^+ - W_\mu^+ W_\nu^-), \qquad (6.58)$$

where $F_{\mu\nu}$ and $\mathcal{Z}_{\mu\nu}$ are defined by formulae (6.45). This expression is invariant under the transformations (6.55), (6.56), (6.57). Thus, we obtain from (6.54) and (6.58) terms of the Lagrangian containing electromagnetic interactions:

$$
\begin{aligned}
-\frac{1}{4} F_{\mu\nu}^a F_{\mu\nu}^a &= -\frac{1}{2}|W_{\mu\nu}^-|^2 - \frac{1}{4}(F_{\mu\nu}^3)^2 \\
&= -\frac{1}{2}|\mathcal{D}_\mu W_\nu^- + ig\cos\theta_W Z_\mu W_\nu^- - (\mu \leftrightarrow \nu)|^2 \\
&\quad -\frac{1}{4}[F_{\mu\nu}\sin\theta_W + \mathcal{Z}_{\mu\nu}\cos\theta_W \\
&\quad + ig(W_\mu W_\nu^\dagger - W_\mu^+ W_\nu^-)]^2. \qquad (6.59)
\end{aligned}
$$

Thus, the Lagrangian of the fields describing deviations from the non-trivial vacuum indeed has an explicit $U(1)_{\text{e.m.}}$ gauge symmetry, and the interaction of the fields with the electromagnetic field A_μ is contained in the terms (6.59) in the Lagrangian.

We note that the simplest, most "minimal" way of ensuring $U(1)$ gauge invariance involves replacing the conventional derivative ∂_μ in the initial Lagrangian for the complex fields by the covariant derivative $\mathcal{D}_\mu = \partial_\mu - ieA_\mu$. In this way, we have constructed a gauge-invariant Lagrangian describing the interaction of the complex scalar field with the electromagnetic field (Section 2.7). Application of an analogous procedure to the Lagrangian of a massive vector field (Section 2.3) would lead to a minimal interaction of the complex vector field C_μ with the electromagnetic field A_μ:

$$\mathcal{L} = -\frac{1}{2}|\mathcal{D}_\mu C_\nu - \mathcal{D}_\nu C_\mu|^2 + m_C^2 C_\mu^* C_\mu - \frac{1}{4}F_{\mu\nu}^2.$$

Unlike this Lagrangian, the interaction with the electromagnetic field which emerges in the theory with $SU(2) \times U(1)$ broken down to $U(1)_{\text{e.m.}}$, is non-minimal: the contribution (6.59) contains terms of the form

$$F_{\mu\nu} W_\mu^- W_\nu^+$$
$$\mathcal{D}_\mu W_\nu^- \cdot W_\mu^+ Z_\mu. \tag{6.60}$$

Their structure, as well as the values of the corresponding interaction constants (numerical factors in front of the terms of (6.60)) are uniquely fixed. The interactions of W- and Z-bosons between themselves and with the Higgs field χ are also fixed, as is the self-interaction of the Higgs field. Thus, the model contains various interactions (nonlinear terms in the Lagrangian) between the fields $W_\mu^\pm, Z_\mu, A_\mu, \chi$; however, all the coupling constants and masses of the fields are expressed in terms of a small set of parameters, containing only dimensionless constants g, g' and λ and a single dimensionful parameter v.

Problem 11. *Determine the interaction of the fields $W_\mu^\pm$ with the field Z_μ and the Higgs field χ, which occurs owing to the presence of the term $(D_\mu \varphi)^\dagger D_\mu \varphi$ in the original Lagrangian. Show that this interaction is invariant under gauge transformations from $U(1)_{\text{e.m.}}$.*

The electric charges of the fields $W_\mu^\pm$, the connection of the Z-boson and photon fields with the original fields A_μ^a and B_μ (formula (6.47)) and the expression for the unit of electric charge (6.53) could be found from the transformation rule for the fields under the action of the group $U(1)_{\text{e.m.}}$, without writing down the Lagrangian of the interaction explicitly. For this, we recall that the ground state (6.33) is invariant under gauge transformations from the group $SU(2)$ with the gauge function

$$\omega(x) = \mathrm{e}^{i \frac{\tau^3}{2} \alpha(x)} \tag{6.61}$$

which at the same time are supplemented by a transformation from the group $U(1)$ with gauge function

$$\Omega(x) = \mathrm{e}^{i \alpha(x)}. \tag{6.62}$$

It is these gauge transformations (of the form $\omega(x)\Omega(x)$) that form the unbroken subgroup $U(1)_{\text{e.m.}}$. Under these gauge transformations, the Higgs field $\chi(x)$ does not transform (i.e. $\chi(x)$ is an electrically neutral field), while the fields A_μ^a and B_μ transform according to the general rule

$$\hat{A}_\mu \to \hat{A}_\mu' = \omega \hat{A}_\mu \omega^{-1} + \omega \partial_\mu \omega^{-1} \tag{6.63}$$

$$\hat{B}_\mu \to \hat{B}_\mu' = \hat{B}_\mu + \Omega \partial_\mu \Omega^{-1}, \tag{6.64}$$

where, as usual, $\hat{A}_\mu$ and $\hat{B}_\mu$ are linear combinations of the generators of the groups $SU(2)$ and $U(1)$, respectively (i.e. they belong to the corresponding algebras),

$$\hat{A}_\mu = -ig\frac{\tau^a}{2}A_\mu^a$$

$$\hat{B}_\mu = -ig'B_\mu.$$

Using the explicit form of the function $\omega(x)$ (formula (6.61)), from (6.63) we obtain

$$
\begin{aligned}
\hat{A}'_\mu &= -\frac{ig}{2}\left(\tau^1 A_\mu'^1 + \tau^2 A_\mu'^2 - \frac{ig}{2}\tau^3 A_\mu'^3\right)\\
&= -\frac{ig}{2}e^{i\frac{\tau^3}{2}\alpha}(\tau^1 A_\mu^1 + \tau^2 A_\mu^2)e^{-i\frac{\tau^3}{2}\alpha} - ig\frac{\tau^3}{2}A_\mu^3 - i\frac{\tau^3}{2}\partial_\mu\alpha\\
&= -\frac{ig}{2}[(\cos\alpha A_\mu^1 + \sin\alpha A_\mu^2)\tau^1 + (\cos\alpha A_\mu^2 - \sin\alpha A_\mu^1)\tau^2\\
&\quad -ig\frac{\tau^3}{2}\left(A_\mu^3 + \frac{1}{g}\partial_\mu\alpha\right).
\end{aligned}
$$

Hence we obtain the transformation rule for the fields $A_\mu^1, A_\mu^2, A_\mu^3$ under transformations from $U(1)_{\text{e.m.}}$:

$$
\begin{aligned}
A_\mu'^1 &= A_\mu^1\cos\alpha + A_\mu^2\sin\alpha, \qquad &(6.65)\\
A_\mu'^2 &= A_\mu^2\cos\alpha - A_\mu^1\sin\alpha
\end{aligned}
$$

and also

$$A_\mu'^3 = A_\mu^3 + \frac{1}{g}\partial_\mu\alpha. \qquad (6.66)$$

Moreover, using the explicit form for $\Omega(x)$ (formula (6.62)), from (6.64) we obtain the transformation rule for the field B_μ under transformations from $U(1)_{\text{e.m.}}$:

$$B_\mu' = B_\mu + \frac{1}{g'}\partial_\mu\alpha. \qquad (6.67)$$

Thus, the electromagnetic gauge group acts on the original fields according to the formulae (6.65), (6.66), (6.67).

Formulae (6.65) are equivalent to (6.55), i.e. the fields $W_\mu^\pm$ have electric charges (± 1). From the fields A_μ^3 and B_μ, one can construct the field Z_μ, which is invariant under electromagnetic gauge transformations. From (6.66) and (6.67) it is clear that this field has the form (up to normalization)

$$Z_\mu \propto gA_\mu^3 - g'B_\mu.$$

The orthogonal linear combination transforms non-trivially under electromagnetic gauge transformations; it is the electromagnetic field and has the form (again up to normalization)

$$A_\mu \propto g' A_\mu^3 + g B_\mu.$$

Taking into account the normalization condition (6.40), ensuring that the kinetic term in the free Lagrangian for the fields Z_μ and A_μ has the standard normalization, we obtain

$$A_\mu = \frac{1}{\sqrt{g^2 + g'^2}} (g' A_\mu^3 + g B_\mu), \qquad (6.68)$$

which agrees with (6.39). For Z_μ we obtain the expression (6.38).

The unit of electric charge e is obtained by comparing the standard transformation rule for the electromagnetic field

$$A_\mu \to A'_\mu = A_\mu + \frac{1}{e} \partial_\mu \alpha$$

with the transformation rule for the field (6.68), emanating from (6.64) and (6.63). The latter has the form

$$
\begin{aligned}
A'_\mu &= \frac{1}{\sqrt{g^2 + g'^2}} (g' A_\mu'^3 + g B'_\mu) \\
&= A_\mu + \frac{1}{\sqrt{g^2 + g'^2}} \left(\frac{g'}{g} + \frac{g}{g'} \right) \partial_\mu \alpha.
\end{aligned}
$$

Thus

$$\frac{1}{e} = \frac{1}{\sqrt{g^2 + g'^2}} \left(\frac{g'}{g} + \frac{g}{g'} \right),$$

from which expression (6.53) for the unit of electric charge follows.

In conclusion, we note that the group approach introduced at the end of this section can be easily generalized to more complicated models with partial breaking of gauge symmetry. It enables us to find a number of properties of physical fields by studying their transformation rules under the action of an unbroken gauge subgroup.

Supplementary Problems for Part I

Problem 1. *Mixing of fields.* Consider the theory of two real scalar fields φ_1, φ_2 with Lagrangian

$$\mathcal{L} = \frac{1}{2}(\partial_\mu \varphi_1)^2 + \frac{1}{2}(\partial_\mu \varphi_2)^2$$
$$- \frac{m_{11}^2}{2}\varphi_1^2 - m_{12}^2 \varphi_1 \varphi_2 - \frac{m_{22}^2}{2}\varphi_2^2 \qquad \text{(S1.1)}$$
$$- \frac{\lambda_{11}}{4}\varphi_1^4 - \frac{\lambda_{12}}{2}\varphi_1^2 \varphi_2^2 - \frac{\lambda_{22}}{4}\varphi_2^4.$$

Note that the mass term in this Lagrangian can be written in matrix form

$$\varphi^T M \varphi,$$

where

$$\varphi = \begin{pmatrix} \varphi_1 \\ \varphi_2 \end{pmatrix}, \quad \varphi^T = (\varphi_1, \varphi_2)$$

$$M = \begin{pmatrix} m_{11}^2 & m_{12}^2 \\ m_{12}^2 & m_{22}^2 \end{pmatrix}$$

(the matrix M is called the matrix of mass squares, or the mass matrix). The Lagrangian (S1.1) is invariant under the discrete symmetry $(\varphi_1 \rightarrow -\varphi_1, \varphi_2 \rightarrow -\varphi_2)$.

1. What constraints on m_{ij}^2 and λ_{ij} are imposed by the requirement that the classical energy is bounded from below?

2. Find the range of values of m_{11}^2, m_{12}^2 and m_{22}^2 for which the discrete symmetry is not spontaneously broken.

3. In the case of unbroken symmetry, find the spectrum of small perturbations about the ground state.

127

Problem 2.   *Dilatation symmetry.*

1. Consider the theory of a single real scalar field in four-dimensional space–time, described by the action

$$S = \int d^4x \left[\frac{1}{2} \partial_\mu \varphi \partial_\mu \varphi - \frac{\lambda}{4} \varphi^4 \right].$$

   Show that the action is invariant under the dilatations

$$\varphi(x) \rightarrow \varphi'(x) = \alpha \varphi(\alpha x),$$

   where α is a real parameter. Find the corresponding conserved current. Choose the energy–momentum tensor T^μ_ν such that its trace is zero on the field equations, $T^\mu_\mu = 0$.

2. Find the dilatation symmetry in Yang–Mills theory without matter in four-dimensional space–time. Construct the conserved current. Show that $T^\mu_\mu = 0$ if the field equations are satisfied.

3. Find the most general form of the potential $V(\varphi)$ in the theory of a single real scalar field in d-dimensional space–time ($d \geq 3$) for which the action

$$S = \int d^dx \left[\frac{1}{2} \partial_\mu \varphi \partial_\mu \varphi - V(\varphi) \right]$$

   is invariant under the dilatation symmetry, analogous (but not identical) to that described in 1).

4. Find the analogue of dilatation symmetry in the Liouville model in two-dimensional space–time with the action

$$S = \int d^2x \left[\frac{1}{2} \partial_\mu \varphi \partial_\mu \varphi - a e^{b\varphi} \right],$$

   where $\mu = 0, 1$; a, b are constants, and φ is a real scalar field.

5. For the models of 3) and 4), construct the conserved current, and show that $T^\mu_\mu = 0$.

Problem 3.   *Shift symmetry.* Consider the theory of a single scalar field with Lagrangian

$$\mathcal{L} = \frac{1}{2} \partial_\mu \varphi \partial_\mu \varphi.$$

The action is invariant under the transformations

$$\varphi(x) \rightarrow \varphi'(x) = \varphi(x) + c,$$

where c is a real parameter of the transformation.

1. Find the conserved current.

2. Is the symmetry spontaneously broken? If it is, then is Goldstone's theorem satisfied?

Problem 4. *Goldstone's theorem and currents.* Consider a model with n real scalar fields φ^a and an $SO(n)$-symmetric Lagrangian

$$\mathcal{L} = \frac{1}{2}\partial_\mu \varphi^a \partial_\mu \varphi^a - V(\varphi^a \varphi^a).$$

Suppose the potential is chosen such that the global $SO(n)$ symmetry is spontaneously broken; choose the ground state in the form

$$\varphi^a = \delta^{an}\varphi_0,$$

i.e.

$$\begin{pmatrix} \varphi^1 \\ \vdots \\ \varphi^n \end{pmatrix} = \begin{pmatrix} 0 \\ \vdots \\ \varphi_0 \end{pmatrix}.$$

1. Find the unbroken subgroup (residual symmetry group).

2. Find the Nambu–Goldstone fields.

3. Find the conserved currents j_μ^a. Show that these currents can be divided into two categories: a) currents corresponding to the unbroken subgroup; b) currents corresponding to broken generators. Show that currents of type b) can be represented in the form

$$j_\mu^i = \partial_\mu \theta^i + \dots, \tag{S1.2}$$

where the θ^i are massless Nambu–Goldstone fields, and the dots denote terms of quadratic and higher order in the perturbations about the ground state. On the other hand, currents of type a) are quadratic (and of higher order) in the perturbations.

4. Note that if equation (S1.2) is satisfied, then the masslessness of the fields follows from conservation of the currents j_μ^i:

$$0 = \partial_\mu j_\mu^i = \partial_\mu \partial_\mu \theta^i + \dots$$

so that when interactions of perturbations are neglected (i.e. to linear order in the fields) the fields θ^i satisfy the massless Klein–Gordon equation $\partial_\mu \partial_\mu \theta^i = 0$. On the basis of these considerations, give a proof of Goldstone's theorem (in the general case), alternative to that presented in the text.

Problem 5.   *Weak explicit symmetry breaking and the masses of pseudo-Goldstone bosons.*   Consider the theory of two real scalar fields with Lagrangian

$$\mathcal{L} = \mathcal{L}_0 + \mathcal{L}_1,$$

where

$$\mathcal{L}_0 = \frac{1}{2}\partial_\mu\varphi^1\partial_\mu\varphi^1 + \frac{1}{2}\partial_\mu\varphi^2\partial_\mu\varphi^2 + \frac{\mu^2}{2}\left[(\varphi^1)^2 + (\varphi^2)^2)\right] - \frac{\lambda}{4}\left[(\varphi^1)^2 + (\varphi^2)^2\right]^2,$$

and

$$\mathcal{L}_1 = \varepsilon U(\varphi^1),$$

where ε is a small parameter, and U depends non-trivially on the component φ^1 only. The part $\mathcal{L}_0$ of the full Lagrangian is invariant under global $SO(2)$ symmetry.

1. Find the ground state, the conserved current and the Nambu–Goldstone mode for $\varepsilon = 0$.

2. Find the lightest mode and its mass for $\varepsilon \neq 0$ to leading order in ε (this mode is called the pseudo-Goldstone mode).

3. Find the connection between the four-divergence of the current constructed in 1) and the pseudo-Goldstone mode to the lowest order in the fields of perturbations about the ground state, and to the lowest non-trivial order in ε.

Problem 6.   *Model of **n**-field.*   Consider a model with n real scalar fields $f^a(x)$ subject to the constraint

$$f^a f^a = 1$$

at every point x. In other words, the field takes values on an $(n-1)$-dimensional sphere of unit radius. Choose the Lagrangian invariant under global $SO(n)$ symmetry:

$$\mathcal{L} = \frac{1}{2g^2}\partial_\mu f^a \partial_\mu f^a,$$

where g is a real parameter.

1. Find the dimension of the parameter g in d-dimensional space–time.

2. Derive the field equations.

3. Find the energy-momentum tensor and the conserved currents corresponding to $SO(n)$ symmetry.

4. Find the ground state and show that it breaks the $SO(n)$.

5. Find explicitly the spectrum of small perturbations about the ground state. Find the unbroken symmetry group. Show that Goldstone's theorem is valid.

Problem 7. *Topologically massive gauge theories in three-dimensional space–time (Deser, Jackiw and Templeton 1982).* Consider the theory of a single real vector field in three-dimensional space–time, $\mu = 0, 1, 2$. Choose the action in the form

$$S = \int d^3x \left(-\frac{1}{4} F_{\mu\nu} F^{\mu\nu} + g \varepsilon^{\mu\nu\lambda} A_\mu \partial_\nu A_\lambda \right),$$

where $\varepsilon^{\mu\nu\lambda}$ is a completely antisymmetric tensor, $\varepsilon^{012} = 1$, g is a real constant, and $F_{\mu\nu} = \partial_\mu A_\nu - \partial_\nu A_\mu$.

1. Find the dimension of the constant g.

2. Show that the action is invariant under the following gauge transformations which decrease at infinity:

$$A_\mu(x) \rightarrow A_\mu(x) + \partial_\mu \alpha(x)$$

   ($\alpha(x)$ decreases rapidly as $x \to \infty$).

3. Find the field equations and show that they are gauge invariant.

4. Find the spectrum of physical excitations of the field (i.e. those which cannot be removed by gauge transformations). In particular, find the number of physical degrees of freedom, the structure of the field in these modes and the dependence of the frequency ω on the wave vector $\mathbf{k}$.

5. The term $\varepsilon^{\mu\nu\lambda} A_\mu \partial_\nu A_\lambda$ is called the Chern–Simons Lagrangian. Generalize this expression to the case of non-Abelian gauge fields so that the non-Abelian Chern–Simons Lagrangian contains at most one derivative and is invariant, up to a total derivative, under non-Abelian gauge transformations.

Problem 8.   *Bosonic sector of the Georgi–Glashow model.* Consider the model with gauge group $SU(2)$ and triplet of real scalar fields φ^a (Georgi and Glashow 1972). Let g be the gauge-coupling constant and suppose that the potential of the scalar fields has the form

$$V = -\frac{\mu^2}{2}\varphi^a\varphi^a + \frac{\lambda}{4}(\varphi^a\varphi^a)^2.$$

1. Find the ground state and the residual (unbroken) gauge group.

2. Find the spectrum of all small perturbations about the ground state; classify these perturbations with respect to the unbroken gauge group.

Problem 9. $SU(5)$ *model (Georgi and Glashow 1974).* Consider the theory with the gauge group $SU(5)$.

1. Choose a representation of the scalar fields and the scalar potential such that $SU(5)$ is broken down to $SU(3) \times SU(2) \times U(1)$, where $SU(3)$ and $SU(2)$ are embedded in $SU(5)$ as follows:

$$\left(\begin{array}{c|c} SU(3) & 0 \\ \hline 0 & SU(2) \end{array}\right),$$

   and the group $U(1)$ is diagonal in $SU(5)$.

2. Find the masses of the vector bosons and their representations under the unbroken gauge group.

3. A scalar field in which representation of $SU(5)$ needs to be added in order to ensure subsequent breaking down to $SU(3) \times U(1)$, and in such a way that $SU(2) \times U(1)$ is broken down to $U(1)$ analogously to the Standard Model? Choose the full scalar potential for the breaking $SU(5) \to SU(3) \times U(1)$.

Problem 10.   *Flat directions.* Consider a model with two complex scalar fields φ_1 and φ_2 with Lagrangian

$$\mathcal{L} = \partial_\mu\varphi_1^*\partial_\mu\varphi_1 + \partial_\mu\varphi_2^*\partial_\mu\varphi_2 - \lambda(\varphi_1^*\varphi_1 - \varphi_2^*\varphi_2 - v^2)^2.$$

1. Find the global symmetry group for this Lagrangian.

2. Find the set of classical vacua in the model. Find the unbroken subgroup for each vacuum.

3. Find the spectrum of all small perturbations about each vacuum. Which vacua are physically equivalent, and which are not? Is Goldstone's theorem valid? Is the number of massless perturbations equal to the number of broken generators? Why?

4. By introducing corresponding gauge fields, construct a theory in which the symmetry group found in 1) is a gauge group. Find the spectrum of small perturbations about each vacuum in the gauge theory obtained. Does the spectrum still contain massless scalar excitations?

Find the spectrum of all small perturbations about each vacuum. Which vacua are physically equivalent, and which are not? Is Goldstone's theorem valid? Is the number of massless perturbations equal to the number of broken generators? Why?

4. By introducing corresponding gauge fields, construct a theory in which the symmetry group found in (1) is a gauge group. Find the spectrum of small perturbations about each vacuum in the gauge theory obtained. Does the spectrum still contain massless scalar excitations?

Part II

Chapter 7

The Simplest Topological Solitons

Until now we have considered small linear perturbations of fields about the ground state (classical vacuum) and have been interested mainly in the mass spectrum. In quantum field theory these elementary excitations correspond to *point-like* particles. Here and in subsequent chapters, we shall consider solitons; these are solutions of the classical field equations, which, in their own right, without quantization, are similar to particles. They are lumps of fields (and, hence, of energy) of finite size; more precisely, the fields decrease rapidly from the center of a lump.[1] The existence and stability of solitons is due, in the first place, to the nonlinearity of the field equations. In quantum theory, solitons correspond to extended particles, which, roughly speaking, are composed of the elementary particles in each specific model.

Among the various types of solitons, the class of topological solitons is of particular interest. The meaning of this term is explained below, and here, let us simply say that it is the topological solitons which we shall mainly consider in this and subsequent chapters. Usually, we shall study static solitons, i.e. solutions which in some reference frame do not depend on time.

In particle physics, the use of the soliton concept is rather limited, although it is sometimes very fruitful. At the same time, solitons are to be found very frequently in condensed matter physics.

---

[1] In the mathematical literature, the term "soliton" is usually employed in a narrower sense. We shall use it for any particle-like stable solutions.

7.1   Kink

The simplest topological object, the kink, arises in the theory of a single
real scalar field in two-dimensional space–time. The action for the model
is chosen in the form

$$S = \int d^2x \left[\frac{1}{2}(\partial_\nu \varphi)^2 - V(\varphi) \right],$$ (7.1)

where $\nu = 0, 1$;

$$V(\varphi) = -\frac{\mu^2}{2}\varphi^2 + \frac{\lambda}{4}\varphi^4 + \frac{\mu^2}{4\lambda},$$ (7.2)

or, what amounts to the same thing,

$$V(\varphi) = \frac{\lambda}{4}(\varphi^2 - v^2)^2$$

$$v = \frac{\mu}{\sqrt{\lambda}}.$$

The action is invariant under the discrete transformation $\varphi \to -\varphi$; this
symmetry is spontaneously broken, since the classical vacua are

$$\varphi^{(v)} = \pm v.$$

We recall that in two-dimensional space–time the field φ is dimensionless,
and (as before, we use the system of units $\hbar = c = 1$) the dimensions of
the parameters are as follows:

$$[\mu] = M, \quad [\lambda] = M^2, \quad [v] = 1.$$

Linear excitations about one of the classical vacua have mass

$$m = \sqrt{2}\mu.$$

In what follows, we shall consider weakly coupled theory, i.e. with a small
parameter λ. Since λ is dimensionful, this statement about weak coupling
is written more precisely as follows:

$$\lambda \ll \mu^2$$

(μ^2 is the only parameter of dimension M^2 with which one can compare
λ), or

$$v \gg 1.$$ (7.3)

In order to clarify the meaning of the inequality (7.3), it is appropriate
to use a simple consideration, which arises in quantum theory. Let us
consider a particle with mass m and characteristic spatial momentum $p \lesssim$

m. Its Compton wavelength is of the order of $1/m$. At the classical level, it corresponds to a linear wave $\Delta\varphi(x,t) = \varphi(x,t) - v$ (where we consider perturbations about the vacuum $\varphi = v$). We estimate the amplitude of $\Delta\varphi$ from the requirement that the classical energy for a (linear) perturbation $\Delta\varphi(x,t)$ should be of the order of m. For a linear perturbation we write

$$E = \int \left[\frac{1}{2}(\Delta\dot{\varphi})^2 + \frac{1}{2}(\Delta\varphi')^2 + \frac{m^2}{2}(\Delta\varphi)^2 \right] dx, \qquad (7.4)$$

where the dot and the prime denote differentiation with respect to t and x, respectively. Since the characteristic wave vector p and frequency ω are of order m, we have

$$\Delta\dot{\varphi} \sim \Delta\varphi' \sim m\Delta\varphi.$$

Moreover, the wave packets with typical wave vector p have typical size of the order of $1/p$; therefore, in (7.4) the domain of integration over dx is of order $1/p \sim 1/m$. Thus,

$$E \sim m(\Delta\varphi)^2.$$

By requiring $E \sim m$ we obtain that the amplitude is of order

$$\Delta\varphi \sim 1.$$

For the perturbations about the vacuum value $\varphi^{(v)} = v$ to be indeed small (linear), we require $\Delta\varphi \ll v$, which is precisely the inequality (7.3).

A kink is a static solution of the field equations $\varphi_k(x)$, interpolating between the vacuum $\varphi = -v$ and the vacuum $\varphi = +v$, as x runs from $x = -\infty$ to $x = +\infty$ (Figure 7.1). The fact that configurations similar to that shown in Figure 7.1 may have a finite energy can be seen from the expression for the static energy (i.e. the energy for the fields, which are independent of time):

$$E = \int dx \left[\frac{1}{2} \left(\frac{d\varphi}{dx} \right)^2 + \frac{\lambda}{4}(\varphi^2 - v^2)^2 \right].$$

When $x \to \pm\infty$, both the first and the second terms in the integrand decrease; if this decrease is sufficiently rapid, then the energy is finite.

Since $\dot{\varphi} = 0$, the field equation has the form

$$\varphi'' - \frac{\partial V(\varphi)}{\partial \varphi} = 0, \qquad (7.5)$$

where, as before, the prime denotes $\frac{d}{dx}$. Before we write down the solution of equation (7.5) in explicit form, we note an analogy which is useful also in

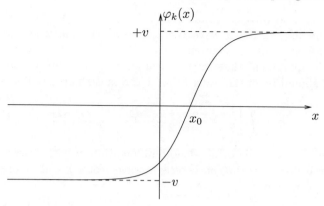

Figure 7.1.

certain other situations. Formally, equation (7.5) has the form of Newton's law for a particle with coordinate φ, moving in time x in the potential

$$U(\varphi) = -V(\varphi)$$

(see Figure 7.2). Here, the solution $\varphi_k(x)$ corresponds to a trajectory beginning in the infinite past at the point $\varphi = -v$ and ending in the infinite future at the point $\varphi = +v$. This trajectory indeed requires infinite time if the energy of the particle is equal to zero (in that case the velocity tends to zero near each bump).

The solution of equation (7.5) is easy to find explicitly. The first integral of that equation has the form

$$\frac{1}{2}\varphi'^2 - V(\varphi) = \varepsilon_0.$$

For the kink $\varphi' \to 0$, $V(\varphi) \to 0$, as $x \to \pm\infty$, thus $\varepsilon_0 = 0$ (we remark that ε_0 is the energy of the particle in the aforementioned analogy). Thus

$$\frac{d\varphi}{dx} = +\sqrt{2V} \equiv \sqrt{\frac{\lambda}{2}(v^2 - \varphi^2)}$$

(the choice of sign in front of the square root corresponds to motion from left to right, i.e. from $\varphi = -v$ to $\varphi = v$). Hence,

$$\varphi = v \tanh\left(\sqrt{\frac{\lambda}{2}}v(x - x_0)\right),$$

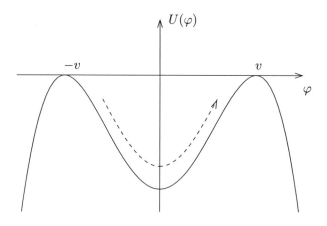

Figure 7.2.

where x_0 is a constant of integration, having the meaning of the position of the center of the kink. For $x_0 = 0$, the configuration of the kink

$$\varphi_k(x) = v \tanh\left(\sqrt{\frac{\lambda}{2}} vx\right) \qquad (7.6)$$

is symmetric under the substitution $\varphi \to -\varphi$, $x \to -x$. It is sometimes convenient to write the configuration of the kink in a different notation:

$$\varphi_k(x) = \frac{\mu}{\sqrt{\lambda}} \tanh\left(\frac{\mu}{\sqrt{2}} x\right). \qquad (7.7)$$

We shall now discuss certain properties of the solution (7.6) (or (7.7)) which are, in fact, of a very general nature.

1. The size of the kink is of order

$$r_k \sim \mu^{-1},$$

or

$$r_k \sim m^{-1}, \qquad (7.8)$$

i.e. of the order of the Compton wavelength of an elementary excitation. Indeed, the solution of (7.7) deviates from the vacuum value only in a region whose size is of the order of r_k. In other words, the energy density of the kink

$$\varepsilon(x) = \frac{1}{2}(\varphi_k')^2 + V(\varphi_k) = \frac{\lambda}{2} v^4 \frac{1}{\cosh^4\left(\frac{\mu}{\sqrt{2}} x\right)}$$

is significantly different from zero only in the region $|x| \lesssim r_k$.

2. The static energy of the kink is equal to

$$E_k = \int_{-\infty}^{\infty} dx \varepsilon(x) = \frac{2}{3} m v^2,$$

where $m = \sqrt{2}\mu$ is the mass of an elementary excitation. The static energy of the kink can be identified with the mass of this particle-like classical excitation of the field; thus

$$M_k = \frac{2}{3} m v^2. \tag{7.9}$$

It is important to note that the size of the kink is much greater than its Compton wavelength $\lambda = 1/M_k$: it follows from equations (7.8) and (7.9) (see also (7.3)) that

$$\frac{r_k}{\lambda_k} \sim v^2 \gg 1.$$

Thus, the classical size of a kink is much greater than its quantum size; even in quantum theory, a kink is essentially a classical object. We also note that the mass of a kink is much greater than the mass m of an elementary excitation.

3. For large $|x|$, the field of the kink differs very little from the vacuum, therefore $|\varphi_k(x)| - v$ must satisfy the Klein–Gordon equation with mass m in this region. Indeed, for large x (to be specific, we shall consider $x > 0$), from (7.7) we have

$$\varphi_k = v - 2v e^{-mx},$$

so that the difference $(\varphi_k - v)$ indeed satisfies the static Klein–Gordon equation with mass m (in two-dimensional space–time) and decreases exponentially.

4. The solution (7.7) is not invariant under spatial translations and Lorentz transformations. However, all these transformations must take a solution of the field equations to another solution. Applying these transformations, we obtain the family of solutions

$$\varphi_k(x - x_0; t; v) = \frac{\mu}{\sqrt{\lambda}} \tanh\left(\frac{\mu}{\sqrt{2}} \frac{(x - x_0) - ut}{\sqrt{1 - u^2}}\right),$$

describing moving kinks. Physically, the parameter u represents the velocity of a soliton; x_0 is the position of the center of the kink at time $t = 0$.

Problem 1. *Calculate the classical energy and the classical momentum of a moving kink. Show that for a kink the relativistic equations between energy, momentum, mass and velocity hold.*

Let us now consider the stability of a solution (7.7) under small perturbations.[2] The approach we shall study below is of a very general nature and is used to study the stability of various static solutions.

Let $\varphi_k(x)$ be a static solution of the field equations (to be specific, we shall consider the $(1 + 1)$-dimensional case). Let us consider small perturbations $\phi(x, t)$ about it, so that the original field has the form

$$\varphi(x, t) = \varphi_k(x) + \phi(x, t). \tag{7.10}$$

The field $\varphi(x, t)$ must satisfy the classical equations

$$-\partial_\mu \partial_\mu \varphi - \frac{\partial V}{\partial \varphi} = 0 \tag{7.11}$$

or

$$-\partial_\mu \partial_\mu (\varphi_k + \phi) - \frac{\partial V}{\partial \varphi}(\varphi_k) - \frac{\partial^2 V}{\partial \varphi^2}(\varphi_k)\phi + \cdots = 0, \tag{7.12}$$

where the dots denote terms of higher order in the perturbations $\phi(x, t)$, and the derivatives of the potential are taken at $\varphi = \varphi_k$; the latter depend explicitly on x, since φ_k depends on x. The field φ_k satisfies equation (7.11), thus, the terms which do not contain ϕ cancel out in (7.12) and, to linear order in ϕ, we obtain

$$-\partial_\mu \partial_\mu \phi - \frac{\partial^2 V}{\partial \varphi}(\varphi_k)\phi = 0. \tag{7.13}$$

In particular, for perturbations around the kink (7.7) in the model (7.1), (7.2), we have

$$\frac{\partial^2 V}{\partial \varphi^2}(\varphi_k) = \mu^2 \left[3 \tanh^2 \left(\frac{\mu}{\sqrt{2}} x \right) - 1 \right]. \tag{7.14}$$

Since, in the general case, $\frac{\partial^2 V}{\partial \varphi^2}(\varphi_k)$ depends only on x, the variables in equation (7.13) can be separated and a solution can be sought in the form

$$\phi(x, t) = e^{i\omega t} f_\omega(x), \tag{7.15}$$

where f_ω satisfies the equation

$$\omega^2 f_\omega + f_\omega'' - \frac{\partial^2 V}{\partial \varphi^2}(\varphi_k) f_\omega = 0 \tag{7.16}$$

---

[2]In mathematics, this is called Lyapunov stability.

or

$$-f_\omega'' + U(x)f_\omega = \omega^2 f_\omega,$$ (7.17)

where

$$U(x) = \frac{\partial^2 V}{\partial \varphi^2}(\varphi_k).$$ (7.18)

Equation (7.17) is an equation for the eigenvalues ω^2 which, formally, coincides with the stationary Schrödinger equation in the potential $U(x)$. A general solution of (7.13) is a linear combination of solutions of (7.17). For the solutions $f_\omega(x)$, we require that they should be smooth and that they should not increase as $|x| \to \infty$, so that the field (7.1) is smooth and bounded as $|x| \to \infty$.

The problem of the stability of the static solution $\varphi_k(x)$ reduces to the problem of the existence or non-existence of negative eigenvalues of the operator

$$-\frac{d^2}{dx^2} + U(x).$$ (7.19)

If there are no negative eigenvalues, then all the ω are real and the small perturbations (7.15) do not grow with time; they either oscillate ($\omega^2 > 0$) or do not depend on time at all. In this case the solution is stable. If there exist negative eigenvalues (even just one), $\omega^2 = -\Omega^2$, then the perturbations (7.15) grow exponentially with time

$$\phi \sim e^{\Omega t},$$

i.e. the field (7.10) becomes exponentially more distant from the solution φ_k over time; in this case, the solution is unstable.[3]

These considerations can be generalized, in an evident manner, to the case of more than one spatial dimension and a system with a more complicated set of fields.

We note that the operator (7.19) with potential (7.18) always has a zero eigenvalue, and the eigenfunction (zero mode) has the form

$$f_0(x) = \frac{\partial \varphi_k(x)}{\partial x}.$$ (7.20)

Indeed, $\varphi_k(x)$ satisfies the equation

$$-\varphi_k'' + \frac{\partial V}{\partial \varphi_k}(\varphi_k) = 0.$$

Differentiating this equation with respect to x, we obtain

$$-(\varphi_k')'' + \frac{\partial^2 V}{\partial \varphi^2}(\varphi_k)\varphi_k' = 0,$$

---

[3]In this case, the eigenvalues Ω are called Lyapunov exponents.

i.e. the function (7.20) indeed satisfies equation (7.16) with $\omega = 0$. Furthermore, $\varphi_k(x)$ tends to a constant as $|x| \to \infty$, and so $\varphi_k'(x)$ decreases as $|x| \to \infty$, so that (7.20) satisfies the requirement of boundedness at spatial infinity, as required. The existence of this mode has to do with the breaking of translational invariance in space: $\varphi_k(x + a)$ is also a static solution of the field equations, but it does not coincide with $\varphi_k(x)$. For small a, the function $\varphi_k(x + a)$ can be represented in the form

$$\varphi_k(x + a) = \varphi_k(x) + a f_0(x),$$

where $a f_0(x)$ is a small deviation from the solution $\varphi_k(x)$. Since $\varphi_k(x + a)$ satisfies the field equations, the small deviation $a f_0(x)$ satisfies the linearized equation (7.12); it does not depend on time, decreases as $|x| \to \infty$ and is therefore a zero mode.

If the number of spatial dimensions d is greater than one, then there are d translational zero modes. In this case the classical solution may also break the rotational symmetry, and rotational zero modes will occur. Finally, in models with internal global symmetry, classical solutions may also break the internal symmetry; to this breaking there will also correspond zero modes.

Problem 2. *Let* $\mathcal{L}(\partial_\mu \Phi_A; \Phi_A) = F^{AB} \partial_\mu \Phi_A \partial_\mu \Phi_B - V(\Phi)$ *be the Lagrangian of a model with scalar fields* Φ_A. *Suppose it is invariant under the global transformations* $\Phi_A \to \Phi_A^{(\varepsilon)}$, *where* ε *is an infinitesimal parameter of the transformation. Suppose also that the ground state does not break this symmetry, while* $\Phi_{k,A}(\mathbf{x})$, *a static solution of field equations of the soliton type, breaks this symmetry (i.e.* $\Phi_{k,A}^{(\varepsilon)}(\mathbf{x}) \neq \Phi_{k,A}(\mathbf{x})$*).*

1. *Formulate in general form, a condition for the stability of the solution* $\Phi_{k,A}(x)$ *under small perturbations.*

2. *Show that in the corresponding eigenvalue problem, there exists a zero mode, corresponding to the symmetry described; find its form.*

Let us now come back to a discussion of the stability of a kink. The spectrum of the operator (7.19) with $U(x) = \frac{\partial^2 V}{\partial \varphi^2}(\varphi_k)$, given by formula (7.14) can be found in explicit form; it does not contain negative eigenvalues. The kink is stable under small perturbations of the field.

Problem 3. *Find the spectrum of small perturbations about the kink, i.e. the spectrum of the eigenvalues and eigenfunctions of the operator (7.19) with the potential (7.14).*

Let us discuss the sense in which the kink is a topological soliton. For this, let us consider possible field configurations, $\varphi(x)$, with finite energy (we fix the moment in time). For finiteness of the energy, we require, in particular, that as $x \to +\infty$ the field should tend to one of the vacua $+v$ or $-v$; the same is true for $x \to -\infty$. Thus, any configuration of fields defines a mapping of the two "points" $x = +\infty$ and $x = -\infty$, which represent spatial infinity in one-dimensional space, to the set of the classical vacua which also consists of the two points $+v$ and $-v$. Here, any changes in the fields in time (classical evolution according to the field equations, the effect of localized external sources, etc.) which do not lead to the occurrence of configurations with infinite energy, and therefore do not affect the values of the field at spatial infinity, do not change the mapping assigned in this way. In this connection, the mapping is said to be a topological characteristic of the field configuration.

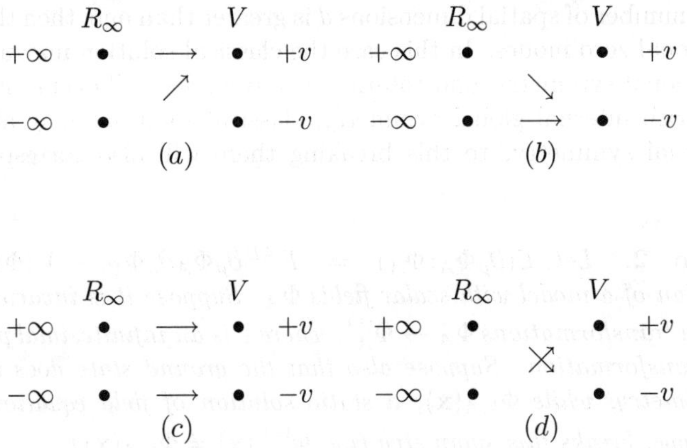

Figure 7.3.

It is clear that the two points of the set $R_\infty \{x = +\infty, x = -\infty\}$ can be mapped to the two points of the set $V\{\varphi = +v, \varphi = -v\}$ in four different ways, as shown in Figure 7.3. This means that all field configurations with finite energy can be divided into four non-intersecting subsets (sectors), and the evolution does not take a field out of its sector. Figure 7.3(a) corresponds to the vacuum sector containing the vacuum $\varphi = +v$, since $\varphi(+\infty) = \varphi(-\infty) = v$; Figure 7.3(b) corresponds to the vacuum sector containing the vacuum $\varphi = -v$. The kink belongs to the sector corresponding to Figure 7.3(c). Finally, the sector of Figure 7.3(d) contains an antikink, namely, a configuration of the form

$$\varphi(x) = -\varphi_k(x) = \varphi_k(-x).$$

Let us suppose that in the sector corresponding to Figure 7.3(c), we manage to find a field realizing a minimum of the static classical energy among fields of that sector. This field is necessarily different from the vacuum (it belongs to a different sector) and is a solution of the static field equations. Such a field is energetically favorable among fields of that sector and, therefore, stable. In the model considered in this section the only such field is the kink configuration (since the kink is the unique static solution in the sector 7.3(c)). Thus, the existence of the kink is associated with a non-trivial topology of mappings of spatial infinity to the set of classical vacua. We note that topological considerations, generalizing those studied here, may not guarantee the existence of solitons in other models, but are very often extremely useful in the search for solitons. We shall encounter considerations of a similar type in subsequent chapters.

Finally, let us introduce the concept of topological current on the example of the model of this section. We define

$$k^\mu = \frac{1}{2v}\varepsilon^{\mu\nu}\partial_\nu\varphi,$$

where $\mu, \nu = 0, 1$, and $\varepsilon^{\mu\nu}$ is an antisymmetric tensor.

The current k^μ is trivially conserved (the field equations need not be satisfied) and is distinguished in this from the Noether currents (conserved only on the field equations):

$$\partial_\mu k^\mu = \frac{1}{2v}\varepsilon^{\mu\nu}\partial_\mu\partial_\nu\varphi \equiv 0.$$

The topological charge

$$Q_T = \int_{-\infty}^{+\infty} k^0 dx^1 = \int_{-\infty}^{+\infty} \frac{1}{2v}\partial_1\varphi dx^1 = \frac{1}{2v}(\varphi(+\infty) - \varphi(-\infty))$$

is equal to zero for the vacuum and $+1$ and -1 for the kink and the antikink, respectively. Thus, the kink realizes a minimum of the energy for a topological charge $Q_T = 1$.

In the model in this section, the introduction of the topological charge does not bring a great benefit; however, as we shall see in Section 7.4, in other models, the use of the topological charge opens up ways of obtaining explicit formulae for soliton configurations.

To conclude this section, we make the following remark, concerning models of the above type in space–time with more than two dimensions (the physically interesting case is the four-dimensional case). In this case, the scalar field equation, as before, will have solution $\varphi_k(x^1)$, which will depend only on one coordinate. Physically, this solution represents an infinite planar "domain wall," on one side of which the field takes the value $(-v)$, and on the other side $(+v)$. The expression (7.9) must now be interpreted as the energy of the wall per unit area. Thus, four-dimensional models with a discrete set of degenerate ground states predict the existence of extended objects, whose energy is concentrated near two-dimensional surfaces, domain walls. This prediction is very significant for cosmology (Zel'dovich *et al.* 1974).

Problem 4. *Sine-Gordon, Bäcklund transformation and breathers. Consider a model of a real scalar field in* $(1 + 1)$*-dimensions, with Lagrangian*

$$\mathcal{L} = \frac{1}{2}\partial_\mu\varphi\partial_\mu\varphi + m^2v^2[\cos(\varphi/v) - 1]. \qquad (7.21)$$

1. *Find the set of vacua in this model.*

2. *Find a soliton, analogous to the kink, which interpolates between neighboring vacua.*

3. *Introduce the variables* $\phi = \varphi/v$, $\xi = mx$, *and* $\tau = mt$, *together with the variables*

$$U = \frac{1}{2}(\xi + \tau)$$
$$V = \frac{1}{2}(\xi - \tau).$$

Let ϕ_0 *be a solution of the classical field equations (depending, generally speaking, on* ξ *and* τ*). Consider the system of equations of first order in derivatives (equations of the Bäcklund transformation)*

$$\frac{1}{2}\frac{\partial}{\partial U}(\phi - \phi_0) = \alpha\sin\left[\frac{1}{2}(\phi + \phi_0)\right] \qquad (7.22)$$
$$\frac{1}{2}\frac{\partial}{\partial V}(\phi + \phi_0) = \frac{1}{\alpha}\sin\left[\frac{1}{2}(\phi - \phi_0)\right].$$

3a. *Show that the solution* ϕ *of this system also satisfies the sine-Gordon equation, i.e. the field equation obtained from the Lagrangian (7.21). This solution of the field equation* ϕ *(which, of course, depends on the choice of the original solution* ϕ_0*) is called the Bäcklund transformation of the solution* ϕ_0.

3b. *By solving the system (7.22) for* $\phi_0 = 0$, *show that the Bäcklund transformation of the classical vacuum* $(\phi_0 = 0)$ *is the sine-Gordon soliton (generally speaking, moving) found in point 1 of the problem.*

3c. *By applying, to the solution* $\phi_0 = 0$, *the Bäcklund transformation twice (with complex and different parameters* α*), find real time-dependent solutions of the sine-Gordon equation. Find among these* solutions a class which corresponds to the classical scattering of solitons, and another class of periodic solutions (breathers). Check, by explicit substitution in the sine-Gordon equation, that these are indeed solutions.

4. Find the topological charges of all possible configurations in the sine-Gordon system, and how they are related to the mapping of infinity to the set of classical vacua. Give examples of configurations with all possible topological charges. To which topological sector does the breather belong?

Problem 5.   Consider the model of a complex scalar field with Lagrangian

$$\mathcal{L} = \partial_\mu \varphi \partial_\mu \varphi^* + \mu^2 \varphi \varphi^* - \frac{\lambda}{2}(\varphi \varphi^*)^2$$

in $(1+1)$-dimensions. Show that the kink considered in this section $\varphi = \varphi^* = \varphi_k(x)$ is a solution of the classical field equations in this model. Is this solution stable against small perturbations in this model?

7.2   Scale transformations and theorems on the absence of solitons

In a number of models in $(d+1)$-dimensional space–time with $d > 1$ it is possible to show that there are no non-trivial static solutions of the field equations by applying scale arguments (Derrick 1964). These arguments, which are studied in this section, apply not only to stable solutions of the soliton type but also to unstable static solutions (these may also be of interest, as we shall see in the following chapters).

Let us consider first the theory of n scalar fields φ^a, $a = 1, \ldots, n$, in $(d+1)$-dimensional space–time. We shall write the Lagrangian in a quite general form

$$\mathcal{L} = \frac{1}{2} F_{ab}(\varphi) \partial_\mu \varphi^a \partial_\mu \varphi^b - V(\varphi), \tag{7.23}$$

where $F_{ab}(\varphi)$ and $V(\varphi)$ are certain functions of the scalar fields φ^a. We shall argue ab absurdo. Let us assume that $\varphi_c^a(\mathbf{x})$ is a static solution of the classical field equations with finite energy. It is an extremum of the energy functional

$$E[\varphi] = \int d^d x \left[\frac{1}{2} F_{ab}(\varphi) \partial_i \varphi^a \partial_i \varphi^b + V(\varphi) \right]. \tag{7.24}$$

We shall assume that for all φ the matrix $F_{ab}(\varphi)$ defines a positive-definite quadratic form, i.e. all the eigenvalues of this matrix are positive for all φ. Then

$$F_{ab} \partial_i \varphi^a \partial_i \varphi^b \geq 0, \tag{7.25}$$

where equality holds only for fields which do not depend on $\mathbf{x}$. In addition, we shall suppose that $V(\varphi)$ is bounded from below and choose the zero-point energy level such that the value of V at the absolute minimum $V(\varphi)$ (classical vacuum) is equal to zero:

$$V(\varphi^{(v)}) = 0.$$

Then

$$V(\varphi) \geq 0, \tag{7.26}$$

and equality holds only for the classical vacuum. Then, the classical vacuum, a homogeneous field realizing the absolute minimum of $V(\varphi)$, will have zero energy, and any other configuration of fields will have positive energy.

Problem 6. *Calculate the energy–momentum tensor for the model (7.23) and show that expression (7.24) is indeed the energy functional for time-independent fields.*

Problem 7. *Show that in the case of static fields, the field equations for the Lagrangian (7.23) are at the same time equations for extremality of the energy functional (7.24).*

Problem 8. *Show that if the matrix $F_{ab}(\varphi)$ has negative eigenvalues for some choice of φ^a, and $F_{ab}(\varphi)$ are smooth functions of the fields φ^a, then the energy (7.24) is not bounded from below.*

If $\varphi_c^a(\mathbf{x})$ is a static solution of the field equations with finite energy, then the energy functional must be extremal for $\varphi^a = \varphi_c^a$ with respect to any variations of the field which vanish at spatial infinity. Let us consider a field configuration of the form (the index a is omitted)

$$\varphi_\lambda(\mathbf{x}) = \varphi_c(\lambda \mathbf{x}). \tag{7.27}$$

For small λ, the difference

$$\varphi_\lambda(\mathbf{x}) - \varphi_c(\mathbf{x}) \equiv \varphi_c(\lambda \mathbf{x}) - \varphi_c(\mathbf{x})$$

is a small variation of the field. It vanishes at spatial infinity, since $\varphi_c(\mathbf{x})$ tends to a constant as $|\mathbf{x}| \to \infty$ (otherwise the gradient contribution to the energy would diverge). Consequently, the energy functional calculated on the configurations (7.27),

$$E(\lambda) = E[\varphi_\lambda(\mathbf{x})],$$

must have at extremum an $\lambda = 1$ (calculated on the one-parameter family of field configurations (7.27) the energy functional is a function of the parameter λ only),

$$\left.\frac{dE}{d\lambda}\right|_{\lambda=1} = 0. \tag{7.28}$$

We shall see that in a number of cases this cannot hold.

Let us calculate the energy for the configuration (7.27):

$$E(\lambda) = \int d^d x \left[\frac{1}{2} F_{ab}(\varphi_c(\lambda \mathbf{x})) \left(\frac{\partial}{\partial x^i} \varphi_c^a(\lambda \mathbf{x}) \right) \left(\frac{\partial}{\partial x^i} \varphi_c^b(\lambda \mathbf{x}) \right) + V(\varphi_c(\lambda \mathbf{x})) \right].$$

In this integral, we make the change of variables

$$\mathbf{y} = \lambda \mathbf{x},$$

so that $d^d x = \lambda^{-d} d^d y$; $\frac{\partial}{\partial x^i} = \lambda \frac{\partial}{\partial y^i}$. We obtain

$$E(\lambda) = \lambda^{-d} \int d^d y \left[\frac{1}{2} F_{ab}(\varphi_c(\mathbf{y})) \cdot \lambda^2 \left(\frac{\partial}{\partial y^i} \varphi_c^a(\mathbf{y}) \right) \left(\frac{\partial}{\partial y^i} \varphi_c^b(\mathbf{y}) \right) + V(\varphi_c(\mathbf{y})) \right]$$

or

$$E(\lambda) = \lambda^{2-d} \Gamma + \lambda^{-d} \Pi, \tag{7.29}$$

where

$$\Gamma = \int d^d x \frac{1}{2} F_{ab}(\varphi_c) \partial_i \varphi_c^a \partial_i \varphi_c^b \tag{7.30}$$

$$\Pi = \int d^d x V(\varphi_c).$$

We stress that Γ and Π are expressed solely in terms of the original solution $\varphi_c^a(\mathbf{x})$; here Γ and Π are the gradient and the potential terms, respectively, in the energy of this configuration. By virtue of conditions (7.25), (7.26), we have

$$\Gamma > 0$$
$$\Pi > 0.$$

Since Γ and Π do not depend on λ, we know the function $E(\lambda)$ explicitly: it is given by formula (7.29). The condition for its extremality (7.28) gives

$$(2 - d)\Gamma - d\Pi = 0. \tag{7.31}$$

Together with the positivity of Γ and Π, this condition leads to serious constraints on the existence of classical solutions in scalar theories, as follows.

1. $d > 2$: condition (7.31) is satisfied only if

$$\Gamma = \Pi = 0.$$

This means that $\partial_i \varphi_c^a = 0$ and φ_c^a is the absolute minimum of the potential $V(\varphi)$, i.e. the only solution is the classical vacuum.

2. $d = 2$: condition (7.31) gives

$$\Pi = 0.$$

If the potential $V(\varphi)$ is non-trivial, then this condition also means that the only static solution is the classical vacuum. The only class of $(2 + 1)$-dimensional scalar models where the existence of non-trivial classical solutions is possible is that of models with

$$V(\varphi) = 0 \text{ for all } \varphi,$$

i.e. there is no potential term in the Lagrangian (in this case, the kinetic term must have a complicated structure). Such a situation is realized, for example, in the model of **n**-field, which will be considered in Section 7.4.

3. For $d = 1$: condition (7.31) gives the virial theorem

$$\Gamma = \Pi$$

and does not impose constraints on the choice of model.

The physical reason for the absence of static solitons in $(d + 1)$-dimensional scalar theories with $d > 2$ (and $d = 2$ for $V(\varphi) \neq 0$) is the following. If $\varphi^a(\mathbf{x})$ is some configuration of scalar fields, then, as can be seen from (7.29), the energy of an adjacent configuration $\varphi^a(\lambda \mathbf{x})$ is less than the energy of the original field, at $\lambda > 1$. The configuration $\varphi^a(\lambda \mathbf{x})$ differs in size from $\varphi^a(\mathbf{x})$: if r is the characteristic size of the configuration $\varphi^a(\mathbf{x})$, then the characteristic size of the configuration $\varphi^a(\lambda \mathbf{x})$ is equal to $\lambda^{-1} r$, i.e. it is a factor of λ^{-1} less ($\lambda > 1$). In other words, it is energetically favorable that a particle-like configuration becomes unboundedly shrunken.

We note that this difficulty can be avoided in scalar theories with $d > 2$ only at the cost of adding terms with higher derivatives to the Lagrangian. For example, if one adds a term of fourth order in derivatives to the Lagrangian (and hence, to the static energy), the scale argument studied earlier is modified; instead of (7.31), one obtains

$$(4 - d)\Gamma^4 + (2 - d)\Gamma^2 - d\Pi = 0, \tag{7.32}$$

where Γ^2 is the contribution to the energy of the field $\varphi_c(\mathbf{x})$ from terms with two derivatives (of the type of (7.30)), while Γ^4 is the contribution to the energy of the configuration $\varphi_c^a(\mathbf{x})$ from terms with four derivatives (of the type $\int d^d x (\partial_i \varphi)^4$). For $d = 3$, condition (7.32) can be satisfied for positive Π, Γ^2, Γ^4, i.e. a soliton may exist. Such a situation is realized in the Skyrme model, considered in one of the Supplementary Problems for Part II.

Let us now consider gauge theories. For ease in writing down the formulae, we shall restrict ourselves to the case of a simple (matrix) gauge group G. Let A_μ be the gauge field, φ a scalar field transforming according to a (generally speaking, reducible) unitary representation T of the group G. The Lagrangian of the gauge theory has the form

$$\mathcal{L} = \frac{1}{2g^2} \mathrm{Tr}\, F_{\mu\nu}^2 + (D_\mu \varphi)^\dagger (D_\mu \varphi) - V(\varphi),$$

where

$$
\begin{aligned}
F_{\mu\nu} &= \partial_\mu A_\nu - \partial_\nu A_\mu + [A_\mu, A_\nu] \\
D_\mu \varphi &= [\partial_\mu + T(A_\mu)]\varphi
\end{aligned}
$$

(we recall that A_ν and $F_{\mu\nu}$ are anti-Hermitian matrices; we shall use the matrix form of the gauge fields).

As an example, let us consider fields which do not depend on time in the gauge $A_0 = 0$. (Note that the time independence of the fields is not a gauge-invariant concept: if some configuration does not depend on time, by performing a gauge transformation over it with a time-dependent gauge function, we obtain a field which depends on time). For the configurations in which we are interested,

$$
\begin{aligned}
F_{0i} &= 0 & (7.33) \\
D_0 \varphi &= 0. & (7.34)
\end{aligned}
$$

We note that these equations are gauge invariant; they could be used as the basis for a definition of the class of fields considered.

In fact, the class of fields satisfying the conditions (7.33) and (7.34) does not exhaust all the interesting cases of static solutions (an example of a static solution which does not satisfy (7.33) and (7.34) is provided by a magnetic monopole with an electric charge, a dyon). However, we shall limit our analysis to this class of fields for definiteness in our subsequent study.

The energy functional for the fields considered here has the form

$$E[A_i, \varphi] = \int d^d x \left[-\frac{1}{2g^2} \mathrm{Tr}\, F_{ij} F_{ij} + (D_i \varphi)^\dagger (D_i \varphi) + V(\varphi) \right],$$

where all three terms are positive (as before, we assume that $V(\varphi)$ is non-negative and equal to zero only for the classical vacuum).

Suppose $\mathbf{A}_c(\mathbf{x})$ and $\varphi_c(\mathbf{x})$ is a classical solution. We again apply a scale transformation, where we choose the transformation of the field $\mathbf{A}$ such that F_{ij} and $D_i\varphi$ transform homogeneously. This requirement leads us to consider the following family of fields:

$$\varphi_\lambda(\mathbf{x}) = \varphi_c(\lambda\mathbf{x}) \qquad (7.35)$$
$$\mathbf{A}_\lambda(\mathbf{x}) = \lambda\mathbf{A}_c(\lambda\mathbf{x}).$$

Then, the covariant derivative with respect to $\mathbf{x}$ for the new configuration is equal to

$$\mathbf{D}_{\mathbf{x}}^{(\lambda)}\varphi_\lambda(\mathbf{x}) = \left[\frac{\partial}{\partial\mathbf{x}} + T(\mathbf{A}_\lambda(\mathbf{x}))\right]\varphi_\lambda(\mathbf{x}) = \lambda\mathbf{D}_{\mathbf{y}}\varphi_c(\mathbf{y}),$$

where $\mathbf{y} = \lambda\mathbf{x}$,

$$\mathbf{D}_{\mathbf{y}}\varphi_c(\mathbf{y}) = \left[\frac{\partial}{\partial\mathbf{y}} + T(\mathbf{A}_c(\mathbf{y}))\right]\varphi_c(\mathbf{y})$$

is the covariant derivative with respect to $\mathbf{y}$ for the original configuration. The strength tensor for the new field is equal to

$$F_{ij}^{(\lambda)}(\mathbf{x}) = \frac{\partial}{\partial x^i}A_\lambda^j(\mathbf{x}) - \frac{\partial}{\partial x^j}A_\lambda^i(\mathbf{x}) + [A_\lambda^i(\mathbf{x}), A_\lambda^j(\mathbf{x})] = \lambda^2 F_{ij}^{(c)}(\mathbf{y}),$$

where

$$F_{ij}^{(c)} = \frac{\partial}{\partial y^i}A_c^j(\mathbf{y}) - \frac{\partial}{\partial y^j}A_c^i(\mathbf{y}) + [A_c^i(\mathbf{y}), A_c^j(\mathbf{y})]$$

is the strength tensor of the original configuration with coordinates $\mathbf{y}$. Substituting these expressions in the energy functional for the configuration (7.35) we obtain

$$E(\lambda) = \lambda^{4-d}G + \lambda^{2-d}\Gamma + \lambda^{-d}\Pi,$$

where

$$G = \int d^dy\left(-\frac{1}{2g^2}\operatorname{Tr}F_{ij}^{(c)}(\mathbf{y})F_{ij}^{(c)}(\mathbf{y})\right)$$

$$\Gamma = \int d^dy(D_{\mathbf{y}}\varphi_c)^\dagger(D_{\mathbf{y}}\varphi_c)$$

$$\Pi = \int d^dy V(\varphi_c).$$

G, Γ, Π are, respectively, the contributions of the gauge field, the covariant derivative and the potential of the scalar field to the energy of the classical solution $(\varphi_c, \mathbf{A}_c)$: all these quantities are positive. The extremality condition on $E(\lambda)$ at $\lambda = 1$ gives

$$(4 - d)G + (2 - d)\Gamma - d\Pi = 0. \qquad (7.36)$$

This condition is far weaker than (7.31); it does not prohibit the existence of non-trivial classical solutions for $d = 2$ and $d = 3$; we shall learn about solitons in gauge theories in two and three spatial dimensions in Section 7.3 and in the following chapters. The case $d = 4$ (instantons) is also interesting; here, condition (7.36) requires that scalar fields be completely absent from the theory (or the value of scalar fields would be the vacuum value everywhere in space).

Thus, the scale argument for gauge theories with scalar fields does not work for $d \leq 3$, while for gauge theories without scalar fields it does not work for $d = 4$. It prohibits the existence of non-trivial static classical solutions

1. in theories with scalar fields for $d \geq 4$

2. in purely gauge theories (without scalar fields) for $d \neq 4$ (in particular, in a purely gauge theory in the physically interesting $(3 + 1)$-dimensional space–time, there are no solitons).

7.3 The vortex

The vortex is the simplest soliton in gauge theory with scalars (Abrikosov 1966, Nielsen and Olesen 1973). It arises in a model with gauge group $U(1)$ and the Higgs mechanism in $(2 + 1)$-dimensional space–time. Thus, the model contains a one-component gauge field $A_\mu(x)$ and a complex scalar field $\varphi(x)$, transforming under gauge transformations as follows:

$$\varphi(x) \rightarrow e^{i\alpha(x)}\varphi(x)$$
$$A_\mu(x) \rightarrow A_\mu(x) + \frac{1}{e}\partial_\mu\alpha(x).$$

Here and subsequently in this section, $\mu, \nu = 0, 1, 2$. The Lagrangian of the model has the form

$$\mathcal{L} = -\frac{1}{4}F_{\mu\nu}^2 + (D_\mu\varphi)^*(D_\mu\varphi) - V(\varphi), \qquad (7.37)$$

where, as usual,

$$F_{\mu\nu} = \partial_\mu A_\nu - \partial_\nu A_\mu$$
$$D_\mu\varphi = (\partial_\mu - ieA_\mu)\varphi,$$

and we choose the potential in a form which ensures that the Higgs mechanism comes into play:

$$V(\varphi) = -\mu^2 \varphi^* \varphi + \frac{\lambda}{2}(\varphi^* \varphi)^2 + \frac{\mu^2}{2\lambda},$$

or, equivalently,

$$V(\varphi) = \frac{\lambda}{2}(\varphi^* \varphi - v^2)^2,$$

where

$$v = \frac{\mu}{\sqrt{\lambda}}.$$

We recall that the ground state in this model can be chosen such that $A_\mu = 0$, $\varphi = v$; then the vector field has the mass

$$m_V = \sqrt{2}ev, \tag{7.38}$$

and the scalar field in the unitary gauge is written in the form $\varphi = v + \eta(x)/\sqrt{2}$, where the real field $\eta(x)$ has mass

$$m_H = \sqrt{2\lambda}v = \sqrt{2}\mu. \tag{7.39}$$

Clearly, formulae (7.38) and (7.39) are relevant for small (linear) perturbations about the vacuum $\varphi = v$.

To find a soliton in this model, we first consider all possible non-singular field configurations which do not depend on time in the gauge $A_0 = 0$. (We recall that in this gauge there is a residual invariance under time-independent gauge transformations). We shall require the field energy to be finite, but we shall not yet fix the residual gauge freedom. For the configurations in which we are interested the energy functional has the form

$$E[A_i(\mathbf{x}), \varphi(\mathbf{x})] = \int \left[\frac{1}{4}F_{ij}F_{ij} + (D_i\varphi)^* D_i\varphi + V(\varphi)\right] d^2x. \tag{7.40}$$

A necessary (but not sufficient) condition for the energy to be finite is that $V(\varphi)$ should tend to zero as $|\mathbf{x}| \to \infty$, i.e.

$$|\varphi| \to v \text{ for } |\mathbf{x}| \to \infty.$$

Let us fix a large circle (radius R) with center at the origin. For sufficiently large R, the modulus of the field on this circle is equal to v; however, the

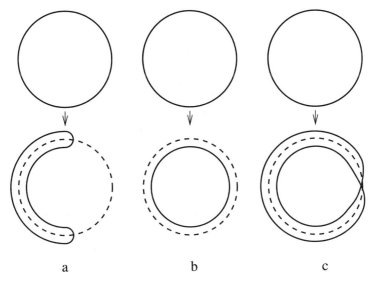

Figure 7.4.

phase of the field φ may depend on the polar angle in the two-dimensional space, θ. Thus, on our circle,

$$\varphi = e^{if(\theta)}v.$$

The function $ve^{if(\theta)}$ defines a mapping of the circle of radius R in space to a circle of radius v in the plane of complex φ. The mapping of circle to circle can be characterized by an integer $n = 0, \pm 1, \pm 2, \ldots$, the "winding number." This can be illustrated as follows: take a rubber ring and place it on a rigid hoop. The way of doing this, which corresponds to $n = 0$, is to contract the whole ring to a single point on the hoop, or to place the ring as shown in Figure 7.4(a). The mapping with $n = +1$ is the mapping shown in Figure 7.4(b); the mapping with $n = -1$ is obtained by placing the rubber ring "upside down." The mapping with $n = 2$ is obtained by making the rubber ring into a figure "eight," folding it as shown in Figure 7.4(c) and then placing it on the hoop. The construction of the mappings with higher n is evident. In analytical form, mappings with different n can be chosen, for example, as

$$\varphi(\theta) = e^{in\theta}v. \tag{7.41}$$

Of course, the phase of the field φ is not gauge invariant. However, the winding number is invariant under gauge transformations which are smooth in the whole of two-dimensional space. For example, one might try to decrease the winding number by unity by taking as gauge function

$$e^{i\alpha(x)} = e^{-i\theta};$$

however, such a gauge transformation is singular at the origin of the coordinate system; in particular, the transformed vector-potential

$$A_i + \frac{1}{e}\partial_i\alpha(x)$$

grows as $1/r$ as $r \to 0$. It is intuitively clear that this example is of a general nature; we shall discuss the properties of mappings of this type in more detail in Chapter 8.

The winding number, which characterizes the field configuration $\varphi(x)$, does not depend on the choice of curve around the origin of the coordinate system, if this curve is in a sufficiently distant region, where $|\varphi| = v$. Indeed, the winding number is discrete and does not change when the curve changes continuously. It is clear from these considerations that the winding number does not change even when the field evolves smoothly in time, provided this evolution does not affect the field at spatial infinity. Thus, the winding number is a topological number characterizing the field configuration and is an integral of motion. As in the example with the kink, the set of all fields with finite energy divides into disjoint sets (sectors); in this case there are infinitely many sectors, which are characterized by an integer $n = 0, \pm 1, \ldots$.

In fact, formula (7.41) exhausts all possible field asymptotics, modulo gauge transformations which are smooth throughout space. To see this, we note, in the first place, that the winding number can be written in explicit form as

$$n = \frac{1}{2\pi i v^2} \oint dx^i \varphi^* \partial_i \varphi,$$

where the integral is taken over a remote circle. If $\varphi = e^{if(\theta)}v$, then $n = \frac{1}{2\pi}[f(2\pi) - f(0)]$. Thus, two fields with the same winding number differ asymptotically by a phase factor with winding number zero: if asymptotically

$$\varphi_1(x) = e^{if_1(\theta)}v$$
$$\varphi_2(x) = e^{if_2(\theta)}v,$$

where $f_1(2\pi) - f_1(0) = f_2(2\pi) - f_2(0)$, then

$$\varphi_2(x) = e^{if_{21}(x)}\varphi_1(x),$$

where $f_{21} = f_2 - f_1$, and

$$f_{21}(2\pi) - f_{21}(0) = 0.$$

Since f_{21} is a single-valued function of θ, it is possible to construct a smooth gauge function

$$\alpha(r, \theta) = g(r) f_{21}(\theta), \qquad (7.42)$$

where $g(r)$ is some arbitrarily chosen smooth function obeying $g(r \to 0) = 0$, $g(r \to \infty) = 1$. Upon gauge transformations with gauge function (7.42), the field φ_1 becomes

$$\varphi_1'(x) = e^{i\alpha(r,\theta)} \varphi_1(x),$$

and the field φ_1' has the same asymptotics as the field φ_2. Thus, using gauge transformations which are smooth throughout space one can achieve that asymptotically the fields have a fixed (for each n) form; this form can be fixed by formula (7.41). In each sector of space of the fields, we now know the asymptotic form of the scalar field.

Let us search for solitons in the sector with $n = 1$. Asymptotically, as $|\mathbf{x}| \to \infty$, the scalar field has the form

$$\varphi = e^{i\theta} v. \qquad (7.43)$$

For the energy to be finite, the covariant derivative $D_i \varphi$ must decrease more rapidly than $1/r$ (otherwise $\int d^2 x |D_i \varphi|^2$ would diverge as $|\mathbf{x}| \to \infty$). The conventional derivative does not have this property, since

$$\partial_i \varphi = e^{i\theta} v i \partial_i \theta = (e^{i\theta} v) \left(-\frac{i}{r} \varepsilon_{ij} n_j \right),$$

where $n_i = \frac{x_i}{r}$ is a unit vector in the direction $\mathbf{x}$. Such a slow decrease in the conventional derivative needs to be compensated by the field A_i with asymptotics

$$A_i = -\frac{1}{er} \varepsilon_{ij} n_j. \qquad (7.44)$$

This asymptotics is a pure gauge, $A_i = \frac{1}{e} \partial_i \theta$, thus the tensor F_{ij} decreases more rapidly than $1/r^2$, although the field A_i decreases as $1/r$.

To obtain a soliton configuration, we need to find a smooth solution of the field equations with asymptotics (7.43), (7.44). To find a corresponding Ansatz we use an approach which works in the case of more complex systems as well. We note that the asymptotics (7.43), (7.44) are invariant under spatial rotations, complemented by global phase transformations of the field φ, i.e.

$$\varphi(\theta) = e^{-i\alpha} \varphi(\theta + \alpha)$$

(the invariance of A_i is evident: rotation of the vector n_i leads to rotation of A_i as a spatial vector). We now write down the most general form of fields, invariant under these generalized spatial rotations:

$$\varphi(r, \theta) = v e^{i\theta} F(r)$$

$$A_i(r, \theta) = -\frac{1}{er}\varepsilon_{ij} n_j A(r) + n_i B(r),$$

where $F(r)$, $A(r)$ and $B(r)$ are functions of radius which have to be found from the field equations. We note further that the contribution from $B(r)$ is a pure gauge,

$$n_i B(r) = \partial_i \left(\int^r B(r) dr \right),$$

therefore, we can set $B(r) = 0$ (this does not contradict the asymptotics (7.44)). Thus, a solution can be sought in the form

$$\varphi(r, \theta) = v e^{i\theta} F(r) \qquad (7.45)$$

$$A_i(r, \theta) = -\frac{1}{er}\varepsilon_{ij} n_j A(r).$$

One not entirely trivial property of this Ansatz is that it "passes through" the field equations, i.e. all field equations are reduced to two equations for $F(r)$ and $A(r)$ (this statement and the results of the next two problems can be generalized to arbitrary models and arbitrary symmetries; this generalization is known as Coleman's theorem).

Indeed, substituting the expression (7.45) in the field equations following from the Lagrangian (7.37), we obtain two ordinary differential equations in the two functions $A(r)$ and $F(r)$:

$$-\frac{d}{dr}\left(\frac{1}{r}\frac{dA}{dr}\right) - 2e^2 v^2 \frac{F^2}{r}(1 - A) = 0, \qquad (7.46)$$

$$-\frac{d}{dr}\left(r\frac{dF}{dr}\right) + \lambda v^2 r F(F^2 - 1) + \frac{F}{r}(1 - A)^2 = 0.$$

It is important that the number of equations is the same as the number of unknown functions, i.e. the system (7.46) is self-consistent.

Problem 9. *Show that for fields of the form (7.45), all the field equations reduce to the two ordinary differential equations (7.46) in $F(r)$ and $A(r)$.*

Problem 10. *Write down the energy functional for fields of the form (7.45) in the form of a simple integral over dr. Find conditions for extremality of this integral and show that they reduce to the equations (7.46).*

It follows from (7.43) and (7.44) that the functions $F(r)$ and $A(r)$ have the following asymptotics:

$$F(r) \to 1, \quad A(r) \to 1, \quad \text{as } r \to \infty. \tag{7.47}$$

Moreover, the requirement of smoothness of the fields at $r = 0$ imposes the condition

$$F(r) \to 0, \quad A(r) \to 0 \quad \text{as } r \to 0 \tag{7.48}$$

(more precisely, $F(r) = O(r)$, $A(r) = O(r^2)$ as $r \to 0$). Unfortunately, it is impossible to find an explicit solution to the equations for $F(r)$ and $A(r)$ with boundary conditions (7.47) and (7.48); the form of the functions $F(r)$ and $A(r)$ can be found numerically.

We can convince ourselves of the existence of solutions of equations (7.46) with the behavior of (7.47) and (7.48), using the following highly reliable, but non-rigorous argument. For definiteness, consider the case $m_H < 2m_V$. We show initially that there is a two-parameter family of solutions of equations (7.46) satisfying the condition (7.47); at this point, we shall not impose the requirement (7.48). For large r, we write $A(r) = 1 - a(r)$, $F = 1 - f(r)$, where we require $a(r) \to 0$ and $f(r) \to 0$ as $r \to \infty$. The first equation of (7.46) is linearized in an evident manner and has the form

$$r \frac{d}{dr} \left(\frac{1}{r} \frac{da}{dr} \right) - m_V^2 a = 0.$$

It has a one-parameter family of solutions, tending to zero as $r \to \infty$,

$$a(r) = C_a \sqrt{r} e^{-m_V r}, \tag{7.49}$$

where C_a is an arbitrary constant. The second equation of (7.46) is also linearized,

$$\frac{1}{r} \frac{d}{dr} \left(r \frac{df}{dr} \right) - m_H^2 f = 0,$$

and we now obtain a family of solutions decreasing as $r \to \infty$:

$$f(r) = C_f \frac{e^{-m_H r}}{\sqrt{r}}, \tag{7.50}$$

where C_f is another arbitrary constant.[4] Thus, we actually have a two-parameter family of solutions, decreasing as $r \to \infty$.

---

[4]The requirement $m_H < 2m_V$ has been imposed to ensure that the third term in the second equation of (7.46) is small in comparison with the second for the solutions (7.49) and (7.50). For $m_H > 2m_V$ a more accurate analysis is required, but the conclusion about the two-parameter family of solutions remains valid.

Let us now study another family of solutions of equations (7.46), namely, those which satisfy the conditions (7.48) (without the requirement (7.47)). For small r we write

$$
\begin{aligned}
F(r) &= \alpha_f r + \beta_f r^3 + \cdots, \\
A(r) &= \alpha_a r^2 + \beta_a r^4 + \cdots,
\end{aligned}
$$

where $\alpha_f, \ldots, \beta_a$ are constants which are as yet unknown. Substituting these expressions in (7.46) and equating terms with the equal powers of r, we obtain that α_f and α_a are arbitrary, while β_f and β_a are expressed in terms of the latter as

$$
\beta_a = -\frac{m_V^2}{8}\alpha_f^2,
$$

$$
\beta_f = -\frac{m_H^2}{16}\alpha_f - \frac{1}{8}\alpha_a\alpha_f
$$

(here, it is important that to leading (zeroth) order in r, the second equation of (7.46) is satisfied identically). Thus, the second family of solutions is also two-parametric, being characterized by the two parameters α_f and α_a.

The solution in which we are interested satisfies both conditions (7.47) and (7.48), i.e. it must belong to both the first and the second family. In other words, some solution from the first family, characterized by certain values C_a and C_f, must be matched to a solution from the second family for some α_a, α_f. The condition for matching of two solutions at some (no matter which) point r_0 is the equality of the functions $F(r_0)$ and $A(r_0)$ and their derivatives $F'(r_0)$ and $A'(r_0)$ at that point. This requirement gives four algebraic equations for four unknown parameters C_a, C_f, α_a and α_f, i.e. the number of parameters is the same as the number of equations for them. Such a system usually has a solution (or a discrete set of solutions), and this is a strong argument supporting the existence of solutions of interest to us.

Of course, this argument is not applicable solely to vortices. Although it is rather heuristic, it leads to correct results in all known cases. We note that when this argument is applied to systems of *linear* equations, account must be taken of the fact that the overall multiplicative constant is in fact not a parameter of a solution.

Problem 11. *Find $F(r)$ and $A(r)$ numerically for $m_H = m_V = 1$.*

Problem 12. *Consider the case $m_H \gg m_V$. Show that the region in which $|\varphi|$ differs significantly from v, i.e. $(|\varphi| - v) \sim v$ has size of the order of $1/m_H$, while the region where $A(r)$ is significantly different from unity* *has size of the order of $1/m_V$. Thus, a soliton has a small scalar core and a relatively large vector core. Find the asymptotics of the function $A(r)$ outside the vector core ($r \gg 1/m_V$). Show that outside the scalar core ($r \gg 1/m_H$, but not necessarily $r \gg 1/m_V$) the field $B_i = (A_i + \frac{1}{er}\varepsilon_{ij}n_j)$ satisfies the equation of a free massive vector field (we note that $B_i \to 0$ as $r \to \infty$). Find an explicit form for $A(r)$ outside the scalar core, i.e. at $r \gg 1/m_H$ (but not necessarily $r \gg 1/m_V$). Find the mass of the soliton with logarithmic accuracy in m_H/m_V, i.e. show that $M_{\mathrm{sol}} = C(m_V, e)(\ln \frac{m_H}{m_V} + O(1))$ and calculate the coefficient C in front of the logarithm.*

Although it is not possible to find a soliton solution explicitly, the mass of the soliton and its size can be estimated based on scaling considerations. We shall do this, assuming that $\frac{m_H}{m_V} \sim 1$. In the energy functional (7.40), we change the variables

$$\begin{aligned}
\varphi(\mathbf{x}) &= v\phi(\mathbf{y}) \\
A_i(\mathbf{x}) &= \frac{m_V}{e}C_i(\mathbf{y}),
\end{aligned}$$

where

$$\mathbf{y} = m_V\mathbf{x}. \tag{7.51}$$

This change of variables is chosen so that the three terms in the energy density have the same order for $\phi \sim 1$, $C_i \sim 1$, $y \sim 1$;

$$\begin{aligned}
F_{ij}^2 &= \left(\frac{m_V^2}{e}\right)^2 C_{ij}^2 \\
|D_i\varphi|^2 &= (m_V v)^2|\mathcal{D}_i\phi|^2 = \left(\frac{m_V^2}{e}\right)^2 \frac{1}{2}|\mathcal{D}_i\phi|^2 \\
\frac{\lambda}{2}(|\varphi|^2 - v^2)^2 &= \frac{\lambda v^4}{2}(|\phi|^2 - 1)^2 = \left(\frac{m_V^2}{e}\right)^2 \frac{m_H^2}{8m_V^2}(|\phi|^2 - 1)^2,
\end{aligned}$$

where $C_{ij} = \frac{\partial C_j}{\partial y^i} - \frac{\partial C_i}{\partial y^j}$, $\mathcal{D}_i\phi = (\frac{\partial}{\partial y^i} - iC_i)\phi$. Thus, the energy functional has the following form in terms of the new variables:

$$E = \frac{m_V^2}{e^2} \int d^2y \left[\frac{1}{4}C_{ij}^2 + \frac{1}{2}(\mathcal{D}_i\phi)^2 + \frac{1}{8}\frac{m_H^2}{m_V^2}(|\phi|^2 - 1)^2\right]. \tag{7.52}$$

For $m_H \sim m_V$, the expression in the integrand does not contain small or large parameters; thus, a minimum of the energy functional is attained at

$$\phi \sim 1, \quad C_i \sim 1,$$

and the characteristic size of the soliton in the new variable is of the order of unity

$$y \sim 1.$$

The last equation, together with (7.51), means that the size of a soliton in space is of the order

$$r_{\text{sol}} \sim \frac{1}{m_V}.$$

Finally, the mass of the soliton (value of the energy functional (7.52) at the minimum) is of the order

$$M_{\text{sol}} \sim \frac{m_V^2}{e^2}.$$

More precisely, the dependence of the mass on the parameters of the model follows from (7.52),

$$M_{\text{sol}} = \frac{m_V^2}{e^2} M\left(\frac{m_H}{m_V}\right), \qquad (7.53)$$

where the function $M\left(\frac{m_H}{m_V}\right)$ can be found numerically.

To end this section, we note that the soliton described in this section can also be viewed as a static solution of the field equations in the model (7.37) in $(3+1)$-dimensional space–time, which does not depend on x^3 and has $A_3 = 0$. Such a solution describes an infinitely long straight object, a vortex or a string. Here, in the general case the topological number n is proportional to the vortex magnetic flux

$$n = \frac{e}{2\pi} \int H_3 d^2 x, \qquad (7.54)$$

where the integral is evaluated over some plane orthogonal to the vortex (or system of parallel vortices), for example, the plane (x^1, x^2). Indeed, the requirement that the covariant derivative of the field $\varphi = e^{i\theta n}$ should decrease rapidly leads to asymptotics of the field A_i of the form (compare with (7.44))

$$A_i = \frac{n}{e} \partial_i \theta, \quad i = 1, 2.$$

Hence, n is expressed in terms of an integral over a remote path in the plane (x^1, x^2):

$$n = \frac{e}{2\pi} \oint A_i dx^i,$$

which leads to (7.54) by Stokes's theorem. The fact that the topological number is integer valued means that the vortex magnetic flux is quantized. Expression (7.53) represents the mass of the vortex per unit length.

The physical systems in which vortices actually occur are type-2 superconductors. The energy functional (7.40) is the Ginzburg–Landau Hamiltonian.

Vortices of the type described are considered in particle theory, although it is still not known whether or not they exist in nature (there are none in the Standard Model). In the cosmological context they are called cosmic strings.

7.4   Soliton in a model of n-field in (2 + 1)-dimensional space–time

The model of **n**-field gives an example of a situation when in theories with scalar fields only the potential term in the Lagrangian is absent, and the existence of a soliton in (2 + 1)-dimensional space–time is not forbidden by the scale arguments of Section 7.2. Another interesting feature of this model is the presence of topological properties, somewhat different from those considered in the previous examples (Sections 7.1 and 7.3).

The model contains three real scalar fields φ^a, $a = 1, 2, 3$, on which a nonlinear condition is imposed at each point x:

$$\varphi^a(x)\varphi^a(x) = 1 \tag{7.55}$$

(sometimes the fields are denoted by n^a, this is how the model obtained its name; we shall reserve the notation **n** for a unit radius vector). Thus, the fields belong to the sphere S^2 of unit radius; only two variables for the three fields φ^a are independent. We choose the Lagrangian of the model in the form

$$\mathcal{L} = \frac{1}{2g^2}\partial_\mu\varphi^a\partial_\mu\varphi^a, \tag{7.56}$$

where g is some constant, $\mu, \nu - 0, 1, 2$. We note that the Lagrangian and condition (7.55) are invariant under global transformations from the group $O(3)$, under which the fields φ^a transform as components of a three-dimensional vector.

Although the Lagrangian (7.56) is quadratic in the fields, the field equations are nonlinear, since the nonlinear condition (7.55) is imposed upon the fields. To obtain these equations we use the standard Lagrange method. We add the relation (7.55) to the action with a Lagrange multiplier, i.e. we write

$$\tilde{S} = \int d^3x \frac{1}{2g^2}\partial_\mu\varphi^a\partial_\mu\varphi^a + \frac{1}{2g^2}\int d^3x\lambda(x)(\varphi^a(x)\varphi^a(x) - 1).$$

Here $\lambda(x)$ is a Lagrange multiplier; the factor $1/2g^2$ in the second term is introduced for convenience. The field equations are obtained by varying $\tilde{S}$ with respect to the fields $\varphi^a(x)$, assuming them to be independent, and choosing the Lagrange multiplier $\lambda(x)$ such that the constraint (7.55) is satisfied.

From the variation of $\tilde{S}$ we obtain

$$-\partial_\mu\partial_\mu\varphi^a(x) + \lambda(x)\varphi^a(x) = 0. \qquad (7.57)$$

We find the Lagrange multiplier by multiplying (7.57) by φ^a and using (7.55);

$$-\varphi^a(x)\partial_\mu\partial_\mu\varphi^a(x) + \lambda(x) = 0,$$

i.e.

$$\lambda(x) = \varphi^a\partial_\mu\partial_\mu\varphi^a. \qquad (7.58)$$

Thus, we finally obtain from (7.57) and (7.58) an equation which does not contain the Lagrange multiplier:

$$-\partial_\mu\partial_\mu\varphi^a + (\varphi^b\partial_\mu\partial_\mu\varphi^b)\varphi^a = 0. \qquad (7.59)$$

This equation is nonlinear, as was to be expected.

We shall study the static (not dependent upon time) configurations of fields with a finite energy. The energy functional for these has the form

$$E = \frac{1}{2g^2}\int \partial_i\varphi^a\partial_i\varphi^a, d^2x. \qquad (7.60)$$

Let us first consider the ground state, which is a field configuration with least energy. It is clear that the least value of the energy is zero, which is realized for homogeneous (not depending on $\mathbf{x}$) fields. As usual, taking into account the global $O(3)$ symmetry, we can choose as the ground state any constant vector, and the global $O(3)$ symmetry is broken. Let us choose as the ground state the configuration

$$\varphi^a = -\delta^{a3}, \qquad (7.61)$$

which corresponds to the south pole of the sphere S^2.

Let us now discuss static configurations of fields with finite energy. The finiteness of energy means that $\varphi^a(x)$ tends to a constant as $|\mathbf{x}| \to \infty$, where this constant does not depend on the angle in the plane (x^1, x^2) (otherwise $\nabla\varphi^a \sim 1/r$ and the integral in (7.60) diverges). As far as topological properties of such configurations are concerned, all spatially infinite "points" can be identified, and the space becomes topologically equivalent to the two-dimensional sphere S^2. Every configuration of fields $\varphi^a(\mathbf{x})$ with finite energy determines a mapping of the sphere S^2 (space with infinity identified) to S^2 (set of values of the field). As in the case of mappings from S^1 to S^1, the mapping from S^2 to S^2 is characterized by a topological number $n = 0, \pm 1, \pm 2, \ldots$, called the degree of the mapping. The construction of mappings with different n reiterates the construction studied in Section 7.3, except that instead of a ring, one has to consider a sphere. Thus, the set of configurations of fields φ^a divides into disjoint subsets (sectors): the sector with $n = 0$ contains the vacuum, while in the sector with $n = 1$ the topological soliton can be sought.

It is useful to derive an explicit formula for n as a functional of $\varphi^a(x)$. For this, we consider a mapping of the region near the point **x** to the region near the point $\vec{\varphi}(\mathbf{x})$, shown in Figure 7.5 (we shall sometimes consider the triplet φ^a as the vector $\vec{\varphi}$ of unit length in the three-dimensional space of the fields).

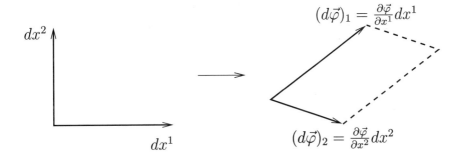

Figure 7.5.

The area of the region obtained by this mapping is equal to

$$d\vec{\sigma} = (d\vec{\varphi})_1 \times (d\vec{\varphi})_2,$$

and the vector $d\vec{\sigma}$ may be either parallel or antiparallel to the vector $\vec{\varphi}$ ($d\sigma$ is a region on the sphere S^2, and $\vec{\varphi}$ is orthogonal to that sphere). If the mapping $S^2_{\text{space}} \rightarrow S^2_{\text{field}}$ has degree n, then the sphere S^2_{field} is covered n times, i.e.

$$n = \frac{1}{4\pi}(\text{surface area, swept by mapping}).$$

Here, the area of an element of the surface should be taken with the plus sign if the orientation of $(d\vec{\varphi})_1$ and $(d\vec{\varphi})_2$ is the same as that of $(d\vec{x}^1)$ and $(d\vec{x}^2)$, and with minus sign otherwise (this latter case is shown in Figure 7.5). The sign is derived correctly if we write

$$n = \frac{1}{4\pi} \int \vec{\varphi} d\vec{\sigma},$$

which gives

$$n = \frac{1}{4\pi} \int d^2x\, \vec{\varphi} \cdot \left[\frac{\partial\vec{\varphi}}{\partial x^1} \times \frac{\partial\vec{\varphi}}{\partial x^2} \right]$$

$$= \frac{1}{8\pi} \int d^2 x \varepsilon^{abc} \varepsilon_{ij} \varphi^a \partial_i \varphi^b \partial_j \varphi^c. \tag{7.62}$$

This topological number does not change under smooth variations of the fields φ^a, which do not affect spatial infinity. Unlike the topological numbers we met in Sections 7.1 and 7.3, it is associated not with the properties of the field at spatial infinity, but with the fields in the whole space.

Problem 13. *Show by explicit calculation that the topological number (7.62) does not change under local variations of the field φ^a, i.e. under replacement of φ^a by $\varphi^a + \delta\varphi^a$, where $\delta\varphi^a$ is an infinitesimal variation of the field which decreases rapidly at spatial infinity.*

In what follows, we shall suppose, without loss of generality, that the field $\varphi^a(\mathbf{x})$ tends to the vacuum value $(-\delta^{a3})$ at spatial infinity.

Problem 14. *Consider the configuration of fields of the form*

$$\begin{aligned}
\varphi^\alpha(\mathbf{x}) &= n_\alpha \sin f(r) \\
\varphi^3(\mathbf{x}) &= \cos f(r),
\end{aligned}$$

where $\alpha = 1, 2$; $n_\alpha = \frac{x_a}{r}$ is the unit radius vector in two-dimensional space, $f(r)$ is a real function, $f(r) \to \pi$ as $r \to \infty$, such that the field φ^a tends to the vacuum value at spatial infinity. Since $\varphi^\alpha \varphi^\alpha + (\varphi^3)^2 = 1$, this configuration is actually a possible $\mathbf{n}$-field configuration.

1. *Find the boundary condition for $f(r)$ at $r = 0$ which ensures that the field $\varphi^a(\mathbf{x})$ is smooth. Show that all boundary conditions are characterized by an integer.*

2. *Calculate the topological number (7.62) for the given configurations. Find its connection with the integer from the previous point.*

As previously mentioned, the soliton has to be sought in the sector with topological number 1. In order to find an explicit solution, we use a technique (Belavin and Polyakov 1975) which also has an analogy in certain more complicated models. Let us consider the quantity

$$F_i^a = \partial_i \varphi^a \pm \varepsilon^{abc} \varepsilon_{ij} \varphi^b \partial_j \varphi^c$$

(later, it will become clear that the sign $+$ should be chosen for positive topological numbers, and the sign $-$ otherwise). Evidently, we have

$$\int F_i^a F_i^a d^2 x \geq 0, \tag{7.63}$$

where equality holds only if

$$\partial_i \varphi^a \pm \varepsilon^{abc} \varepsilon_{ij} \varphi^b \partial_j \varphi^c = 0. \tag{7.64}$$

On the other hand,

$$F_i^a F_i^a = \partial_i \varphi^a \partial_i \varphi^a \pm 2\varepsilon^{abc} \varepsilon_{ij} \partial_i \varphi^a \varphi^b \partial_j \varphi^c + \varepsilon^{abc} \varepsilon_{ij} \varepsilon^{ade} \varepsilon_{ik} \varphi^b \partial_j \varphi^c \varphi^d \partial_k \varphi^e. \tag{7.65}$$

The last term here is equal to

$$
\begin{aligned}
\delta_{jk}(\delta^{bd}\delta^{ce} - \delta^{be}\delta^{cd})\varphi^b \varphi^d \partial_j \varphi^c \partial_k \varphi^e &= \partial_j \varphi^c \partial_j \varphi^c - (\varphi^b \partial_j \varphi^b)(\varphi^c \partial_j \varphi^c) \\
&= \partial_j \varphi^c \partial_j \varphi^c,
\end{aligned}
$$

where we use the relation $\varphi^a \varphi^a = 1$ and its consequence

$$\varphi^a \partial_i \varphi^a = 0.$$

Renaming and permuting the indices in (7.65), we obtain

$$F_i^a F_i^a = 2\partial_i \varphi^a \partial_i \varphi^a \mp 2\varepsilon^{abc} \varepsilon_{ij} \varphi^a \partial_i \varphi^b \partial_j \varphi^c.$$

Thus, the inequality (7.63) gives

$$\int d^2x \, \partial_i \varphi^a \partial_i \varphi^a \geq \pm \int d^2x \varepsilon^{abc} \varepsilon_{ij} \varphi^a \partial_i \varphi^b \partial_j \varphi^c,$$

or

$$E \geq \frac{4\pi}{g^2}|n|. \tag{7.66}$$

Hence, we note that in the sector with topological number n, the energy is bounded from below by the value $\frac{4\pi}{g^2}|n|$, where the absolute minimum of the energy in each sector is attained if the field satisfies equation (7.64). A soliton is the absolute minimum of the energy in the sector with $n = 1$; this can be found by solving equation (7.64) (with the choice of sign $+$). It is important to note that, unlike the original field equation (7.59), equation (7.64) is a first-order equation and is easier to solve.

Problem 15. *Show that any solution of equation (7.64) is also a solution of equation (7.59).*
    We now find an explicit expression for soliton fields. For this, we use an Ansatz which is invariant under $SO(2)$ spatial rotations, complemented by $SO(2)$ rotations around the third axis in the space of the fields

$$
\begin{aligned}
\varphi^\alpha(\mathbf{x}) &= n^\alpha \sin f(r) \\
\varphi^3(\mathbf{x}) &= \cos f(r),
\end{aligned}
\tag{7.67}
$$

where $n^\alpha = \frac{x^\alpha}{r}$, $\alpha = 1, 2$. The condition $\varphi^a \varphi^a = \varphi^\alpha \varphi^\alpha + (\varphi^3)^2 = 1$ is automatically satisfied. The derivatives of the fields are equal to

$$\partial_i \varphi^\alpha = \frac{1}{r}(\delta^{i\alpha} - n^i n^\alpha) \sin f + n^i n^\alpha f' \cos f$$

$$\partial_i \varphi^3 = -n^i f' \sin f.$$

Moreover,

$$\varepsilon_{ij} \varepsilon^{3\alpha\beta} \varphi^\alpha \partial_j \varphi^\beta = \varepsilon_{ij} \varepsilon_{\alpha\beta} n_\alpha \sin f \left[\frac{1}{r}(\delta^{i\alpha} - n^i n^\alpha) \sin f + n^i n^\alpha f' \cos f \right]$$

$$= \varepsilon_{ij} \varepsilon_{\alpha\beta} \delta^{j\beta} n_\alpha \frac{1}{r} \sin^2 f = n_i \frac{1}{r} \sin^2 f.$$

Equation (7.64) with $a = 3$ takes the form

$$-n_i f' \sin f + n_i \frac{1}{r} \sin^2 f = 0,$$

or

$$f' = \frac{1}{r} \sin f. \tag{7.68}$$

Equation (7.64) with $a = 1, 2$ reduces to equation (7.68).

Problem 16. *Show that equation (7.64) with $a = 1, 2$ for fields of the form (7.67) reduces to equation (7.68).*

The solution of equation (7.68) with boundary condition which ensures that $\varphi^a = -\delta^{a3}$ as $r \to \infty$,

$$f(\infty) = \pi,$$

has the form

$$f = 2 \arctan \frac{r}{r_0}$$

so that

$$\varphi^\alpha = 2 \frac{x_\alpha r_0}{r_0^2 + r^2}$$

$$\varphi^3 = \frac{r_0^2 - r^2}{r_0^2 + r^2},$$

where r_0 is an arbitrary constant (soliton size). The fact that the soliton size may be arbitrary, actually follows already from the scale considerations of Section 7.2. From the results of Problem 14 of this section, it follows that the topological number of this soliton is actually equal to 1, and hence it follows from (7.66) that its energy (mass) does not depend on size and is equal to

$$M_{\text{sol}} = \frac{4\pi}{g^2}. \tag{7.69}$$

Problem 17.   *Calculate the soliton energy explicitly and verify that equation (7.69) holds.*

Problem 18.   *Let us introduce the variables $W_\alpha = 2\frac{\varphi^\alpha}{1-\varphi^3}$, $\alpha = 1, 2$ and also the complex variable $W = W_1 + iW_2$. Show that equations (7.64) are Cauchy–Riemann conditions, guaranteeing that W is a function of the complex variable $z = x^1 + ix^2$. Using this property, find a general n-soliton solution (i.e. a solution with topological number n). Show that the one-soliton solution found earlier is the unique solution modulo $O(3)$ rotations and spatial translations.*

Chapter 8

Elements of Homotopy Theory

8.1 Homotopy of mappings

In Chapter 7, we met examples of mappings from one manifold to another. There, the global properties of these mappings, which remain unchanged under continuous changes of the mappings, were important. In this chapter, we briefly discuss certain topological (i.e. global) properties of mappings in quite general form, together with some specific results which are useful for physics.

Let X, Y be topological spaces (we shall often simply refer to spaces), i.e. sets in which the concept of the proximity of two points is defined. For us, the important cases will be those in which the topological spaces are domains of dimension n or lower, of Euclidean space R^n. A mapping $f : X \to Y$ is continuous if it takes any nearby points in X to nearby points in Y. In what follows we shall consider only continuous mappings, unless otherwise stated. The mappings $f : X \to Y$ and $g : X \to Y$ are said to be *homotopic* if one can be continuously deformed to the other, i.e. if there exists a family of mappings h_t, depending continuously on the parameter $t \in [0, 1]$, such that

$$h_0 = f, \quad h_1 = g.$$

In other words, if I is the interval $[0, 1]$, then there exists a continuous mapping H of the direct product $X \times I$ to Y, such that

$$H(x, 0) = f(x), \quad H(x, 1) = g(x)$$

(here, $H(x, t) = h_t(x)$ for all $x \in X$).

This definition can be formulated slightly differently: suppose $C(X,Y)$ is the set of continuous mappings from X to Y, then f is homotopic to g if f and g belong to the same connected component of $C(X,Y)$.

Problem 1. *Show that homotopy is an equivalence relation in $C(X,Y)$.*

The homotopy relation divides $C(X,Y)$ into equivalence classes. The central problem, for us, is to describe the set of equivalence classes (homotopy classes), which we denote by $\{X,Y\}$.

Example. If X consists of a single point, then $C(X,Y)$ coincides with Y and $\{X,Y\}$ coincides with the set of connected components of the space Y.

A mapping $f : X \to Y$ is said to be *homotopic to zero* if it is homotopic to a mapping taking the whole space X to a single point of Y. If Y is connected, then all such mappings are homotopic to each other. Their equivalence class is called the *zero homotopy class*. In what follows, we shall consider connected topological spaces X, Y, unless otherwise stated.

Example. Let $X = S^1$, $Y = R^2 \setminus \{0\}$ (Y is the plane with a point removed). The mapping $f : S^1 \to R^2 \setminus \{0\}$, shown in Figure 8.1, is homotopic to zero, the mapping g is not homotopic to zero.

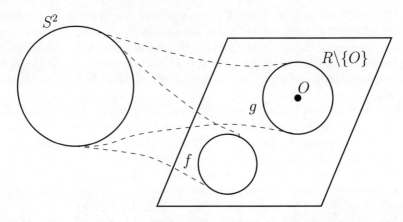

Figure 8.1.

Problem 2. *Let Y be a convex space in R^n, i.e. a set which for any two of its points also contains the straight interval connecting them. Show that a mapping of any space X to Y is always homotopic to zero.*

Problem 3. *Let f be a mapping of the sphere S^n to the topological space X. Let D^{n+1} be the ball with boundary S^n (written as $\partial D^{n+1} = S^n$). Show* *that f is homotopic to zero if and only if it can be continuously extended into the entire ball D^{n+1}, i.e. there exists a continuous mapping $g : D^{n+1} \to X$, such that $g|_{\partial D^{n+1}} = f$.*

The well-known concept of a simply connected space (two paths with the same beginning and end can always be continuously deformed to each other) can be formulated as follows in the language of homotopy theory: the space Y is simply connected if any mapping $S^1 \to Y$ is homotopic to zero.

Let us consider a mapping to a direct product

$$f : X \to Y \times Z.$$

It can be viewed as a pair of mappings

$$f_1 : X \to Y, \quad f_2 : X \to Z,$$

such that $f(x)$ is the pair $(f_1(x), f_2(x))$. For continuous deformation of the mapping f, the deformations of the mappings f_1 and f_2 are continuous. Consequently, there exists a one-to-one correspondence between $\{X, Y \times Z\}$ and $\{X, Y\} \times \{X, Z\}$. The classification of mappings to a direct product reduces to the classification of mappings to each of its factors.

The homotopy of mappings is closely related to the concept of the *homotopy of topological spaces.* Two spaces Y_1 and Y_2 are said to be equivalent (homotopic) if there exist mappings

$$h_1 : Y_1 \to Y_2 \text{ and } h_2 : Y_2 \to Y_1$$

such that $h_1 h_2 : Y_2 \to Y_2$ and $h_2 h_1 : Y_1 \to Y_1$ are homotopic to the identity (the mapping $e : Y \to Y$ is said to be the identity if $e(y) = y$ for all $y \in Y$). If Y_1 and Y_2 are homotopic and X is any topological space, then there exists a one-to-one correspondence between $\{X, Y_1\}$ and $\{X, Y_2\}$.

Problem 4. *Let Y_1 and Y_2 be homotopic and h_1, h_2 the mappings of the previous paragraph. Each mapping $f : X \to Y_1$ induces a mapping $g : X \to Y_2$ via the formula $g = h_1 f$. 1) Show that in this way a mapping from $\{X, Y_1\}$ to $\{X, Y_2\}$ is induced, i.e. if f and f' belong to the same class in $\{X, Y_1\}$, then $h_1 f$ and $h_1 f'$ belong to the same class in $\{X, Y_2\}$. 2) Show that this mapping from $\{X, Y_1\}$ to $\{X, Y_2\}$ is one to one.*

Problem 5. *Show that S^1 and $R^2 \backslash \{0\}$ are homotopically equivalent. For this, construct the mappings $h_1 : S^1 \to R^2 \backslash \{0\}$ and $h_2 : R^2 \backslash \{0\} \to S^1$, given in the definition of homotopic equivalence.*

Generally, the sphere S^{n-1} and the n-dimensional Euclidean space with a point removed, $R^n \backslash \{0\}$, are homotopically equivalent.

Another example of homotopically equivalent spaces is given by the sphere with a point removed (let us say, the north pole) $S^n \setminus \{$n.p.$\}$ and R^n. Finally, we recall one further example from Section 7.4: the space R^n with infinity identified is homotopically equivalent to the sphere S^n (here, stereographic projection plays the role of the mapping h_1).

Thus, the homotopy classes $\{X, Y\}$ can be analyzed by considering mappings of X to a space Y', homotopically equivalent to Y. The choice of the space Y' is a matter of convenience.

Conversely, if X_1 and X_2 are homotopically equivalent spaces, then the sets $\{X_1, Y\}$ and $\{X_2, Y\}$ are in one-to-one correspondence for any topological space Y.

Problem 6. *Prove the last assertion.*

8.2   The fundamental group

In this section, we shall discuss mappings of a circle S^1 to topological spaces. These mappings can be viewed as mappings f from the interval $[0, 1]$ to X, such that $f(0) = f(1)$. We shall consider connected spaces X. We fix the point x_0 to which the initial and final points of the interval $[0, 1]$ are mapped: $f(0) = f(1) = x_0$. Mappings with this property form a subset of the set $C(S^1, X)$ of all mappings from S^1 to X; for all these mappings, one can also define the concept of homotopy by complete analogy with Section 8.1 (in addition to the usual continuity requirement, the condition $h_t(0) = h_t(1) = x_0$ for each t is imposed on the family of mappings h_t occurring in the definition of homotopy; we shall see that this constraint is not significant). The set of homotopy classes of mappings f from the interval $[0, 1]$ to X such that $f(0) = f(1) = x_0$ will be denoted by $\pi_1(X, x_0)$.

A mapping from the interval $[0, 1]$ to X is called a path in X. Thus, the mappings in which we are interested are paths in X, beginning and ending at the point x_0. $\pi_1(X, x_0)$ is the set of homotopy classes of closed paths starting and ending at x_0.

We introduce a *group structure* in $\pi_1(X, x_0)$, as follows. Suppose f and g are two paths in X, starting and ending at the point x_0. We construct the path $f * g$ to be the path which first runs along g and then along f, as shown in Figure 8.2. The mapping $f * g$ from the unit interval $[0, 1]$ to X can be written as

$$(f * g) = \begin{cases} g(2\xi), & 0 \leq \xi \leq 1/2 \\ f(2\xi - 1), & 1/2 \leq \xi \leq 1, \end{cases}$$

where $\xi \in [0, 1]$. Since $g(1) = f(0) = x_0$, the two halves of the formula "match" at the point $\xi = 1/2$, so that the mapping $f * g$ defined by this equation is indeed continuous.

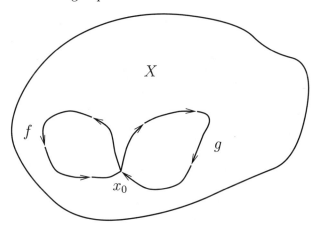

Figure 8.2.

The inverse mapping to f is the path going in the reverse direction:

$$f^{-1}(\xi) = f(1 - \xi)$$

(Figure 8.3). The operation $*$ and the taking of the inverse mapping induce operations in $\pi_1(X, x_0)$.

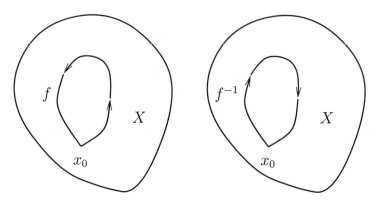

Figure 8.3.

Problem 7. *Suppose we have mappings from* $[0, 1]$ *to* X *starting and ending at* x_0, *such that* f' *is homotopic to* f *and* g' *is homotopic to* g.

Show that $f' * g'$ is homotopic to $f * g$ and f'^{-1} is homotopic to f^{-1}. Thus, the operations defined above induce operations in $\pi_1(X, x_0)$.

Let us choose as the unit element in $\pi_1(X, x_0)$ the homotopy class containing the mapping for which the whole of the interval $[0, 1]$ is mapped to the single point x_0 (zero homotopy class of mappings from S^1 to X). The operations described above, $*$ and the taking of the inverse element, are group operations in $\pi_1(X, x_0)$.

Problem 8.   *Verify that $\pi_1(X, x_0)$ is a group under the operations $*$ and the taking of an inverse element.*

The group $\pi_1(X, x_0)$ is called the *fundamental group*.

If the space X is connected, then $\pi_1(X, x_0)$ and $\pi_1(X, x_0')$ are isomorphic for all x_0 and x_0'. The isomorphism is induced as follows. Let us choose, once and for all, a path joining x_0 and x_0'. Every path f beginning and ending at the point x_0 corresponds to a path f' starting and ending at the point x_0', as illustrated in Figure 8.4. (Here, the path from x_0' to x_0 is traversed twice–in the direct and the reverse directions). This correspondence between paths induces a correspondence between $\pi_1(X, x_0)$ and $\pi_1(X, x_0')$. One proves that this correspondence is one-to-one and preserves the group properties.

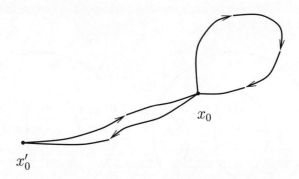

Figure 8.4.

This isomorphism depends, generally speaking, on the choice of path from x_0 to x_0'. It can also be shown that, in the case of a commutative fundamental group, the isomorphism *does not depend* on this path. Then, one can speak about the group $\pi_1(X)$, without specifying the point x_0. In the general case of a non-commutative fundamental group, we can speak about the group $\pi_1(X)$ only as an abstract group, i.e. without specifying which particular element corresponds to which specific path.

Problem 9. *Based on the discussions in Section 7.3, show that*

$$\pi_1(S^1) = Z,$$

where Z is the group of integers under addition.

Problem 10. *Give an example of a topological space X, whose fundamental group is non-commutative.*

From the result of the last problem, it follows that, generally speaking, the fundamental group is non-commutative. One can prove, however, that the fundamental group of any *Lie group* is commutative.

8.3 Homotopy groups

In the previous section, we saw that the set of homotopy classes $\pi_1(X)$ of mappings of a circle S^1 to a topological space X has a group structure. This construction can be generalized to mappings of the spheres S^n of higher dimensions $n \geq 2$. Here, the homotopy groups $\pi_n(X)$ turn out to be Abelian.

Let us consider mappings $f : S^n \to X$ for which the south pole of the sphere is mapped to a fixed point x_0. Such mappings are called *spheroids* (multidimensional analogues of a path). Two spheroids are said to be homotopic if they can be continuously deformed, the one to the other, in such a way that the south pole is always mapped to the point $x_0 \in X$. The set of homotopy classes (with respect to the homotopy defined in this way) is called an *nth homotopy group* and is denoted by $\pi_n(X, x_0)$. We shall see that homotopy groups do not depend on the choice of x_0 for connected spaces X, but for the present, we shall hold the point x_0 fixed.

For what follows, it is useful to note that the sphere S^n is homotopically equivalent to an n-dimensional cube I^n with the boundary identified. Thus, a spheroid is a mapping from the cube I^n to X, under which the whole of the boundary of the cube is mapped to x_0.

We define the *sum* of two spheroids as follows (for $n \geq 2$ the group $\pi_n(X)$ is Abelian and the group operation in it is called the sum). Let $f, g : I^n \to X$ be two spheroids (f and g map the boundary of the cube to x_0). We define their sum $h = f + g$ by the formula

$$h(x^1, x^j) = \begin{cases} f(2x^1, x^j) & \text{for } 0 \leq x^1 \leq 1/2 \\ g(2x^1 - 1, x^j) & \text{for } 1/2 \leq x^1 \leq 1 \, . \end{cases} \tag{8.1}$$

Here, $j = 2, 3, \ldots, n$, $x^1, \ldots, x^n$ are the coordinates in the cube I^n and take values from 0 to 1. Since all points $(1, x^j)$ and $(0, x^j)$ belong to the boundary of the cube, we have

$$f(1, x^j) = g(0, x^j) = x_0 \tag{8.2}$$

and (8.1) defines a continuous mapping. The sum (8.1) is illustrated in Figure 8.5; half of the cube is mapped using the mapping f, the other half using the mapping g.

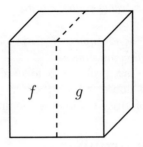

Figure 8.5.

Summation of spheroids induces summation of the homotopy classes $\pi_n(X, x_0)$: if f is homotopic to f', and $\tilde{f}_t$ is a family of mappings connecting f and f' (i.e. $\tilde{f}_0 = f$, $\tilde{f}_1 = f'$), g is homotopic to g' and $\tilde{g}_t$ is a corresponding family of mappings, then $\tilde{f}_t + \tilde{g}_t$ is a family of mappings connecting $(f + g)$ and $(f' + g')$.

The zero (group unit) in $\pi_n(X)$ is the class containing the mapping of the whole cube to x_0.

The inverse spheroid is the mapping given by the formula

$$f^{-1}(x^1, x^j) = f(1 - x^1, x^j).$$

The taking of the inverse mapping also induces an operation in $\pi_n(X, x_0)$.

Problem 11. *Show that the mapping $f + f^{-1}$ is homotopic to zero.*

Thus, the addition operation in $\pi_n(X, x_0)$ is indeed a group operation (its associativity is obvious). For $n \geq 2$, this operation is commutative, since it can be shown that $(f+g)$ is homotopic to $(g+f)$. The corresponding family of mappings is constructed as follows. We first deform $f+g$, as shown in Figure 8.6 (the dimensions $x^3, \dots, x^n$ are not shown in the figure). Here, the shaded areas are mapped entirely to x_0. The remaining deformation is evident (see Figure 8.7). Thus, $\pi_n(X, x_0)$ is an Abelian group for $n \geq 2$.

As we have already mentioned, the groups $\pi_n(X, x_0)$ and $\pi_n(X, x_0')$ are isomorphic for all $x_0, x_0' \in X$, if the topological space X is connected. The isomorphism is constructed as follows. Suppose the path α in X, connecting the points x_0 and x_0', is chosen once and for all. The spheroid f will be viewed as a mapping of the ball D^n, under which the whole of the boundary of the latter is mapped to the point x_0 (the ball is clearly equivalent to a cube). Suppose the spheroid f maps the boundary of the

Figure 8.6.

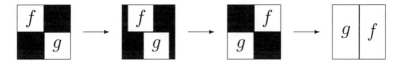

Figure 8.7.

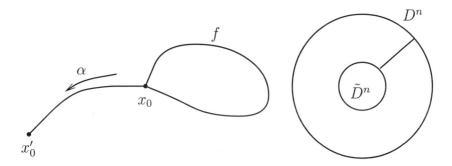

Figure 8.8.

ball to the point x_0, then f belongs to one of the homotopy classes in $\pi_n(X, x_0)$. Let us construct a mapping f' of the ball D^n to X, such that the boundary of the ball is mapped to x_0'. For this we take a ball $\tilde{D}^n$ of smaller size inside D^n and map it using $f \in X$. Here, its boundary $\partial \tilde{D}^n$ is mapped to the point x_0. The mapping of the remaining part of the ball D^n is performed in such a way that each radial interval is mapped to the path α (see Figure 8.8). Since the boundary of the sphere $\tilde{D}^n$ is mapped to x_0, this can be done; the mapping obtained is continuous, and the boundary of the sphere D^n is mapped to x_0'. One can show that the correspondence between spheroids with designated point x_0 and spheroids with designated point x_0' constructed in this way induces an isomorphism of the groups $\pi_n(X, x_0)$ and $\pi_n(X, x_0')$. Thus, the Abelian group $\pi_n(X, x_0)$ does not depend on the choice of point x_0 (for connected X); it is called the nth homotopy group of the space X and is denoted by $\pi_n(X)$.[1]

To end this section, we present some simple results, relating to the homotopy groups of spheres.

The *homotopy groups* $\pi_n(S^m)$ are *trivial* (consist of the zero element only) for $n < m$; we write

$$\pi_n(S^m) = 0 \text{ for } n < m.$$

Indeed, let us consider a continuous mapping[2] $S^n \to S^m$ for $n < m$. There is at least one point in S^m, to which no point of S^n is mapped.[3] This point can be removed from S^m, S^m with this point removed is homotopically equivalent to R^m, and in R^m any spheroid is homotopic to zero (this is evident from the fact that R^m is homotopically equivalent to a single point).

The *homotopy groups* $\pi_n(S^n)$ are *isomorphic to Z, the group of all integers under addition.* We have already met this statement in Chapter 7 (for $n = 1$ and $n = 2$). The corresponding topological number is called the degree of the mapping f, $\deg f$. An analytic expression for $\deg f$ can be found which generalizes the expression for the degree of the mapping from $S^2 \to S^2$ of Section 7.4. Let S and S' be two n-dimensional spheres, and suppose the mapping $f : S \to S'$ is defined by the relationship between the coordinates $x^1, \ldots, x^n$ of a point on the sphere S and the coordinates $y^1, \ldots, y^n$ on the sphere S':

$$\begin{aligned}
y^1 &= f_1(x^1, \ldots, x^n) \\
y^2 &= f_2(x^1, \ldots, x^n) \\
&\vdots \\
y^n &= f_n(x^1, \ldots, x^n).
\end{aligned} \qquad (8.3)$$

One can prove that for almost all points $y \in S'$ there are no roots of the equation $f(x) = y$ such that the Jacobian $J(x) = \det\left(\frac{\partial f_i}{\partial x^j}\right)$ is equal to zero (Sard's theorem). Points where the Jacobian is non-zero are said to

---

[1] As in the case of the fundamental group, this isomorphism depends on the path α. If this isomorphism does not depend on the path, then the space is said to be n-simple. For $n > 1$, n-simplicity is no longer associated with commutativity of the fundamental group.

[2] Here and in what follows, we shall not distinguish between continuous and smooth mappings; this does not lead to confusion.

[3] Strictly speaking, this is not true (a counterexample can be constructed using Peano's curve). To make the proof rigorous, we can use the free-point lemma. *Let U be an open subset of the space R^p and $\varphi : U \to \text{Int } D^q$ a continuous mapping such that the set $V = \varphi^{-1}(d^q) \subset U$, where d^q is some closed ball in $\text{Int } D^q$, is compact. If $q > p$ then there exists a continuous mapping $\psi : U \to \text{Int } D^q$, which coincides with φ outside V and is such that its image does not cover the whole of the ball d^q.* For a proof of the lemma, see the book "Course in Homotopic Topology" by A.T. Fomenko and D.B. Fuks.

be regular in S and S'; thus, almost all points of S and S' are regular ("almost all" means all, with the exception of a set of measure zero).

This situation is illustrated in Figure 8.9, which shows a mapping $f :$ $S^1 \to S^1$ with degree 1. All points of the circle S, apart from x_0 and x_1 are regular; all points of the circle S', apart from the points y_0, y_1, are regular.

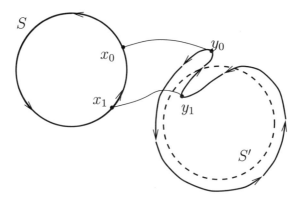

Figure 8.9.

The degree of the mapping from S to S' coincides with the algebraic number of solutions of the equation

$$f(x) = y$$

at a regular point y. This algebraic number is defined as the sum

$$\sum_{\text{roots } f(x)=y} \text{sign } J(x_i). \qquad (8.4)$$

Thus

$$\deg f = \sum_{\text{roots } f(x)=y} \text{sign } J(x_i), \qquad (8.5)$$

for a regular point y. One can prove that the right-hand side of this equation does not depend on y for regular points y and indeed determines the topological number of the mapping from S^n to S^n.

From Figure 8.9, the equation $f(x) = y$ for a regular point y may have either one or three solutions; in either case, the sum (8.4) for regular points is equal to one.

Formula (8.5) is an analytic formula for the degree of a mapping. To bring it to a more convenient form, we use the equation for the delta function

$$\delta(f(x) - y) = \sum_i \frac{1}{|J(x_i)|} \delta(x - x_i(y)),$$

which holds for a regular point y; here, the summation is over all roots of the equation $f(x) = y$. Hence,

$$\deg f = \int dx\, J(x)\delta(f(x) - y)$$

for any regular point y. Integrating this equation with an arbitrary weight $\mu(y)$, we obtain (taking into account that almost all points of S' are regular)

$$\deg f = \frac{1}{\int dy\mu(y)} \int dx\, J(x)\mu(f(x)). \qquad (8.6)$$

This is the desired analytic expression for the degree of the mapping. The choice of the weight μ is a matter of convenience for each specific problem.

Problem 12. *Verify that* $\deg f$ *does not change under smooth deformations of the mapping* f, *i.e. under smooth variations of the functions* f_i *(see (8.3)).*

Problem 13. *Show that the analytic expression for the degree of the mapping* $S^2 \to S^2$ *introduced in Section 7.4, indeed has the structure (8.6). Find the corresponding weight* $\mu(y)$.

8.4　Fiber bundles and homotopy groups

When calculating homotopy groups (and not only for problems of that type), the following construction is very useful.

Let E and B be topological spaces. Suppose we have a mapping

$$p : E \to B.$$

To this correspond disjoint subspaces of E; for each point $b \in B$, one can construct the space of its inverse images in E, i.e. of points $x \in E$ such that $p(x) = b$. We denote the space of inverse images of the point b by F_b. If all F_b are topologically equivalent to one another, then, we have a *fiber bundle*, the space E is called the *bundle space*, B is the *base space*, the space F, to which all F_b are equivalent is a *fiber* of the fiber bundle, p is the *bundle projection*. Here $F_b = p^{-1}(b)$ is the fiber over the point b. The whole construction is denoted by (E, B, F, p) or (E, B, F).

The simplest example of a fiber bundle is the surface of a cylinder, where p is its orthogonal projection onto the base. The base space of this fiber bundle is the circle, a typical fiber is an interval. This example is an example of a *trivial* fiber bundle. The bundle (E, B, F, p) is said to be trivial if $E = B \times F$, and the projection p maps the point $(b, f) \in E$ to the point $b \in B$. Clearly, $p^{-1}(b) = F$ for each point b of a trivial fiber bundle.

A less simple example is that of a *tangent bundle*. Suppose B is a manifold in R^n. At each point $b \in B$, we construct all possible tangent vectors; these form a space, equivalent to R^k, where k is the dimension of the manifold B. The space of the tangent bundle E is the set of all tangent vectors at all points of the manifold B, where the vectors are assumed to be "attached" to points of the manifold B (if b and b' are different points of the manifold B, then the vectors tangent to B at the points b and b' are considered as different). The base space of the tangent bundle is the manifold B, and the projection p takes any vector to the point of B to which it is "attached". Clearly, a typical fiber is R^k.

Problem 14. *Let B be the two-dimensional sphere S^2 and F_b the set of non-zero tangent vectors at the point b in B, such that F_b is topologically equivalent to R^2 with the origin of the coordinate system removed, $F_b = R^2 \setminus \{0\}$. Let us construct the fiber bundle with the same projection p as in the tangent bundle (i.e. E consists of all non-zero tangent vectors to S^2). Show that this fiber bundle is non-trivial (hint: use the "hedgehog theorem": there does not exist a continuous configuration of non-zero tangent vectors to the sphere S^2).*

In what follows, we shall be concerned with "*locally trivial fiber bundles*," i.e. such that for each point $b \in B$, there exists a neighborhood $U \subset B$ of b such that $p^{-1}(U) \approx U \times F$; moreover, there exists a homeomorphism $\varphi : p^{-1}(U) \to U \times F$, such that $p = f \circ \varphi$, where $f : U \times F \to U$ is the natural projection.

Furthermore, we shall assume the necessary smoothness properties for all mappings, when we are dealing with manifolds.

A number of important examples of fiber bundles emerge if one has a space E upon which a compact Lie group G acts (we recall that the group G acts on E if to any $g \in G$ there corresponds an invertible mapping $\varphi(g) : E \to E$, such that

$$
\begin{aligned}
\varphi(g_1 g_2) &= \varphi(g_1)\varphi(g_2) \\
\varphi(e) &= \text{identity mapping} \\
\varphi(g^{-1}) &= [\varphi(g)]^{-1}.
\end{aligned}
$$

G is also said to act *transitively* on E if for all points $x_1, x_2 \in E$, there exists g such that $x_2 = \varphi(g)x_1$; in this case, E is said to be a homogeneous space, see Section 3.1).

Thus, suppose the group G acts on the space E. Suppose also that the stationary subgroup of any point $x \in E$ is trivial (consists solely of the unit element $e \in G$; we recall that in the general case, the stationary subgroup of the point $x \in E$ consists of elements $h \in G$ such that $\varphi(h)x = x$, see Section 3.1). In this case, G is said to act *freely* on E. The *orbit* of the point $x \in E$ is the set of elements of the form $\varphi(g)x$, where g runs through the whole group G. If G acts freely on E, then the orbit of the point x is equivalent to G. Indeed, suppose y belongs to the orbit of the point x; then $y = \varphi(g)x$. This equation defines a one-to-one correspondence between y and g (points of the orbit and of the group): if $\varphi(g)x = \varphi(g')x$, then $\varphi(g^{-1}g')x = x$, i.e. $g^{-1}g' = e$ and $g' = g$. Thus, we obtain a fiber bundle with bundle space E, base space the set of orbits in E and fiber G. The projection p for this fiber bundle is the mapping of the point x to the orbit which contains that point. Such a fiber bundle is called a *major fiber bundle*.

Finally, the notion of a homogeneous space also leads to a fiber bundle. Namely, if G is a compact Lie group with subgroup H, and G/H is a homogeneous space (we recall that any homogeneous space is equivalent to G/H, where H is the stationary subgroup of any point of the homogeneous space, see Section 3.1). Then, we can define the projection $p : G \to G/H$ which takes the point $g \in G$ to its coset, an element of G/H. The fiber of the fiber bundle obtained in this way is the group H: indeed, if u is some coset in G/H (for example, containing the element $g \in G$), then its inverse image $p^{-1}(u)$ is the set of elements of G belonging to u, i.e. elements of the form gh, where $h \in H$ (for definiteness, we shall consider right cosets). As mentioned in Section 3.1, the set of such elements is equivalent to H. Thus, it makes sense to consider the fiber bundle $(G, G/H, H)$.

We now state without proof a number of assertions which enable one to compute many important homotopy groups. In what follows, we shall consider connected spaces and will be interested in the groups $\pi_1, \pi_2, \ldots$.

1. If $E = B \times F$, then

$$\pi_k(E) = \pi_k(B) + \pi_k(F),$$

where the $+$ sign denotes the direct sum. This assertion follows from the remarks made in Section 8.1.

2. Suppose we are given a fiber bundle (E, B, F, p) and

$$\pi_k(B) = \pi_{k+1}(B) = 0$$

(zero here denotes the group with one element, the unit element). Then

$$\pi_k(E) = \pi_k(F).$$

This isomorphism is induced as follows. We choose the point $b \in B$, to which there corresponds a subset in E of the form $p^{-1}(b)$ (the fiber over b), where $p^{-1}(b)$ is equivalent to F. Thus, every mapping of the sphere S^k to F induces a mapping from S^k to $p^{-1}(b)$, i.e. a mapping from S^k to E. Furthermore, when $\pi_k(B) = 0$, any image of the sphere S^k (spheroid) in E can be deformed to a spheroid in $p^{-1}(b)$. Moreover, when $\pi_{k+1}(B) = 0$, any spheroids homotopic in E are homotopic in F.

3. Suppose the following holds for the fiber bundle (E, B, F, p):

$$\pi_{k-1}(F) = \pi_k(F) = 0.$$

Then

$$\pi_k(E) = \pi_k(B).$$

This isomorphism is induced by the projection p.

4. If for the fiber bundle (E, B, F, p) we have

$$\pi_{k-1}(E) = \pi_k(E) = 0,$$

then

$$\pi_k(B) = \pi_{k-1}(F).$$

This isomorphism is constructed as follows. Let $f : S^{k-1} \to F$ be a mapping of the $(k-1)$-dimensional sphere to a fiber. This can be used to construct a mapping $\tilde{f} : S^{k-1} \to E$, in a way described in point 2. The mapping $\tilde{f}$ is contractible in E (since $\pi_{k-1}(E) = 0$), i.e. there exists an extension $\tilde{f}_I$ of the mapping $\tilde{f}$ to the ball D^k. We project the mapping $\tilde{f}_I : D^k \to E$ onto the base space, i.e. we consider the mapping $p\tilde{f}_I : D^k \to B$. We note that on the boundary of the ball (i.e. on the original sphere S^{k-1}), $\tilde{f}$ maps S^{k-1} to the fiber over some point $b \in B$, i.e. $p\tilde{f}$ maps S^{k-1} to the point b. Consequently, $p\tilde{f}$ is a mapping of the ball D^k to B, where the boundary of the ball is mapped to a single point, and the boundary of the ball can be identified if we consider the mapping $p\tilde{f}_I$. Since the ball D^k with identified boundary is equivalent to S^k, in this way we have constructed a mapping from S^k to B from the mapping from S^{k-1} to F. This correspondence between mappings induces a correspondence between homotopy classes, which is one-to-one when $\pi_k(E) = 0$.

5. If G is a Lie group and Ξ a discrete normal subgroup of G, then

$$\pi_k(G) = \pi_k(G/\Xi), \quad k \geq 2.$$

Let us give some examples of calculations of the homotopy groups of simple Lie groups, based on the above assertions and on the homotopy groups of spheres, which we know.

1.
$$\pi_k(S^1) = 0 \text{ for } k \geq 2$$

$$\pi_k(SO(2)) = \pi_k(U(1)) = 0, \quad k \geq 2.$$

2. The group $SU(2)$ is homeomorphic to S^3 (see Section 3.2), thus

$$\pi_k(SU(2)) = 0 \text{ for } k = 1, 2$$
$$\pi_3(SU(2)) = Z.$$

3. The group $SO(3)$ is isomorphic to $SU(2)/Z_2$ (see Section 3.2), thus

$$\pi_2(SO(3)) = 0$$
$$\pi_3(SO(3)) = Z.$$

Moreover, one can prove that the fundamental group is equal to

$$\pi_1(SO(3)) = Z_2.$$

4. The group $SO(4)$ is isomorphic to $SU(2) \times SU(2)/Z_2$, thus

$$\pi_k(SO(4)) = \pi_k(S^3) + \pi_k(S^3), \quad k \geq 2.$$

In particular,

$$\pi_2(SO(4)) = 0$$
$$\pi_3(SO(4)) = Z + Z.$$

Moreover, one can prove that

$$\pi_1(SO(4)) = Z.$$

5. The sphere S^{n-1} is homeomorphic to the coset space $SO(n)/SO(n-1)$ (see Section 3.1). Thus, one can construct the fiber bundle $(SO(n), S^{n-1}, SO(n-1))$. For $k < n-2$, we have $\pi_k(S^{n-1}) = \pi_{k+1}(S^{n-1}) = 0$, whence

$$\pi_k(SO(n)) = \pi_k(SO(n-1)) \text{ for } k < n-2.$$

In particular

$$\pi_1(SO(n)) = Z_2 \text{ for } n \geq 3.$$

Moreover, because $\pi_2(SO(4)) = 0$, we have

$$\pi_2(SO(n)) = 0 \text{ for all } n$$

(the cases $n = 2$ and $n = 3$ have already been considered).

6. The sphere S^{2n-1} is homeomorphic to $SU(n)/SU(n-1)$ (see Section 3.1). Thus, one can construct the fiber bundle

$$(SU(n), S^{2n-1}, SU(n-1)).$$

Hence

$$\pi_k(SU(n)) = \pi_k(SU(n-1)) \text{ for } k < 2n - 2.$$

In particular

$$\pi_2(SU(n)) = 0 \text{ for all } n$$

and

$$\pi_3(SU(n)) = Z \text{ for } n \geq 2.$$

7. In general, the following holds for any compact group

$$\pi_2(G) = 0.$$

8. Finally, let us consider the fiber bundle $(G, G/H, H)$. Let G be compact and simply connected (for example, $SU(N)$). Then, because $\pi_1(G) = \pi_2(G) = 0$, we have

$$\pi_2(G/H) = \pi_1(H).$$

We note further that if G is connected, then

$$\pi_1(G/H) = \pi_0(H),$$

where $\pi_0(H)$ is the set of connected components of the group H.

8.5 Summary of the results

We present the most important results from calculations of homotopy groups as far as applications are concerned. Most of these have already figured in the previous sections; the remainder are given without proof or discussion.

In what follows, Z_n is the group of integers modulo n under addition (mod n). In particular, Z_2 comprises two elements, 0 and 1, where $0+0 = 0$, $0+1 = 1$, $1+1 = 0$ (the plus here denotes the group operation in Z_2). We recall that Z is the group of integers under addition.

1. Homotopy groups of spheres:

(a) $\pi_k(S^n) = 0$ for $k < n$

(b) $\pi_n(S^n) = Z$.

In particular,

$$\begin{aligned} \pi_1(S^1) &= Z \\ \pi_2(S^2) &= Z \\ \pi_3(S^3) &= Z. \end{aligned}$$

(c) $\pi_n(S^1) = 0$ for $n \geq 2$

(d) $\pi_3(S^2) = Z$

$\pi_4(S^2) = Z_2$

(e) $\pi_n(S^3) = \pi_n(S^2)$ for $n \geq 3$.

In particular,

$$\pi_4(S^3) = Z_2.$$

2. Homotopy groups of Lie groups:

(a) $\pi_1(G)$ is an Abelian group for all Lie groups G.

(b) $\pi_n(G_1 \times G_2) = \pi_n(G_1) + \pi_n(G_2)$ for all n and for all Lie groups G_1 and G_2 (indeed, for any two manifolds, if instead of the direct sum, we write the direct product for $n = 1$)

(c) $\pi_2(G) = 0$ for any compact Lie group G

(d) $\pi_3(G) = Z$ for any simple compact Lie group G

(e) $\pi_1(U(1)) = \pi_1(SO(2)) = Z$

$\pi_n((U(1)) = \pi_n(SO(2)) = 0$ for $n \geq 2$

(f) $\pi_1(SU(n)) = 0$

$\pi_2(SU(n)) = 0$

$\pi_3(SU(n)) = Z$.

These three equations hold for all $n \geq 3$.

(g) $\pi_1(SO(n)) = Z_2$ for all $n \geq 2$

$\pi_2(SO(N)) = 0$ for all n

$\pi_3(SO(3)) = Z$

$\pi_n(SO(4)) = \pi_n(SO(3)) + \pi_n(SO(3))$ for $n \geq 2$.

3. Homotopy groups of homogeneous spaces:

(a) $\pi_1(G/H)$ is the set of connected components of the group H. This is true for connected and simply connected Lie groups G.

(b) $\pi_2(G/H) = \pi_1(H)$.

This equation holds for a simply connected Lie group G. If the group G is not simply connected, this relation is generalized in terms of so-called universal covering groups.

Chapter 9

Magnetic Monopoles

We end the discussion of topological solitons with an important example, the magnetic monopole of 't Hooft and Polyakov. Interest in these solutions is, in the first place, due to the fact that they exist in four-dimensional space–time and are present in all models unifying the strong, weak and electromagnetic interactions in the framework of gauge theory with a compact simple or semi-simple gauge group. These models are called grand unified theories. Thus, the existence of magnetic monopoles is a very general prediction of grand unified theories, essentially independently of the choice of model (although some properties of monopoles, such as their mass, are model dependent). It is not surprising that monopoles are sought experimentally, although these searches have not (yet?) been successful.

9.1 The soliton in a model with gauge group $SU(2)$

The simplest model in which monopole solutions arise is the Georgi–Glashow model with gauge group $SU(2)$ and triplet of real Higgs fields φ^a, $a = 1, 2, 3$. The Lagrangian of the model has the form

$$\mathcal{L} = -\frac{1}{4} F_{\mu\nu}^a F_{\mu\nu}^a + \frac{1}{2} (D_\mu \varphi)^a (D_\mu \varphi)^a - \frac{\lambda}{4} (\varphi^a \varphi^a - v^2)^2. \tag{9.1}$$

Here, $\mu, \nu = 0, 1, 2, 3$ (four-dimensional space–time),

$$D_\mu \varphi^a = \partial_\mu \varphi^a + g \varepsilon^{abc} A_\mu^b \varphi^c$$
$$F_{\mu\nu}^a = \partial_\mu A_\nu^a - \partial_\nu A_\mu^a + g \varepsilon^{abc} A_\mu^b A_\nu^c.$$

We shall sometimes also use the matrix notation

$$A_\mu = \frac{g}{2i} A_\mu^a \tau^a,$$

$$\varphi = \frac{1}{2i} \varphi^a \tau^a,$$

such that the matrices A_μ and φ belong to the $SU(2)$ algebra.

The potential of the scalar field in (9.1) was chosen so that in the model the Higgs mechanism is realized. The ground state φ_0^a can be chosen in the form

$$
\begin{aligned}
\varphi_0^1 &= \varphi_0^2 = 0, \\
\varphi_0^3 &= v.
\end{aligned}
\tag{9.2}
$$

As a result of the Higgs mechanism in the model, among the small excitations about the vacuum, there exist one real massless vector field, two real massive vector fields and one real massive scalar field (see Supplementary Problems for Part I, Problem 8). In a unitary gauge field φ with small perturbations about the ground state, (9.2) has the form

$$
\begin{aligned}
\varphi^1 = \varphi^2 &= 0 \\
\varphi^3 &= v + \eta(x);
\end{aligned}
$$

the massive scalar field is $\eta(x)$; its mass is equal to

$$m_H = \sqrt{2\lambda} v.$$

The massless vector field corresponds to an unbroken gauge subgroup $U(1)$ of rotations (in internal space) about the third axis. We shall call this field the electromagnetic field and the unbroken subgroup $U(1)$ the electromagnetic gauge group. In a unitary gauge, the electromagnetic vector potential coincides with A_μ^3,

$$\mathcal{A}_\mu = A_\mu^3.$$

Finally, the two real massive vector fields in a unitary gauge are described by the vector potentials A_μ^1 and A_μ^2. They have the same mass

$$m_V = gv.$$

Instead of two real fields, it is convenient to consider a single complex field W_μ^+ and its conjugate W_μ^-,

$$W_\mu^\pm = \frac{1}{\sqrt{2}}(A_\mu^1 \pm iA_\mu^2).$$

The electric charge of the field W^+ is equal to g; this is also the result of Supplementary Problem 8 for Part I. The field $\eta(x)$ is electrically neutral.

Thus, the model in question can be used as a prototype for more complicated theories, where a compact (semi-)simple gauge group is broken down to a subgroup containing the electromagnetic group $U(1)$.

Our most immediate task is to show that topological solitons may exist in the model. Let us begin as in Chapter 7 by considering static configurations of fields with finite energy. By static configurations, we shall mean configurations for which

$$A_0^a = 0,$$

and the fields A_i^a and φ^a do not depend on time,

$$
\begin{aligned}
A_i^a &= A_i^a(\mathbf{x}) \\
\varphi^a &= \varphi^a(\mathbf{x}).
\end{aligned}
$$

For such configurations, the energy functional has the form

$$E = \int d^3x \left[\frac{1}{4} F_{ij}^a F_{ij}^a + \frac{1}{2}(D_i\varphi)^a (D_i\varphi)^a + \frac{\lambda}{4}(\varphi^a\varphi^a - v^2)^2 \right]. \tag{9.3}$$

A necessary condition for finiteness of the energy is the requirement

$$\varphi^a\varphi^a = v^2 \text{ for } |\mathbf{x}| = \infty, \tag{9.4}$$

which guarantees the finiteness of the contribution to the energy associated with the scalar potential. Here, the direction of the fields φ^a in internal space may depend on the direction in physical three-dimensional space

$$\varphi^a\big|_{|\mathbf{x}|\to\infty} = \varphi^a(\mathbf{n}),$$

where $\mathbf{n} = \frac{\mathbf{x}}{r}$ is a unit radius vector. Thus, each configuration of fields with finite energy is associated with a mapping of an infinitely remote sphere S_∞^2 in physical space to the sphere S_{vac}^2 in the space of scalar fields, defined by the equation

$$\varphi^a\varphi^a = v^2.$$

We note that S_{vac}^2 is the set of classical vacua of the model.

Since $\pi_2(S^2) = Z$, the mappings $S_\infty^2 \to S_{\text{vac}}^2$ are characterized by an integer topological number $n = 0, \pm 1, \pm 2, \dots$. This number does not change under small deformations of the configurations $\varphi^a(\mathbf{x})$, for which the energy remains finite; for such deformations equation (9.4) is satisfied for each configuration, and the mappings $S_\infty^2 \to S_{\text{vac}}^2$ remain in the same homotopy class. As in the examples in Chapter 7, we are concerned with the set of connected components (topological sectors) in the space of configurations of fields with finite energy, where different connected components correspond to different topological numbers. The sector with topological number zero contains the classical vacuum, the topological soliton should be sought in the sector with $n = 1$; it realizes the absolute minimum of the energy among fields with $n = 1$. It is important that the scale arguments of Section 7.2 do not rule out the existence of a minimum of the energy functional.

In order to choose the Ansatz (form of the fields) for solution of the field equations, we note that the most symmetric form of the field $\varphi^a(\mathbf{n})$, corresponding to a mapping $S^2_\infty \to S^2_{\text{vac}}$ with a unit topological number, is

$$\varphi^a(\mathbf{n}) = n^a v \text{ as } r \to \infty. \tag{9.5}$$

This asymptotics of the field as $|\mathbf{x}| \to \infty$ is invariant under spatial rotations, supplemented by global $SU(2)$ transformations (the former rotate the vector $\mathbf{n}$, the latter act on φ^a as on a vector with vector index a):

$$(\Lambda^{-1})^a_b \varphi^b(\Lambda^i_j n^j) = \varphi^a(n^i), \tag{9.6}$$

where Λ is a matrix from the group $SO(3)$ of three-dimensional rotations.

Let us now find the asymptotics of the vector field $A^a_i(\mathbf{x})$. For finiteness of the soliton energy, the covariant derivative

$$D_i \varphi^a = \partial_i \varphi^a + g \varepsilon^{abc} A^b_i \varphi^c \tag{9.7}$$

is required to decrease at spatial infinity faster than $1/r$. At the same time, the conventional derivative for fields with asymptotics (9.5) decreases as $1/r$:

$$\partial_i \varphi^a = \frac{1}{r}(\delta^{ai} - n^a n^i)v \text{ as } r \to \infty. \tag{9.8}$$

This behavior must be compensated by the second term in (9.7) for which we require that A^a_i decrease as $1/r$. The requirement for cancellation of the term (9.8) leads to the following asymptotics of the field A^a_i:

$$A^a_i(\mathbf{x}) = \frac{1}{gr} \varepsilon^{aij} n^j \text{ as } r \to \infty. \tag{9.9}$$

Indeed, the second term in (9.7) for fields of the form (9.5) and (9.9) is equal to

$$
\begin{aligned}
g\varepsilon^{abc} A^b_i \varphi^c &= g\varepsilon^{abc} \frac{1}{gr} \varepsilon^{bij} n^j \cdot n^c v = \frac{1}{r}\left(-\delta^{ai}\delta^{cj} + \delta^{aj}\delta^{ci}\right) n^j n^c v \\
&= -\frac{1}{r}(\delta^{ai} - n^a n^i)v,
\end{aligned}
$$

which precisely cancels (9.8) in the covariant derivative (9.7).

To obtain a smooth solution one can seek a soliton configuration in the form suggested by the asymptotics (9.5), (9.9),

$$\varphi^a = n^a v(1 - H(r)) \tag{9.10}$$

$$A_i^a = \frac{1}{gr} \varepsilon^{aij} n^j (1 - F(r)),$$

where $H(r)$ and $F(r)$ are unknown functions of the radius. It is not a completely trivial fact that this Ansatz "passes through" the field equations, i.e. all field equations for configurations of the form (9.10) are reduced to two equations for the functions $H(r)$ and $F(r)$. From the asymptotic behavior of the fields (9.5), (9.9), we have the boundary conditions on $F(r)$ and $H(r)$:

$$F(r) = H(r) = 0 \text{ as } r \to \infty. \tag{9.11}$$

Two other boundary conditions arise from the requirement that the fields be smooth at the origin:

$$H(0) = F(0) = 1. \tag{9.12}$$

More precisely, the function $(1 - H(r))$ must decrease at least like r as $r \to 0$:

$$1 - H(r) = O(r) \text{ as } r \to 0,$$

while the function $(1 - F(r))$ must decrease at least like r^2:

$$1 - F(r) = O(r^2) \text{ as } r \to 0$$

(if $(1 - F(r)) \sim r$ as $r \to 0$, the value of A_i^u near $r = 0$ would depend on the direction $\mathbf{n}$).

In the general case, it is not possible to find a solution of the equations for $F(r)$ and $H(r)$ in analytic form, and these functions must be determined numerically.

The fact that the Ansatz (9.10) "passes through" the field equations can be understood as follows. We have already mentioned that if the Lagrangian has a global symmetry, then the most general Ansatz, consistent with this symmetry, always "passes through" the field equations. In our case, this symmetry is spherical symmetry, which is to be understood in the sense of (9.6), i.e. symmetry under $SO(3)$ spatial rotations, supplemented by $SU(2)$ transformations with respect to the index a. The Ansatz $\varphi^a(\mathbf{x}) = n^a v(1 - H(r))$ is clearly the most general form of the Higgs field, which is invariant under generalized spherical symmetry. At the same time, the most general spherically symmetric field A_i^a has the form

$$A_i^a(\mathbf{x}) = n^i n^a a(r) + (\delta^{ai} - n^a n^i) f_1(r) + \varepsilon^{aij} n^j f_2(r), \tag{9.13}$$

where $a(r)$, $f_1(r)$ and $f_2(r)$ are arbitrary functions of the radius (the form of (9.13) is clear from the fact that A_i^a must be a tensor under $SO(3)$ rotations with respect to the indices a and i, constructed from the only available vector n^a and the tensors δ_{ai} and ε_{aij}). The Ansatz (9.10) contains only the third term of (9.13). The possibility of a restriction to this term alone follows from the fact that it is odd under the transformation $\mathbf{x} \to -\mathbf{x}$, whereas the other two terms are even. In other words, the Ansatz (9.10) is the most general form of the fields, invariant under generalized rotations and discrete transformations

$$\varphi^{a\prime}(-\mathbf{x}) = -\varphi^a(\mathbf{x})$$
$$A_i^{a\prime}(-\mathbf{x}) = -A_i^a(\mathbf{x}).$$

It is important that this discrete symmetry is a symmetry of the energy functional of the static fields (and of the Lagrangian).

Problem 1.   *Write down the field equations for the case where $A_0^a = 0$ and the fields φ^a and A_i^a do not depend on time. Show, by direct substitution, that, when the fields are chosen in the form (9.10), all these equations reduce to two equations for the functions $F(r)$ and $H(r)$. Write down these two equations.*

   Show that $F(r)$ and $H(r)$ tend to zero exponentially as $r \to \infty$ in the case where $m_H \sim m_V$.

Problem 2.   *Show that there exists a two-parameter family of solutions of the equations for F and H, satisfying the conditions (9.11) (without the requirement (9.12)). Show that there exists another two-parameter family of solutions of these equations, containing solutions satisfying only the conditions (9.12). (Hint: consider solutions of type*

$$H(r) = 1 + \alpha_H r + \beta_H r^3 + \cdots,$$
$$F(r) = 1 + \alpha_F r^2 + \beta_F r^4 + \cdots$$

for $r \to 0$; show that the constants α_H and α_F are not specified by the equations for H and F, while β_H and β_F are expressed in terms of α_H and α_F). Thus, the argument of Section 7.3 can also be applied to substantiate the existence of a soliton solution in the model of this section.

Problem 3.   *Find the soliton configuration numerically for $m_V = 2m_H$.*

   Let us estimate the energy (mass) and the characteristic size of a soliton for $m_V \sim m_H$ from scaling considerations. We introduce the variables y^i, f^a and B_i^a by the equations

$$y^i = gvx^i$$

$$\varphi^a(\mathbf{x}) = v f^a(\mathbf{y}) \qquad (9.14)$$
$$A_i^a(\mathbf{x}) = v B_i^a(\mathbf{y}).$$

In terms of these, the quantities appearing in the energy functional (9.3) can be written in the form

$$(D_i \varphi^a) = g v^2 \mathcal{D}_i f^a \qquad (9.15)$$
$$F_{ij}^a = g v^2 B_{ij}^a$$
$$(\varphi^a \varphi^a - v^2) = v^2 (f^a f^a - 1)$$
$$d^3 x = \frac{1}{(g v)^3} d^3 y,$$

where

$$\mathcal{D}_i f^a = \frac{\partial}{\partial y^i} f^a + \varepsilon^{abc} B_i^b f^c \qquad (9.16)$$

$$B_{ij}^a = \frac{\partial}{\partial y^i} B_j^a - \frac{\partial}{\partial y^j} B_i^a + \varepsilon^{abc} B_i^b B_j^c. \qquad (9.17)$$

Substituting (9.15) in (9.3) we obtain the following expression for the energy functional

$$E = \frac{v}{g} \int d^3 y \left[\frac{1}{4} B_{ij}^a B_{ij}^a + \frac{1}{2} (\mathcal{D}_i f^a)^2 + \frac{\lambda}{4g^2} (f^a f^a - 1)^2 \right]. \qquad (9.18)$$

We are considering the case $m_V \sim m_H$, thus

$$\frac{\lambda}{4g^2} = \frac{m_H^2}{8 m_V^2} \sim 1.$$

Hence, and from (9.16) and (9.17), it follows that the integral in (9.18) does not contain parameters which are significantly different from unity. The soliton configuraton is its minimum, thus for that we have

$$E \sim \frac{v}{g} \sim \frac{m_V}{\alpha_g},$$

where $\alpha_g = \frac{g^2}{4\pi}$ is the "fine structure constant" (we recall that the electric charge of the field W^+ is equal to g). The characteristic size of a soliton in terms of $\mathbf{y}$ is of the order of unity, thus the physical size is of the order of $(gv)^{-1}$, i.e. of the order

$$r_0 = m_V^{-1}.$$

Therefore, the mass of the soliton (its static energy) and its size are estimated by

$$M \sim \frac{m_V}{\alpha_g}$$

$$r \sim r_0 = m_V^{-1}.$$

As in the preceding examples, in the theory with weak coupling (i.e. for $\alpha_g \ll 1$), the Compton wavelength of the soliton, $\lambda = M^{-1}$, is small in comparison with its size:

$$\frac{\lambda}{r_0} \sim \alpha_g \ll 1.$$

The soliton can be viewed, with a good accuracy, as a classical object.

Problem 4. *Find the number of zero modes for small perturbations of the fields A_μ and φ^a about the soliton solution. Hint: consider symmetries of the classical solution; use the fact that some of these symmetries are gauge transformations; these symmetries (there are infinitely many) should be considered individually.*

9.2 Magnetic charge

Unlike the Abelian Higgs model, considered in Section 7.3, the model (9.1), after symmetry breaking, contains a massless vector field, which we call the electromagnetic field. Far from the center of the soliton, massive fields must decrease exponentially; however, the massless field may decrease slowly. In other words, generally speaking, a soliton may have an electric or magnetic charge (and also an electric or magnetic dipole moment and higher multipole moments). In this section, we show that the soliton solution of Section 9.1 has zero electric charge and non-zero magnetic charge, i.e. it is a magnetic monopole.

Far from the center of a soliton, the square of the Higgs field $\varphi^a \varphi^a$ tends to the vacuum value v^2, and the vector fields must represent small deviations from the vacuum $\mathbf{A}^a = 0$. If the soliton solution satisfied $\varphi^a = \delta^{a3} v$ as $r \to \infty$ (unitary gauge) the electromagnetic field (far from the center) would be described by the vector potential A_μ^3. However, the asymptotics (9.5) does not correspond to the unitary gauge, so the potential A_μ^3 cannot be used to calculate the electromagnetic field for large r. The simplest way to find the electromagnetic field far from the center is to introduce the gauge invariant quantity

$$\mathcal{F}_{\mu\nu} = \frac{1}{v} F_{\mu\nu}^a \varphi^a. \tag{9.19}$$

For small perturbations of the field A_μ^3 about the vacuum in the unitary gauge, we have

$$\mathcal{F}_{\mu\nu} = F_{\mu\nu},$$

thus $\mathcal{F}_{\mu\nu}$ coincides with the electromagnetic field strength. Since the quantity (9.19) is gauge invariant, the definition (9.19) can be used in any gauge, provided there are no massive fields, the massless fields are weak, and $\varphi^a\varphi^a$ is close to v. Now, far from the center of the soliton massive fields should decrease rapidly (exponentially); the electromagnetic field also decreases (but slowly). Thus, expression (9.19) can actually be used to find the electromagnetic field of a soliton far from its center.

Let us calculate the electromagnetic field strength $\mathcal{F}_{\mu\nu}$ for the solution (9.10) far from the center of a monopole. Since $F(r)$ tends to zero exponentially for large r (see the problem in Section 9.1), to calculate any finite number of electromagnetic multipole moments, we can use the asymptotic expression

$$A_i^a = \frac{1}{gr}\varepsilon^{aij}n_j. \qquad (9.20)$$

We recall that $A_0^a = 0$.

Since A_i^a does not depend on time, and $A_0^a = 0$, the field strength F_{0i}^a is equal to zero, i.e. the electric field is equal to zero far from the center of the soliton:

$$\mathcal{E}_i = \mathcal{F}_{0i} = 0.$$

We now calculate the magnetic field

$$\mathcal{H}_i = \frac{1}{v}H_i^a\varphi^a,$$

where

$$H_i^a = -\frac{1}{2}\varepsilon_{ijk}F_{jk}^a.$$

Using the definition of F_{ij}^a, we write

$$H_i^a = -\varepsilon_{ijk}\partial_j A_k^a - \frac{1}{2}g\varepsilon_{ijk}\varepsilon^{abc}A_j^b A_k^c.$$

The calculation of H_i^a for the field (9.20) is performed using the formulae for differentiation:

$$\partial_i r = n_i$$
$$\partial_i n_j = \frac{1}{r}(\delta_{ij} - n_i n_j)$$

and convolution:

$$\delta^{ij}\delta^{ij} = 3$$
$$\varepsilon^{ijk}\varepsilon^{ilm} = \delta^{jl}\delta^{km} - \delta^{jm}\delta^{kl}$$

(we shall not distinguish between Latin superscripts and subscripts). We obtain

$$-\varepsilon_{ijk}\partial_j A_k^a = \frac{2}{gr^2}n_i n_a$$

$$-\frac{1}{2}\varepsilon_{ijk}\varepsilon^{abc}A_j^b A_k^c = -\frac{1}{gr^2}n_i n_a.$$

Thus, we have

$$H_i^a = \frac{1}{gr^2}n_i n_a. \tag{9.21}$$

We note that, as one might expect, H_i^a is directed in space (with respect to the index i) along the radius vector n_i and, in internal space, along the vector φ^a. This latter property is a reflection of the fact that, far from the center of the soliton there is only an electromagnetic field (in the unitary gauge the electromagnetic field strength $F_{\mu\nu}^3$ is directed in internal space along the vacuum field $\varphi_{\mathrm{vac}} = v\delta^{a3}$; the fields $F_{\mu\nu}^a$ and φ^a transform in the same way under gauge transformations, thus, the electromagnetic field strength is directed in internal space along φ^a in any gauge). The first property is a consequence of the spherical symmetry of the configuration (9.10).

From (9.21) and the asymptotics $\varphi^a = vn^a$ (once again true modulo terms decreasing exponentially for large r) we obtain the expression for the magnetic field of the solution (9.10):

$$\mathcal{H}_i = \frac{1}{gr^2}n_i. \tag{9.22}$$

This is the field of a magnetic monopole with magnetic charge

$$g_M = \frac{1}{g}. \tag{9.23}$$

Indeed, the magnetic field (9.22) is directed along the radius vector and its value is equal to $\frac{g_M}{r^2}$.

Problem 5. *By adding new scalar fields with zero expectation values, in all possible representations of the gauge group $SU(2)$, show that the minimum electric charge in the models obtained is equal to $q_{\min} = \frac{1}{2}g$, while the electric charge of any field is a multiple of $q_{\min}$. Thus, the magnetic charge of a monopole is equal to $g_M = \frac{1}{2q_{\min}}$. The fact that the magnetic charge, in theories where there is a single long-ranged vector field, namely the electromagnetic field, is a multiple of $1/2q_{\min}$ is known as the Dirac quantization condition ($q_{\min}$ is the minimum electric charge; the charge of any field is assumed to be a multiple of $q_{\min}$). We note that, in the* *case of a point-like (Dirac) monopole in electrodynamics, the quantization condition for magnetic charge arises at the level of quantum mechanics in the monopole background field; in the case of the 't Hooft–Polyakov monopole, the quantization condition exists already at the level of classical field theory. This last fact has a deep connection with the quantization of the electric charge, which arises automatically, even at the classical level, in models of the type considered in this section.*

Expression (9.22) for the magnetic field of a monopole is valid modulo contributions to $F_{\mu\nu}^a$, decreasing as e^{-mvr} far from the center. The magnetic field contains no components decreasing as an inverse power of the radius for large r, other than the component with behavior r^{-2} described earlier in (9.22). This means that the magnetic dipole moment and higher magnetic multipoles are absent, as are electric multipoles.

The way of calculating the magnetic charge presented above does not hint at its connection with the topological properties of field configurations, i.e. with the fact that a monopole is a topological soliton. To pursue this connection,[1] we again consider the asymptotics of fields at spatial infinity. The field $\varphi^a = n^a v$ is not in the unitary gauge, thus it is not convenient to discuss the gauge vector potential in the original gauge. It would be desirable to convert the field φ^a to the unitary gauge. It is impossible to do so by performing a gauge transformation which is non-singular everywhere at an infinitely remote sphere. Indeed, if there existed a non-singular gauge transformation $\omega(\mathbf{n})$ on S_∞^2, such that

$$(\varphi^\omega)^a = v\delta^{a3}$$

(here and in what follows φ^ω denotes the gauge transformed field), then it would be possible to construct a family of smooth fields $\varphi^a(\tau)$ on S_∞^2, continuous in the parameter $\tau \in [0, 1]$, and interpolating between vn^a and $v\delta^{a3}$: since $\pi_2(SU(2)) = 0$, $\omega(\mathbf{n})$ is homotopically equivalent to a gauge transformation with $\omega(\mathbf{n}) = 1$; the corresponding family $\omega_\tau(\mathbf{n})$ deforming ω to $\omega(\mathbf{n})$, would generate the family $\varphi^a(\tau) = (\varphi^{\omega_\tau})^a$, which is impossible.

Therefore, a gauge transformation taking the field $\varphi^a = n^a v$ to the unitary gauge exists only on part of the infinitely remote sphere S_∞^2, for example, everywhere, apart from some small neighborhood of the south pole. We denote this gauge transformation by $\omega_N(\mathbf{n})$ (it acts on the north). There is another gauge transformation $\omega_S(\mathbf{n})$ which is non-singular everywhere, apart from some small neighborhood of the north pole. (The existence of ω_S and ω_N is guaranteed by the fact that the sphere with a point deleted (the north or south pole) is homotopic to R^2). Here

$$(\varphi^{\omega_N})^a = (\varphi^{\omega_S})^a = v\delta^{a3} \equiv (\varphi_{\text{vac}})^a, \qquad (9.24)$$

---

[1] The following considerations are essentially the same as the approach of Arafune *et al.* (1975).

or

$$(\varphi_{\text{vac}}^{\omega_N^{-1}})^a = (\varphi_{\text{vac}}^{\omega_S^{-1}})^a = n^a v \tag{9.25}$$

everywhere, apart from small neighborhoods of the north and south poles. From (9.24) and (9.25) it follows that

$$\varphi_{\text{vac}}^{(\omega_S \omega_N^{-1})} = \varphi_{\text{vac}},$$

i.e.

$$\Omega(\mathbf{n}) = \omega_S \omega_N^{-1}$$

belongs to the unbroken subgroup $U(1)_{\text{e.m.}}$ of rotations around the third axis.

Let us consider $\Omega(\mathbf{n})$ on the equator of the sphere S_∞^2. There it maps the circle S^1 (equator) to the group $U(1)_{\text{e.m.}}$. Clearly, this mapping must belong to a non-trivial homotopy class of $\pi_1(U(1))$ (we recall that $\pi_1(U(1)) = Z$). Indeed, if it were homotopically zero, it would be smoothly extensible to the northern hemisphere; if we denote this extension by $\tilde{\omega}(\mathbf{n})$ ($\tilde{\Omega}(\mathbf{n}) \in U(1)_{\text{e.m.}}$ for all $\mathbf{n}$), then the gauge function, equal to

$$\omega(\mathbf{n}) = \begin{cases} \omega_S(\mathbf{n}) & \text{on the southern hemisphere} \\ \tilde{\Omega}(\mathbf{n})\omega_N(\mathbf{n}) & \text{on the northern hemisphere,} \end{cases}$$

would be continuous on all S_∞^2 and would transform the field $\varphi^a = n^a v$ to the unitary gauge. We have seen that it is impossible to do this, thus, $\Omega(\mathbf{n})$ on the equator belongs to a non-trivial homotopy class of $\pi_1(U(1))$.

On the equator, we can write $\Omega(\mathbf{n})$ in the form

$$\Omega = e^{if(\varphi)\tau^3}, \tag{9.26}$$

where φ is the polar angle at the equator (we recall that the group $U(1)_{\text{e.m.}}$ consists of matrices of the form $e^{if\tau^3}$). The topological non-triviality of $\Omega(\mathbf{n})$ at the equator implies that

$$f(2\pi) - f(0) = 2\pi n, \tag{9.27}$$

where n is a non-zero integer. It can be shown (we shall do this by explicitly constructing $\omega_S(\mathbf{n})$ and $\omega_N(\mathbf{n})$) that n coincides with the topological number of the field $\varphi^a(\mathbf{n})$ (the degree of the mapping $S_\infty^2 \to S_{\text{vac}}^2$); in our case

$$n = 1.$$

Let us now consider the vector potential A_i^a. Let us move to the unitary gauge everywhere, apart from a small neighborhood of the south pole, using the gauge transformation ω_N. The gauge potential

$$\hat{A}_i^N = \omega_N A_i \omega_N^{-1} + \omega_N \partial_i \omega_N^{-1}$$

is smooth everywhere, apart from a small neighborhood of the south pole. Here, A_i is the vector potential in the original gauge (in our case its components have the form (9.9)); we are using matrix notation. It is clear that, in the unitary gauge, far from the monopole, there exist only electromagnetic components of the vector potential (the other fields are massive), thus

$$\hat{A}_i^N = \frac{g\tau^3}{2i} \mathcal{A}_i^N, \qquad (9.28)$$

where $\mathcal{A}_i^N$ are the magnetic field potentials (real functions depending on **x**).

Furthermore, apart from in a small neighborhood of the north pole, we define the vector potential

$$\hat{A}_i^S = \omega_S A_i \omega_S^{-1} + \omega_S \partial_i \omega_S^{-1}.$$

It also has the third component only

$$\hat{A}_i^S = \frac{g\tau^3}{2i} \mathcal{A}_i^S. \qquad (9.29)$$

Although the potentials $\mathcal{A}_i^N$ and $\mathcal{A}_i^S$ describe the same magnetic field, everywhere away from the north and south poles, they are not equal and are related by the gauge transformation $\Omega(\mathbf{n})$:

$$\hat{A}_i^S = \Omega \hat{A}_i^N \Omega^{-1} + \Omega \partial_i \Omega^{-1}.$$

This is a gauge transformation from the group $U(1)_{\text{e.m.}}$ over the electromagnetic vector potentials (in the unitary gauge), i.e. it is an Abelian transformation. Using (9.26) and (9.28), (9.29), we obtain the connection between real quantities on the equator of the sphere S_∞^2:

$$\mathcal{A}_i^S(\mathbf{n}) = \mathcal{A}_i^N + \frac{2}{g} \partial_i f(\varphi). \qquad (9.30)$$

Finally, we calculate the magnetic field flux through the sphere S_∞^2. We do this in the unitary gauge. The magnetic field on the northern hemisphere can be found in terms of the non-singular vector potentials there, $\mathcal{A}_i^N$. $\mathcal{A}_i^N$ are not useful in the southern hemisphere, since they are singular at the south pole. In the southern hemisphere we use the non-singular vector potentials $\mathcal{A}_i^S$. Therefore

$$\vec{\mathcal{H}} = \begin{cases} -\text{curl}\,\vec{\mathcal{A}}^N & \text{in the northern hemisphere} \\ -\text{curl}\,\vec{\mathcal{A}}^S & \text{in the southern hemisphere.} \end{cases}$$

We note that $\text{curl}\,\vec{\mathcal{A}}^N = \text{curl}\,\vec{\mathcal{A}}^S$ everywhere apart from at the north and south poles, since $\vec{\mathcal{A}}^N$ and $\vec{\mathcal{A}}^S$ are related by the gauge transformation

$\Omega(\mathbf{n})$ from $U(1)$. The magnetic field flux $\vec{\mathcal{H}}$ is non-singular everywhere on the sphere S^2_∞. The magnetic field flux is equal to

$$
\begin{aligned}
\int \vec{\mathcal{H}} d\vec{s} &= -\int_{\substack{\text{upper} \\ \text{hemisphere}}} \operatorname{curl} \vec{A}^N d\vec{s} - \int_{\substack{\text{lower} \\ \text{hemisphere}}} \operatorname{curl} \vec{A}^S d\vec{s} \\
&= -\int_{S^1} \vec{A}^N d\vec{l} + \int_{S^1} \vec{A}^S d\vec{l},
\end{aligned}
$$

where S^1 denotes the equator of the sphere S^2_∞. From (9.30) we have

$$
\int \vec{\mathcal{H}} d\vec{s} = \frac{2}{g} \int_{S^1} \partial_i f(\varphi) d\vec{l}
$$

or, taking into account (9.27),

$$
\int \vec{\mathcal{H}} d\vec{s} = \frac{4\pi}{g} n.
$$

This magnetic field flux corresponds to a monopole with magnetic charge

$$
g_M = \frac{1}{g} n, \tag{9.31}
$$

where n is the topological number appearing in (9.27). In our case $n = 1$ and we reach (9.23).

The considerations presented here are very general and shed light upon the connection between magnetic charge and Higgs field topology. In particular, they show that the magnetic charge in the $SU(2)$ model is a multiple of $\frac{1}{g}$, and the number n in (9.31) is equal to the degree of the mapping from S^2_∞ to S^2_{vac}, characterizing the Higgs field at spatial infinity.

The functions $w^N(\mathbf{n})$ and $w^S(\mathbf{n})$ for the specific case of $\varphi^a = n^a v$ can be constructed in explicit form:

$$
\begin{aligned}
w^N &= e^{-i\frac{\varphi}{2}\tau^3} e^{i\frac{\theta}{2}\tau^2} e^{i\frac{\varphi}{2}\tau^3} \tag{9.32} \\
w^S &= -i\tau^2 e^{-i\frac{\varphi}{2}\tau^3} e^{-i(\frac{\pi-\theta}{2})\tau^2} e^{i\frac{\varphi}{2}\tau^3}, \tag{9.33}
\end{aligned}
$$

where φ and θ are standard angles on the sphere S^2. The fact that the w^N are non-singular everywhere, apart from at the south pole $\theta = \pi$, and w^S is non-singular everywhere apart from at the north pole $\theta = 0$ is evident. At the equator

$$
\Omega(\varphi) = (w_S w_N^{-1}) \left(\theta = \frac{\pi}{2} \right) = e^{i\varphi\tau^3},
$$

which corresponds to (9.26) and $n = 1$.

Problem 6. *Show that the gauge transformations (9.32) and (9.33) indeed convert the field $\varphi^a = n^a v$ to the unitary gauge. Hint: use the matrix formula for the Higgs field. Using these transformations, find an explicit form for the monopole field in the unitary gauge far from the center.*

The fields obtained in this problem correspond to the Dirac point-like monopole in pure electrodynamics, for which one can use either a formulation in terms of a singular vector potential (with unobservable Dirac string) or a formulation with different vector potentials in the northern and southern hemispheres (Wu and Yang 1975) as we did above.

9.3 Generalization to other models

The considerations studied in Sections 9.1 and 9.2 indicate that the existence of magnetic monopoles as topological solitons must be a quite common property of gauge theories with breaking of a compact simple or semi-simple gauge group down to a subgroup containing a factor of the type $U(1)$. In this section we shall substantiate this using topological arguments (Tyupkin, Fateev and Schwarz 1975; Monastyrsky and Perelomov 1975).

Let us consider a theory with a compact simple or semi-simple gauge group G. To be specific, we shall assume that the group G is simply connected (in reality, this assumption does not restrict the generality of the subsequent discussion). Suppose that the theory includes a Higgs field φ, transforming according to some, generally speaking reducible, representation $T(g)$ of the group G. Suppose that the Higgs mechanism is realized in the model and that G is broken down to some subgroup H. This means that the vacuum value of the Higgs field is non-zero and that for some choice of the classical vacuum φ_{vac}, its stationary subgroup is equal to H,

$$T(h)\varphi_{\text{vac}} = \varphi_{\text{vac}} \text{ for all } h \in H$$

(we recall that φ_{vac} does not depend on x^μ). Of course, the vacuum, φ_{vac}, is not unique; any field value of the form

$$\varphi = T(g)\varphi_{\text{vac}}, \tag{9.34}$$

where g does not depend on x^μ, is also a ground state (throughout, we assume that in the ground state $A_\mu = 0$). Let us suppose that all possible vacua have the form (9.34), i.e. that there is no accidental degeneracy of vacua in the model (in fact, this assumption can also be omitted).

Let us consider the set of classical vacua M_{vac}. The assumptions made mean that the group G acts transitively on M_{vac}, where this action is determined by the representation T. Hence, M_{vac} is a homogeneous space

$$M_{\text{vac}} = G/H.$$

Let us now consider static configurations of fields with finite energy in three-dimensional space, where we shall assume that the vacuum energy is equal to zero. The contribution of the scalar potential to the energy will be finite, if the field at spatial infinity tends to some vacuum value, which may depend on the direction $\mathbf{n}$. Thus, we have a mapping from the infinitely remote sphere S^2_∞ to M_{vac}. Since

$$\pi_2(M_{\text{vac}}) \equiv \pi_2(G/H) = \pi_1(H),$$

this mapping may be topologically non-trivial in the case of a group H which is not simply connected, i.e. when

$$\pi_1(H) \neq 0.$$

In this case, the space of configurations with finite energy divides into disjoint subspaces (topological sectors), each of which corresponds to an element of $\pi_1(H)$. A minimum of the energy in a non-trivial topological sector is given by a soliton.

Problem 7. *Let $H = U(1)$. Using considerations analogous to those of Section 9.2, show that solitons just described are magnetic monopoles.*

Thus, monopoles exist in all models with compact simple or semi-simple gauge groups, where the Higgs mechanism leads to symmetry breaking down to a non-simply connected subgroup. Since, in nature, the electromagnetism group $U(1)_{\text{e.m.}}$ is indeed unbroken, monopoles always exist in realistic grand unified theories with a simple or semi-simple gauge group.

9.4   The limit $m_H/m_V \to 0$

We continue the study of monopoles in the $SU(2)$ model of Sections 9.1 and 9.2. In the limiting case, when the self-interaction constant of the Higgs field λ tends to zero, a solution of the field equations can be obtained in explicit form. Physically, this limit (called the Bogomolny–Prasad–Sommerfield limit) corresponds to the situation where the mass of the Higgs boson $m_H = \sqrt{2\lambda}v$ is much less than the mass of a vector boson[2] $m_V = gv$; it follows from formulae (9.14) and (9.18) that λ occurs in a solution in a combination $\lambda/g^2 = m_H^2/2m_V^2$.

---

[2]In some theories, the absence of a scalar potential can be ensured by symmetry considerations; this is the case in a number of supersymmetric theories.

When $\lambda \to 0$, the potential term in the energy functional (9.3)

$$\int d^3x \frac{\lambda}{4}(\varphi^a \varphi^a - v^2)^2$$

tends to zero, if *the condition (9.4), ensuring its finiteness, is satisfied.* In other words, in this limit, a monopole configuration is a minimum of the functional

$$E_{\lambda=0} = \int d^3x \left[\frac{1}{2}(H_i^a)^2 + \frac{1}{2}(D_i\varphi)^a(D_i\varphi)^a \right] \tag{9.35}$$

in the class of fields satisfying the condition

$$\varphi^a \varphi^a = v^2 \quad \text{for } |\mathbf{x}| = \infty \tag{9.36}$$

and having topological number one. In (9.35), $H_i^a = -\frac{1}{2}\varepsilon_{ijk}F_{jk}^a$. To find this minimum, we use a technique (Bogomolny 1976) analogous to that used in the model of **n**-field. We write down the following inequality

$$\int d^3x \frac{1}{2}(H_i^a - D_i\varphi^a)(H_i^a - D_i\varphi^a) \geq 0, \tag{9.37}$$

where equality holds when

$$D_i\varphi^a = H_i^a. \tag{9.38}$$

Expanding the brackets in (9.37), we have

$$E_{\lambda=0} \geq \int H_i^a D_i\varphi^a d^3x.$$

Furthermore, using the definition

$$D_i\varphi^a = \partial_i\varphi^a + g\varepsilon^{abc}A_i^b\varphi^c$$

and integrating by parts, we obtain

$$\int H_i^a D_i\varphi^a d^3x = \int_{S_\infty^2} H_i^a \varphi^a d\Sigma^i - \int \varphi^a(\partial_i H_i^a + g\varepsilon^{abc}A_i^b H_i^c)d^3x. \tag{9.39}$$

We now recall the Bianchi identity

$$\varepsilon_{\mu\nu\lambda\rho}D_\nu F_{\lambda\rho}^a = 0, \tag{9.40}$$

where

$$D_\nu F_{\lambda\rho}^a = \partial_\nu F_{\lambda\rho}^a + g\varepsilon^{abc}A_\nu^b F_{\lambda\rho}^c.$$

The component $\mu = 0$ in (9.40) gives

$$\partial_i H_i^a + g\varepsilon^{abc} A_i^b H_i^c = 0,$$

so that the second term in (9.39) vanishes.

Finally, the first term in (9.39) is proportional to the magnetic charge (see Section 9.2):

$$\int_{S_\infty^2} H_i^a \varphi^a d\Sigma^i = v \int_{S_\infty^2} \mathcal{H}_i d\Sigma^i = \frac{4\pi v}{g}.$$

Thus, we have an inequality for the energy as $\lambda \to 0$:

$$E_{\lambda=0} \geq \frac{4\pi v}{g}, \tag{9.41}$$

where there is equality only if equation (9.38) is satisfied.

A monopole configuration is a minimum of the energy for a unit magnetic charge. Thus, in the limit $\lambda \to 0$, it can be found by solving equation (9.38) (it is called the Bogomolny equation).

Problem 8. *Write down the field equations arising from minimization of the energy functional (9.35). Show that any solution of equations (9.38) is at the same time a solution of these field equations.*

A spherically symmetric (in the sense of generalized symmetry, discussed in Section 9.1) solution of the equations (9.38) can be obtained in explicit form. Let us again write down the spherically symmetric fields, but this time in the form

$$\begin{aligned} \varphi^a &= v n^a h(r) \\ A_i^a &= \varepsilon^{aij} n_j \frac{1}{gr}(1 - F(r)) \end{aligned}$$

(this Ansatz differs from (9.10) only by the replacement $h = 1 - H$). By calculating the "magnetic" field H_i^a and the covariant derivative, we obtain

$$\begin{aligned} H_i^a &\equiv -\frac{1}{2}\varepsilon_{ijk}F_{jk}^a = -\frac{\delta_{ia} - n_i n_a}{rg}F' + \frac{n_i n_a}{gr^2}(1 - F^2) \\ D_i \varphi^a &= v\frac{\delta_{ai} - n_i n_a}{r}Fh + v n_i n_a h'. \end{aligned}$$

From the independence of the tensors $n_i n_a$ and $(\delta_{ai} - n_i n_a)$, it follows that the equations (9.38) reduce to the equations

$$\begin{aligned} F' &= -(gv)hF \tag{9.42} \\ h' &= \frac{1}{gvr^2}(1 - F^2). \end{aligned}$$

A solution of these equations must satisfy the boundary conditions

$$
\begin{aligned}
F(0) &= 1, \quad h(0) = 0 \\
F(r \to \infty) &= 0, \quad h(r \to \infty) = 1.
\end{aligned}
\tag{9.43}
$$

These conditions ensure the non-singularity of the fields φ^a and A_i^a as $r \to \infty$ and $r \to 0$, the finiteness of the energy (9.35) and the satisfaction of the auxiliary equation (9.36). A solution of the equations (9.42) with the boundary conditions (9.43) can be guessed at:

$$
F = \frac{\rho}{\sinh \rho}
$$

$$
h = \coth \rho - \frac{1}{\rho},
$$

where

$$
\rho = gvr = \frac{r}{r_0}
$$

$$
r_0 = (gv)^{-1} = (m_V)^{-1}.
$$

We note that it was possible to find an explicit form for the solution by virtue of the fact that instead of the original second-order equations, we solved the first-order equations (9.38).

As one might expect, the function $F(r)$ tends exponentially to zero as $r \to \infty$:

$$
F(r) \propto e^{-r/r_0}.
$$

At the same time, $h(r)$ tends to one according to the power law:

$$
1 - h(r) \propto \frac{1}{r}.
\tag{9.44}
$$

This behavior has to do with the fact that the mass of the Higgs boson m_H is equal to zero as $\lambda \to 0$, i.e. among the small excitations over the vacuum there is a massless scalar field.

The mass of a monopole (its static energy) in the limit $\lambda \to 0$, as follows from the discussions leading up to (9.41), equals

$$
M = E_{\lambda=0} = \frac{4\pi v}{g} = \frac{m_V}{\alpha_g},
\tag{9.45}
$$

where, as before, $\alpha_g = \frac{g^2}{4\pi}$.

Problem 9. *Show by direct calculation of the energy functional $E_{\lambda=0}$ for the solution found that the mass of the monopole in the limit $\lambda \to 0$ is indeed given by formula (9.45).*

We note that there exist (and these have been found in explicit form) multi-monopole solutions of equations (9.38). At first glance, this assertion contradicts our expectations: monopoles experience a "Coulomb" repulsion due to their magnetic charges; thus, it would seem, the energy of two monopoles must have a minimum when the distance between their centers is infinite. In reality, the repulsion of magnetic charges is precisely compensated due to attraction associated with the long-ranged scalar field (see (9.44)), so that two monopoles are in neutral equilibrium. Of course, this situation only occurs at $\lambda = 0$; for $\lambda > 0$, the Higgs field is massive and interaction of monopoles associated with that field decreases exponentially for $r > m_H^{-1}$.

We note further that for $\lambda > 0$, the contribution of the potential term to the energy is positive, and a modification of the discussions of this section leads to the inequality

$$E_{\lambda>0} > \frac{m_V}{\alpha_g}.$$

Thus, the monopole has least mass in the limit $m_H/m_V \to 0$.

9.5   Dyons

Until now we have considered solutions with zero electric charge. In addition to these, the $SU(2)$ model contains solutions with non-zero magnetic charge and non-zero electric charge, dyons (Julia and Zee 1975). Since a static electric charge can be described by a potential $A_0(\mathbf{x})$, which does not depend on time, a dyon solution can be sought, on the assumption that $A_\mu^a(\mathbf{x})$ and $\varphi(\mathbf{x})$ do not depend on time; here, $A_0^a(\mathbf{x})$ is non-zero. In the spherically symmetric case for the field $A_0^a(\mathbf{x})$, we choose the Ansatz

$$A_0^a = n^a B(r), \tag{9.46}$$

where, from the requirement of regularity at $r = 0$ we obtain the condition

$$B(r = 0) = 0.$$

For the fields $\varphi^a(\mathbf{x})$ and $A_i^a(\mathbf{x})$ as before, we use the Ansatz (9.10) with boundary conditions (9.11) and (9.12). The choice of (9.46) leads to

$$D_0\varphi^a = \partial_0\varphi^a + g\varepsilon^{abc}A_0^b\varphi^c = 0.$$

Thus, the energy of a dyon will be finite if the electric component of the strength tensor F_{0i}^a decreases as $\frac{1}{r^2}$.

Let us determine the asymptotics of the function $B(r)$ for a dyon with (electric) charge q. For this, we calculate F_{0i}^a far from the center of the dyon, where the field A_i^a is given by the formula (9.9). We obtain

$$E_i^a \equiv F_{0i}^a = -n_a n_i B'.$$

Since E_i^a is directed in internal space along φ^a, asymptotically, there is indeed only the electric field which corresponds to the unbroken subgroup $U(1)_{\text{e.m.}}$. This is equal to

$$\mathcal{E}_i = \frac{1}{v} E_i^a \varphi^a = -n_i B'.$$

The electric field is directed along the radius vector, as it must be for a field of an electric charge. If the charge of the dyon is equal to q, then we must have

$$\mathcal{E}_i = n_i \frac{q}{r^2},$$

whence we obtain the asymptotics

$$B = \frac{q}{r} + C, \tag{9.47}$$

where C is a constant depending on q.

Similarly to the monopole case, it is not possible to find a dyon solution in analytic form. The only exception is the limiting case $\lambda \to 0$. Field configurations for the dyon with $\lambda \neq 0$ have been found numerically.

Problem 10. *Find equations for the radial functions $B(r)$, $F(r)$ and $H(r)$ for the dyon. Show that the asymptotics (9.11) and (9.47) satisfy these equations in the limit of large r.*

Chapter 10

Non-Topological Solitons

Until now, we have considered solitons, whose existence is related to the topological properties of field configurations. Furthermore, we have limited ourselves to a discussion of static solutions, which was in fact justified for topological solitons (it is more or less clear that the absolute minimum of the energy in the sector with non-zero topological number is attained, if it is attained at all, at a static configuration). However, neither non-trivial topological properties, nor time independence of the fields are mandatory for solitons, which are conceived as stable (or metastable) particle-like solutions of the field equations for fields with finite energy. In this chapter, we shall present an example of a stable soliton, whose field depends on time and does not have a non-trivial topology (Friedberg *et al.* 1976). This example is of a very general nature; the principle governing the existence and stability of the soliton is the presence of charge on the soliton. Such solitons, following Coleman (1985), are often called Q-balls. Up to now, it is not known whether they play any important role in particle physics, although similar ideas have been used in hadron (quark bag) models. We note that a completely different example of a non-topological soliton is given in one of the Supplementary Problems for Part II.

As the simplest model possessing non-topological solitons, we can choose the theory with two scalar fields: a real field $\varphi(x)$ and a complex field $\xi(x)$. The Lagrangian of the model has the form

$$\mathcal{L} = \frac{1}{2}(\partial_\mu \varphi)^2 - V(\varphi) + (\partial_\mu \xi)^* \partial_\mu \xi - h^2 \varphi^2 \xi^* \xi, \qquad (10.1)$$

where the scalar potential of the field φ is equal to

$$V(\varphi) = \frac{m_\varphi^2}{2}(\varphi - v)^2.$$

215

We shall suppose that

$$m_\varphi, v, h > 0.$$

To be specific, we shall consider this model in $(1 + 1)$-dimensional space–time, although this analysis is easy to generalize to the case of space–time with an arbitrary number of dimensions.

The model has a continuous global $U(1)$ symmetry

$$
\begin{aligned}
\varphi(x) &\rightarrow \varphi(x) \\
\xi(x) &\rightarrow e^{i\alpha}\xi(x) \\
\xi^*(x) &\rightarrow e^{-i\alpha}\xi^*(x).
\end{aligned}
\tag{10.2}
$$

This symmetry gives rise to the Noether current

$$j_\mu = \frac{1}{i}(\xi^*\partial_\mu\xi - \partial_\mu\xi^*\xi).$$

The charge

$$Q = \int \frac{1}{i}(\xi^*\partial_0\xi - \partial_0\xi^*\xi)dx^1 \tag{10.3}$$

is conserved on the field equations (i.e. provided the latter are satisfied).

To find the ground state, as usual, we write down the energy functional

$$E = \int dx^1 \left[\frac{1}{2}(\partial_0\varphi)^2 + \frac{1}{2}(\partial_1\varphi)^2 + V(\varphi) + |\partial_0\xi|^2 + |\partial_1\xi|^2 + h^2\varphi^2|\xi|^2\right]. \tag{10.4}$$

It has a minimum at a homogeneous configuration of the form

$$
\begin{aligned}
\varphi &= v \\
\xi &= 0,
\end{aligned}
\tag{10.5}
$$

which is the ground state in the model. We note that the symmetry (10.2) is not broken.

By considering small perturbations of the fields about the ground state (10.5), we conclude that they describe a massive real scalar field

$$
\begin{aligned}
\eta(x) &= \varphi(x) - v \\
m_\eta &= m_\varphi
\end{aligned}
$$

and a massive complex scalar field ξ with mass

$$m_\xi = hv.$$

Since the charge (10.3) is conserved, it is legitimate to ask about configurations of fields with minimum energy *among all configurations with a given value of* Q. If for some fixed Q a configuration with minimum energy is a localized state, then the model includes an absolutely stable soliton (Q-ball). In other words, the set of field configurations can be divided into disjoint sectors, characterized by the value of Q, each of which contains all the fields with the same value of the charge (in quantum theory these sectors are called superselection sectors). In the classical evolution, fields do not move from one sector into another because of charge conservation. The soliton (Q-ball) realizes the absolute minimum of the energy in a sector for some Q, thus it is stable.

We note that in the model (10.1), the topology of the field configurations is trivial; the existence of a soliton does not rely on topological arguments. Furthermore, the charge (10.3) is non-zero, only if the fields depend on time, thus the soliton (if it exists) is a time-dependent configuration.

Problem 1. *Write down equations for a conditional extremum of the energy functional (10.4) for a fixed charge* Q. *Show that the fields* $\xi(x^1, t)$ *and* $\varphi(x^1, t)$, *satisfying these equations, also satisfy the Lagrange field equations, arising from the Lagrangian (10.1). Hint: assume that the variations* $\delta(\partial_0 \varphi)$ *and* $\delta(\partial_0 \xi)$ *are not dependent on* $\delta \varphi$ *and* $\delta \xi$.

We shall show that a non-topological soliton actually exists in our model for sufficiently large values of Q. For this, we need to show that there exist localized configurations of fields, whose energies are smaller than the energy of non-localized configurations at fixed (and large) Q.

Let us first consider non-localized configurations. If fields are not localized then their amplitudes are small and a theory linearized near the classical vacuum (10.5) can be used. For a linearized theory the energy functional has the form

$$\int dx^1 \left[\frac{1}{2}(\partial_0 \eta)^2 + \frac{1}{2}(\partial_1 \eta)^2 + \frac{m_\varphi^2}{2}\eta^2 + |\partial_0 \xi|^2 + |\partial_1 \xi|^2 + m_\xi^2 |\xi|^2 \right]. \quad (10.6)$$

Clearly, for fixed Q, the functional (10.6) will be least if

$$\eta = 0.$$

To determine the energy for fixed Q it is convenient to use a Fourier representation for solutions of the field equations in the linearized theory:

$$\xi(x) = \frac{1}{\sqrt{2\pi}} \int \left[e^{i(kx^1 - \omega_k t)} a_k + e^{-i(kx^1 - \omega_k t)} b_k^* \right] \frac{dk}{\sqrt{2\omega_k}} \quad (10.7)$$

$$\xi^*(x) = \frac{1}{\sqrt{2\pi}} \int \left[e^{i(kx^1 - \omega_k t)} b_k + e^{-i(kx^1 - \omega_k t)} a_k^* \right] \frac{dk}{\sqrt{2\omega_k}},$$

where

$$\omega_k = \sqrt{k^2 + m_\xi^2}, \quad (10.8)$$

the factors $\frac{1}{\sqrt{2\pi}}$ and $\frac{1}{\sqrt{2\omega_k}}$ are introduced for convenience, and a_k and b_k are arbitrary small amplitudes; integration in (10.7) involves both positive and negative k. In terms of the amplitudes a_k and b_k, the energy (10.6) and the charge (10.3) can be written in the form

$$E = \int dk\,\omega_k (b_k^* b_k + a_k^* a_k) \tag{10.9}$$

$$Q = \int dk\,(b_k^* b_k - a_k^* a_k). \tag{10.10}$$

From formulae (10.9), (10.10) and (10.8), it is clear that if the charge Q is fixed, then

$$E \geq m_\xi Q, \tag{10.11}$$

where we have equality when $a_k = 0$, and b_k is non-zero only for $k = 0$ (to find the amplitudes b_k, one actually has to consider the theory in a large spatial box).

Problem 2.   *Consider the system with Lagrangian (10.1) in a large box of size L, i.e. assuming $-\frac{L}{2} \geq x^1 \geq \frac{L}{2}$. Find the value of the amplitudes b_k, minimizing the energy (10.9) for a fixed (and finite) charge Q. Is the assertion in the text concerning the smallness of the field $\xi(x)$ true at every point of space–time?*

Equation (10.11) has the following interpretation in quantum theory. The field $\xi(x)$ describes particles with mass m_ξ and unit charge (and also antiparticles with charge -1). For the charge to be equal to Q requires at least Q particles; their energy is minimal and equal to $E = m_\xi Q$ if they are all at rest (spatial momentum k is equal to zero).

Next, we show that for sufficiently large Q there exist localized configurations of fields with charge Q, whose energy is less than $m_\xi Q$. Taking into account (10.11), this will imply the existence of a non-topological soliton.[1]

Let us choose a field configuration φ which is static and has the form of a well, as shown in Figure 10.1. Outside a region of size l the field $\varphi(x^1)$ is equal to its value in the ground state, $\varphi = v$. There is a small transitional region (size much less than l) where the field changes from v to zero, and within the well, the field φ is equal to zero. The energy, associated with the field φ itself,

$$E_\varphi = \int \left[\frac{1}{2}(\partial_i\varphi)^2 + V(\varphi) \right] dx^1,$$

[1]Strictly speaking, we would need to show that the minimum energy for fixed charge Q is actually attained, i.e. a situation of the type studied in Section 7.2 does not arise. This can be done based on considerations studied below (see formula (10.19)).

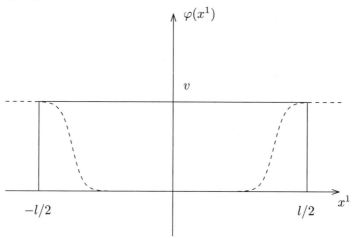

Figure 10.1.

is made up of the energy of the transitional region and the energy of the well. We shall be interested in the dependence of the energy on l, and we can write

$$E_\varphi = \mu + V_0 l,$$

where the first term arises due to the transitional region and does not depend on l, the second term comes from the region inside the well, where the energy density equals

$$V_0 = V(\varphi = 0) = \frac{m_\varphi^2}{2} v^2. \tag{10.12}$$

Let us now consider the field configuration $\xi(x^1, t)$. In the background field of our trial configuration $\varphi(x^1)$, the equation for the field ξ has the form

$$-\partial_\mu \partial_\mu \xi - h^2 \varphi^2(x^1)\xi = 0. \tag{10.13}$$

Let us choose the dependence of ξ on time in the form

$$\xi(x^1, t) = e^{i\Omega t}\xi_\Omega(x^1),$$

then equation (10.13) becomes an eigenvalue equation

$$[-\partial_1^2 + h^2 \varphi^2(x^1)]\xi_\Omega(x^1) = \Omega^2 \xi_\Omega(x^1). \tag{10.14}$$

We shall be interested in the smallest eigenvalue Ω. The potential $h^2 \varphi^2(x^1)$ has the form of a well of size l (and depth m_ξ^2), and it is equal to zero on the bottom of the well. Thus, modulo higher-order corrections in $1/l$, the least eigenvalue is equal to

$$\Omega_0 = \frac{\pi}{l} + O(1/l^2). \qquad (10.15)$$

The corresponding eigenfunction $\xi_{\Omega_0}(x^1)$, the lowest bound state in the potential $h^2\varphi^2(x^1)$, is real and decreases exponentially outside the well.

Thus, for the configuration $\xi(x^1, t)$, we choose the function

$$\xi(x^1, t) = Ne^{i\Omega_0 t}\xi_{\Omega_0}(x^1), \qquad (10.16)$$

where the function ξ_{Ω_0} is assumed to be normalized to one:

$$\int_{-\infty}^{\infty} \xi_{\Omega_0}^2(x^1)dx^1 = 1, \qquad (10.17)$$

and the constant N is defined from the requirement that the charge (10.3) should be equal to the given value Q. We have

$$Q = 2N^2\Omega_0,$$

so that

$$N^2 = \frac{Q}{2\Omega_0}. \qquad (10.18)$$

Finally, we calculate the energy associated with the field $\xi(x^1, t)$,

$$E_\xi = \int dx^1 [\partial_0\xi^*\partial_0\xi + \partial_1\xi^*\partial_1\xi + h^2\varphi^2\xi^*\xi].$$

Integrating by parts in the second term and taking into account (10.16), (10.17) and (10.14), we obtain

$$E_\xi = 2\Omega_0^2 N^2.$$

Using (10.18), we finally have

$$E_\xi = \Omega_0 Q.$$

Thus, the total energy for our chosen configuration of fields $\varphi(x^1)$ and $\xi(x^1, t)$ is equal to

$$E = E_\varphi + E_\xi = \mu + V_0 l + \frac{\pi Q}{l}, \qquad (10.19)$$

where we have used (10.12) and neglected corrections to formula (10.15). Up to now we have not yet fixed the size of the well l and have only assumed that it is sufficiently large. Let us now determine l from the requirement that the energy (10.19) should be least. We have

$$\frac{\partial E}{\partial l} = 0$$

at

$$l = \sqrt{\frac{\pi Q}{V_0}}, \tag{10.20}$$

and the minimum energy value for configurations of the type described is equal to

$$E(Q) = E\left(l = \sqrt{\frac{\pi Q}{V_0}}\right) = \mu + 2\sqrt{\pi Q V_0}. \tag{10.21}$$

For sufficiently large Q this expression is less than $m_\xi Q$, as required.

We stress that the configurations described are indeed localized: the field $\varphi(x^1)$ tends rapidly to its vacuum value for large $|x^1|$, and $\xi(x^1, t)$ tends exponentially to zero for $|x^1| \to \infty$.

For our estimates to be correct, we require the size of the well to be large. Equation (10.20) shows that this is indeed the case for sufficiently large Q.

Let us estimate the value of the charge Q_c, above which the model has a non-topological soliton. For this estimate, we neglect the first term in the soliton energy (10.21) and require the energy to be equal to the energy of a non-localized state:

$$2\sqrt{\pi Q_c V_0} = m_\xi Q_c.$$

Taking into account (10.12), we obtain

$$Q_c = 2\pi \frac{m_\varphi^2}{m_\xi^2} v^2. \tag{10.22}$$

For $Q \sim Q_c$, the size of our constructed configuration is in fact not large,

$$l \sim \sqrt{\frac{Q_c}{m_\varphi^2 v^2}} \sim \frac{1}{m_\xi}.$$

Thus, the above calculation of the energy of this configuration is incorrect, generally speaking, for $Q \sim Q_c$. This means, in particular, that the coefficient $2\pi \frac{m_\varphi^2}{m_\xi^2}$ in formula (10.22) for the critical charge cannot be trusted. The form of the soliton configuration for $Q \sim Q_c$ and the true value of the critical charge can only be obtained numerically; one can only assert that

$$Q_c \sim v^2$$

for large v and $m_\varphi \sim m_\xi$.

We recall that in $(1+1)$-dimensional space–time the characteristic value of the field v plays the role of an inverse coupling constant, i.e. the weak coupling limit corresponds to $v^2 \to \infty$ for fixed m_φ and m_ξ. Thus, non-topological solitons in models with weak coupling have a large charge

$$Q \gtrsim (\text{coupling constant})^{-2}.$$

We note that the mass of the lightest non-topological soliton is also large:

$$M_c = E(Q_c) \sim m_\xi v^2.$$

Similarly to topological solitons, the size of a soliton is much greater than its Compton wavelength: from (10.20) and (10.21) we have the relation

$$\frac{\lambda_c(Q)}{l(Q)} = \frac{1}{E(Q)l(Q)} \lesssim \frac{1}{Q} \ll 1.$$

In this sense, topological and non-topological solitons have analogous properties.

Problem 3.   *Show that in the model (10.1) (for sufficiently large charges) non-topological solitons exist in space–time of any dimension $(d+1)$, $d \geq 1$. Estimate the critical charge and mass of a non-topological soliton for $d = 3$, assuming the dimensionless constant h is small, and $m_\varphi / m_\xi \sim 1$.*

Non-topological solitons of the type described in this chapter arise in many models where there is charge conservation and the corresponding symmetry is not spontaneously broken. Again we note the physical principle governing their occurrence: for sufficiently large Q, the state in which the charge-carrying fields (in our case ξ) are localized in a finite region of space is energetically favorable; in this region, the field giving them the masses (in our case ϕ) takes a value different from the vacuum, so that the masses of the charged fields are equal to zero.

In the language of particles, the fact that this state is energetically favorable can be seen from the following qualitative consideration. For the charge of the state to be equal to Q, there would have to be Q charged particles (corresponding to the field ξ). The field configuration shown in Figure 10.1 is a potential well for these particles. In the well, their mass is equal to zero, which gives an energy gain of the order of Q in comparison with free particles (for fixed l). For sufficiently large Q, this energy gain is greater than the energy needed for the formation of the well (the latter energy is equal to E_φ and does not depend on Q for fixed l).

This consideration is clearly very general. It holds true in models where the role of charged particles is played by fermions, provided the number of fermion types is sufficiently large (Pauli's principle prevents us from applying these arguments in models with a single type of fermion or a small number of fermion species).

Chapter 11

Tunneling and Euclidean Classical Solutions in Quantum Mechanics

Localized solutions of classical field equations, whose existence is a result of the nonlinearity of these equations, are important, not only for the description of particle-like states, or solitons. Such solutions also arise in the study of a completely different class of problems, concerning tunneling processes in quantum field theory (and in quantum mechanics). In theories with a small coupling constant these processes can be studied within a semiclassical approach, where localized solutions of the field equations in *Euclidean space–time*, i.e. solutions of the instanton type, play a key role. They determine the leading semiclassical exponential in the tunneling probability. Another class of solutions, sphalerons, determine the height of the barrier separating classically stable field states. Here and in subsequent chapters, we shall be concerned with this class of problems, initially, for the example of quantum-mechanical systems, and then in the framework of field theory.

We note that instanton effects are important both for an understanding of the properties of the ground state, the vacuum, in gauge theories and for the theoretical study of the processes which took place in the early Universe. The methods which will be studied here and in subsequent chapters have close analogues in condensed matter theory.

11.1 Decay of a metastable state in quantum mechanics of one variable

The simplest system in which the problem of the decay of a metastable state arises is the quantum mechanics of a single variable q, described by the potential $V(q)$ shown in Figure 11.1. We shall be interested in the decay of a *lowest* metastable state, i.e. a state which would be the ground state in the potential well near the point q_0, if the barrier were impenetrable. In the semiclassical situation, the probability of tunneling through the potential barrier is given by the well-known expression

$$\Gamma = Ae^{-S_B}, \tag{11.1}$$

where Γ is the width of the metastable state ($\Gamma = 1/\tau$, τ is the lifetime), A is a pre-exponential factor, and S_B is the leading semiclassical exponent

$$S_B = 2 \int_{q_0}^{q_1} \sqrt{2MV(q)}dq, \tag{11.2}$$

where M is the particle mass, and q_1 is a turning point at which $V(q_1) = 0$, see Figure 11.1; in what follows, for convenience we shall assume that $V(q_0) = 0$ at the local minimum of the potential.

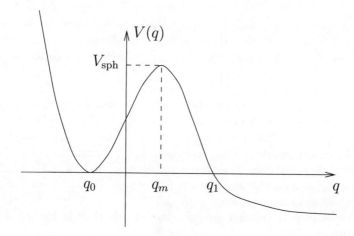

Figure 11.1.

We immediately note that S_B coincides with the Euclidean action, calculated on the classical trajectory of a particle moving "in imaginary time" with energy $E = 0$ from the point $q = q_0$ to the point $q = q_1$ and backwards. To explain this statement, we first write down the conventional action for a classical particle in the potential $V(q)$:

$$S = \int \left[\frac{M}{2} \left(\frac{dq}{dt} \right)^2 - V(q) \right] dt. \tag{11.3}$$

In this action, we shall make the purely formal substitution $t = -i\tau$ and assume that τ is real. Then, the action (11.3) becomes iS_E, where

$$S_E = \int \left[\frac{M}{2} \left(\frac{dq}{d\tau} \right)^2 + V(q) \right] d\tau\tau. \tag{11.4}$$

We shall call the functional S_E the Euclidean action, and τ the Euclidean time.

The origin of the term "Euclidean time" lies in the fact that the Minkowski metric $ds^2 = dt^2 - dx^2$, with the formal substitution $t = -i\tau$ becomes the Euclidean metric, $ds_E^2 = d\tau^2 + dx^2$ modulo an overall sign. We shall see that tunneling processes in field theory are described by the solutions of field equations in Euclidean space–time. We stress that the introduction of Euclidean time is a purely formal device.

The equation of motion following from the action (11.4) has the form

$$M \frac{d^2 q}{d\tau^2} = \frac{\partial V}{\partial q} \equiv -\frac{\partial(-V)}{\partial q}. \tag{11.5}$$

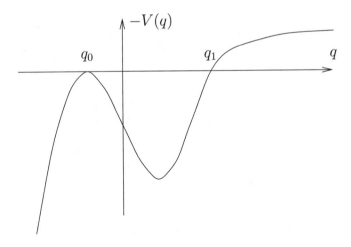

Figure 11.2.

This equation formally coincides with the equation of Newtonian mechanics for particles in the potential $(-V)$, shown in Figure 11.2. The integral of motion of this equation,

$$\frac{M}{2}\left(\frac{dq}{d\tau}\right)^2 - V(q) = \mathcal{E},$$

will be called the Euclidean energy.

Let us consider a solution with zero Euclidean energy, which starts when $\tau \to -\infty$ at the point $q = q_0$, goes to the point q_1 and returns to the point q_0 when $\tau \to +\infty$. We note that the choice of the origin for the Euclidean time τ can be made in such a way that the turning point $q = q_1$ is reached at $\tau = 0$. Such a solution is called a "bounce"; we shall denote it by $q_B(\tau)$. The Euclidean action on this solution is equal to

$$S_E(q_B) = \int_{-\infty}^{+\infty} d\tau \left[\frac{M}{2}\left(\frac{dq_B}{d\tau}\right)^2 + V(q_B)\right] = 2\int_{-\infty}^{0} 2V(q_B(\tau))d\tau, \quad (11.6)$$

where we have used the fact that $\mathcal{E} = 0$ for bounce. The integral (11.6) is converted to the form (11.2) by changing the integration variable, $\tau \to q_B$ using the formula $d\tau = \sqrt{\frac{M}{2V(q_B)}}dq_B$. Thus, we have

$$S_B = S_E[q_B(\tau)]. \quad (11.7)$$

Thus, the semiclassical exponent of the decay of a metastable state, is equal to the Euclidean action on the bounce solution.

Problem 1. *Determine the behavior of the bounce solution near $q = q_0$. Show that the point q_0 is reached asymptotically for $\tau \to -\infty$ or $\tau \to +\infty$. Assume that the potential near $q = q_0$ is quadratic $V(q) = \frac{1}{2}V''(q_0)(q - q_0)^2 + O[(q - q_0)^3]$.*

Problem 2. *Show that the semiclassical exponent of the decay of a highly excited metastable state with energy E in the well with potential $V(q)$ (Figure 11.1) is determined by a truncated Euclidean action on a periodic solution of the Euclidean equation (11.5), with $\mathcal{E} = -E$. (The expression $(S_E[q(\tau)]+\mathcal{E}\tau_0)$, where $S_E[q(\tau)]$ is the Euclidean action over a single period, is called the truncated Euclidean action for the periodic solution $q(\tau)$ with period τ_0 and Euclidean energy $\mathcal{E}$).*

Of course, the result (11.7) is no accident. To explain the connection between the tunneling exponent and the bounce solution and to determine a way of generalizing it to quantum mechanics with many variables and to field theory, we shall derive it rather more systematically. Let us write down the Schrödinger equation for the stationary state such that near $q = q_0$ it coincides with the ground state in the potential well:

$$\left[\frac{1}{2M} \left(-i \frac{\partial}{\partial q} \right)^2 + V(q) \right] \psi = 0.$$

Here, we have neglected the difference of the energy of this state from zero, which is legitimate for calculation of the semiclassical exponential. In the classically forbidden region $q_0 < q < q_1$, the solution is, generally speaking, a linear combination of decreasing and increasing exponentials, where the decreasing exponential is significantly greater than the increasing one, except for in a small neighborhood of the point $q = q_1$. Thus, we shall be interested in the exponentially decreasing component of the wave function which, modulo a pre-exponential factor is equal to

$$\psi(q) = e^{-S(q)},$$

where, as usual, $S(q)$ satisfies the equation

$$\frac{1}{2M} \left(\frac{dS}{dq} \right)^2 - V(q) = 0. \tag{11.8}$$

The semiclassical exponential for the decay probability is determined by the wave function on the boundary of the forbidden region (at the point q_1). Thus

$$\Gamma \propto |\psi(q_1)|^2 = e^{-2S(q_1)}. \tag{11.9}$$

Near $q = q_0$, the wave function is not small, therefore

$$S(q_0) = 0. \tag{11.10}$$

For systems with a single variable q, it is easy to solve equation (11.8) directly (and to obtain expressions (11.1) and (11.2) using (11.9)). One might, however, remark that equation (11.8) is the stationary Hamilton–Jacobi equation for the theory with the action (11.4). As we know from classical mechanics, the classical trajectories are characteristics of the Hamilton–Jacobi equation and solutions of the Hamilton–Jacobi equation can be obtained by integrating the Lagrangian along the classical trajectories. In our case, we are concerned with theories with the action (11.4), therefore, the classical trajectories satisfy the Euclidean equation (11.5). Moreover, the stationary Hamilton–Jacobi equation (11.8) is characterized by zero Euclidean energy $\mathcal{E}$, so that its solution is determined by the trajectory with $\mathcal{E} = 0$. Taking into account (11.10), we obtain a solution of equation (11.8) in the form

$$S(q) = \int_{-\infty}^{\tau(q)} d\tau \left[\frac{M}{2} \left(\frac{dq_B}{d\tau} \right)^2 + V(q_B(\tau)) \right], \tag{11.11}$$

where q_B is the solution of equation (11.5) with $\mathcal{E} = 0$, i.e. precisely the bounce solution, and the integration is performed from $\tau = -\infty$, where $q_B = q_0$, to the value of τ such that $q_B(\tau) = q$.

Problem 3. *Show by direct calculation of the derivative dS/dq that the function $S(q)$, given by formula (11.11), satisfies equation (11.8).*

Assuming that the turning point is reached at $\tau = 0$, we obtain for the exponent in (11.9)

$$S(q_1) = \int_{-\infty}^{0} d\tau \left[\frac{M}{2} \left(\frac{dq_B}{d\tau} \right)^2 + V(q_B) \right].$$

Finally, taking into account the symmetry of the bounce solution with respect to the substitution $\tau \to -\tau$, we obtain $S_B = 2S(q_1)$; thus expression (11.9) indeed coincides with (11.1), (11.2).

Problem 4. *Generalize the above consideration to the case of non-zero energy and hence obtain the result of Problem 2.*

To conclude this section, we mention another important solution of the equations of classical motion. Namely, the point of the maximum of the potential (q_m in Figure 11.1) is an unstable (in conventional time) *static* solution of the equation of Newtonian mechanics. This solution determines the height of the potential barrier $V_{\text{sph}} \equiv V(q_m)$ for a given system. Solutions with this property are called sphalerons. It is useful to know the height of the potential barrier when studying processes taking place at finite energies (for $E > V_{\text{sph}}$, the process of leaving a well may take place without tunneling, with probability of the order of unity in quantum mechanics with one degree of freedom; for $E < V_{\text{sph}}$, tunneling is obligatory, and the probability is exponentially small), and also for processes at finite temperatures. Let us consider the latter case in somewhat more detail. For finite, but not excessively high temperatures ($T \ll V_{\text{sph}}$; we set the Boltzmann constant equal to one, so that the temperature is measured in units of energy), the excited levels in a well are populated according to the Boltzmann distribution

$$P(E) \propto e^{-\frac{E}{T}},$$

where $P(E)$ is the probability that a particle, interacting with a thermostat with temperature T, has energy E. In particular, there is a finite probability that a particle will have energy $E \geq V_{\text{sph}}$, will be incident on the top of the barrier and will leave the well without tunneling. This process is purely classical; it leads to a decay probability Γ (the reciprocal of the time spent by a particle in the well at finite temperature), which is suppressed by the Boltzmann exponential, rather than the tunneling exponential

$$\Gamma_T = e^{-\frac{V_{\text{sph}}}{T}}$$

modulo a pre-exponential factor (Arrhenius formula). Even for $T \ll V_{\text{sph}}$ this value may be significantly greater than the tunneling exponential e^{-S_B}, thus, starting from some temperature, thermal jumps become the dominant decay mechanism. Their rate is determined by the barrier height, i.e. by the sphaleron energy $V_{\text{sph}} \equiv V(q_m)$.

Problem 5. *At intermediate temperatures, the process of tunneling from excited levels with energy $E < V_{\text{sph}}$ may have the greatest probability. Modulo a pre-exponential factor, this probability is given by the product of the probability that a particle has energy E, and the probability of tunneling from a state with energy E:*

$$\Gamma \propto e^{-\frac{E}{T}} e^{-\tilde{S}(E)} = e^{-\frac{E}{T} - \tilde{S}(E)}, \tag{11.12}$$

where this expression is to be maximized with respect to energy. Here, $\tilde{S}(E)$ is the semiclassical exponent for tunneling from a state with energy E. Find the connection between the exponent in (11.12) and the periodic Euclidean solutions considered in Problem 2: show that the exponent is related to the action on the corresponding solution, and that the temperature is related to its period. Draw qualitatively the dependence of the exponent in (11.12) on $\beta = T^{-1}$.

We note that everything studied in this section is applicable to the semiclassical situation only. A necessary condition for this is a large value of S_B, i.e. an exponentially small probability of decay of a metastable ground state.

Problem 6. *Suppose the potential has the form*

$$V(q) = \frac{1}{2}\mu^2 q^2 - \frac{\lambda}{3}q^3.$$

Find the relationship between the mass of a particle M and the parameters μ and λ, for which the standard conditions for the applicability of the semiclassical approximation are satisfied. Find the dependence of S_B on the parameters M, μ and λ and show that in the domain of applicability of the semiclassical approximation, the probability of tunneling from the ground state $q_0 = 0$ is indeed exponentially small. Calculate the energy of a metastable ground state to the first two orders in λ in perturbation theory, and show that perturbation theory is applicable for these values of the parameters.

The results of this problem illustrate a very common situation: if the parameters of the theory are such that standard perturbation theory is applicable, then the decay of a metastable ground state can, usually, be described in the framework of a semiclassical approximation.

11.2 Generalization to the case of many variables

The connection, identified in the previous section, between solutions of the classical Euclidean equations of motion and the semiclassical exponential for the probability of tunneling from the *ground* state admits generalization to the quantum mechanics of systems with many variables $\mathbf{q} = (q^1, \ldots, q^n)$ (and to field theory). First, we formulate a prescription for calculating probability (Banks *et al.* 1973; Banks and Bender 1973) and then we move to justify it.

Let us consider, to be specific, a system with the classical action (in conventional time)

$$S = \int dt \left[\frac{M}{2} \left(\frac{d\mathbf{q}}{dt} \right)^2 - V(\mathbf{q}) \right]. \tag{11.13}$$

We shall assume that the potential has a (sufficiently deep) local minimum at the point $\mathbf{q} = \mathbf{q}_0$; we choose the origin for the energy such that $V(\mathbf{q}_0) = 0$. We shall further assume that there is a region where $V(\mathbf{q}) < 0$, which is separated from the local minimum by the barrier. At the classical level, a particle at rest at the point $\mathbf{q}_0$ is in a state of stable equilibrium. In quantum mechanics, the corresponding ground state in a potential well is metastable and its decay probability Γ (the reciprocal of the lifetime) in the semiclassical situation is equal to

$$\Gamma = A e^{-S_B}, \tag{11.14}$$

where A is a sub-leading pre-exponential factor and S_B is the leading semiclassical exponent. The prescription for finding the leading exponent is the following. To calculate S_B, one has to switch to Euclidean time, i.e. to consider the system with the action

$$S_E = \int d\tau \left[\frac{M}{2} \left(\frac{d\mathbf{q}}{d\tau} \right)^2 + V(\mathbf{q}) \right] \tag{11.15}$$

and to find the bounce solution of the classical equations of motion following from the action (11.15). This bounce solution $q_B(\tau)$ must have the following properties: 1) for $\tau \to -\infty$ and $\tau \to +\infty$, the bounce solution must tend to a local minimum of the potential, $q_B(\tau \to \pm\infty) = q_0$; 2) the solution must have a turning point for some τ (without loss of generality, we may suppose that this is the "moment" $\tau = 0$),

$$\left.\frac{dq_B^i}{d\tau}\right|_{\tau=0} = 0 \text{ for all } i. \tag{11.16}$$

There is usually a finite number of such solutions; the relevant one is such that the Euclidean action (11.15) on it is minimal (if the system has a continuous symmetry like a rotation of the coordinates q, then there may — and usually does — exist a family of bounce solutions differing from one another by a symmetry transformation; the Euclidean action on the solutions from the same family, however, is the same and, as before, we need to minimize the Euclidean action among all families). The exponent in (11.14) is equal to the Euclidean action on the (most favorable) bounce solution,

$$S_B = S_E[q_B(\tau)].$$

We would make a few remarks in connection with this prescription. First of all, it only works for tunneling from a metastable *ground* state. No analogous prescription for tunneling from a fixed excited state in a potential well is known.

Next, the equations of motion, arising from the action (11.15), formally have the form of equations of Newtonian mechanics in the potential $[-V(q)]$,

$$M\frac{d^2 q}{d\tau^2} = -\frac{\partial(-V)}{\partial q}. \tag{11.17}$$

The integral of motion for these equations is the Euclidean energy

$$\frac{M}{2}\left(\frac{dq}{d\tau}\right)^2 - V(q) = \mathcal{E}. \tag{11.18}$$

For bounce, we have $\mathcal{E} = 0$; for precisely this Euclidean energy, a particle moving in the potential $[-V(q)]$, moves up to its top q_0 (or moves down from the top) over an infinite time, i.e. $q_B \to q_0$ for $\tau \to \pm\infty$.

By virtue of (11.16), all velocity components are equal to zero at $\tau = 0$, whence it follows that $V(q_B(\tau = 0)) = 0$ (from conservation of Euclidean energy and $\mathcal{E} = 0$); in other words, at $\tau = 0$, a particle lies on the boundary separating the classically allowed $(V < 0)$ and classically forbidden $(V > 0)$ regions of the original system. Finally, again by virtue of (11.16), the bounce solution is symmetric with respect to the substitution $\tau \to -\tau$, i.e. a particle in the potential $(-V)$ moves down from the hump, stops on the boundary of the region $V \geq 0$ and moves up again to the hump along the original trajectory. Therefore,

$$S_B = 2 \int_{-\infty}^{0} \left[\frac{M}{2} \left(\frac{d\mathbf{q}_B}{d\tau} \right)^2 + V(\mathbf{q}_B) \right] d\tau. \qquad (11.19)$$

Taking into account (11.18) and $\mathcal{E} = 0$, this expression can be transformed to the form

$$S_B = 2 \int_{\mathbf{q}_0}^{\mathbf{q}_1} \sqrt{2MV(\mathbf{q})} \, dl,$$

where $\mathbf{q}_1 = \mathbf{q}_B(\tau = 0)$ is a turning point, the integration is taken along the bounce trajectory, and

$$dl = \sqrt{\left(\frac{d\mathbf{q}}{d\tau} \right)^2} \, d\tau$$

is a length element of the trajectory. We note that the bounce trajectory is sometimes called the most probable escape path from under the barrier.

Let us turn to a justification of the above prescription. As in the previous section, we are interested in a solution of the stationary Schrödinger equation with zero energy

$$\left[\frac{1}{2M} \left(-i \frac{\partial}{\partial \mathbf{q}} \right)^2 + V(\mathbf{q}) \right] \psi = 0 \qquad (11.20)$$

in the classically forbidden region $V(\mathbf{q}) > 0$. This solution should decrease exponentially as the point $\mathbf{q}$ moves away from the local minimum of the potential $\mathbf{q}_0$. At least in some neighborhood of the point $\mathbf{q}_0$, a solution can be sought in the form

$$\psi(\mathbf{q}) = A(\mathbf{q}) e^{-S(\mathbf{q})}, \qquad (11.21)$$

where $A(\mathbf{q})$ is a slowly varying pre-exponential factor. As usual, for $S(\mathbf{q})$ we obtain from (11.20) the equation

$$\frac{1}{2M} \left(\frac{\partial S}{\partial \mathbf{q}} \right)^2 - V(\mathbf{q}) = 0. \qquad (11.22)$$

This equation again looks like the stationary Hamilton–Jacobi equation with zero energy for a system with the action (11.15); its characteristics are solutions of the Euclidean Newton's equations (11.17). The condition

$$S(\mathbf{q}_0) = 0,$$

which ensures that the wave function (11.21) is not small at the point $\mathbf{q}_0$, again has to be imposed on $S(\mathbf{q})$.

Let us discuss in detail the solutions of the Newtonian equations (11.17), i.e. the characteristics of equation (11.22), which are important to us. We are interested in solutions with Euclidean energy $\mathcal{E}$, equal to zero, which start (for $\tau \to -\infty$) from the point $\mathbf{q}_0$. Through each point $\mathbf{q}$ in a sufficiently small, but finite, neighborhood of the point $\mathbf{q}_0$, there passes one and only one characteristic of this type (we suppose that the potential $V(\mathbf{q})$ is quadratic near $\mathbf{q}_0$).

Problem 7. *Prove the above assertion, assuming that near the point* $\mathbf{q} = \mathbf{q}_0$ *the potential has the form*

$$V(\mathbf{q}) = \frac{1}{2}\mu_{ij}(q^i - q_0^i)(q^j - q_0^j) + O[(q - q_0)^3],$$

where μ_{ij} is a positive definite quadratic form.

In this region the solution of equation (11.22) has the form

$$S(\mathbf{q}) = \int_{-\infty}^{\tau(\mathbf{q})} d\tau \left[\frac{M}{2}\left(\frac{d\mathbf{q}}{d\tau}\right)^2 + V(\mathbf{q}(\tau)) \right], \qquad (11.23)$$

where $\mathbf{q}(\tau)$ is a trajectory with zero Euclidean energy, starting at $\mathbf{q}_0$ and passing through the point $\mathbf{q}$ at time $\tau(\mathbf{q})$.

The new circumstance in systems with more than one variable is that Euclidean trajectories of the given type do not, generally speaking, cover the whole region with $V(\mathbf{q}) > 0$ (classically forbidden region of the original system). If the potential does not have a sufficiently large symmetry, the majority of trajectories of interest to us (i.e. those which start at the top of the potential $(-V(\mathbf{q}))$ and have zero Euclidean energy) bend, do not reach the surface $V(\mathbf{q}) = 0$, and cross other similar trajectories (see Figure 11.3, where the trajectories are shown by continuous curves with arrows). Obtaining a solution of the Euclidean Hamilton–Jacobi equation in the form (11.23) is only possible in the region bounded by the envelope of our trajectories, called the caustic or caustic surface. The caustic is shown by the dashed curve in Figure 11.3. Outside the caustic surface it is impossible to find a decreasing solution of the Schrödinger equation (11.20) using the method we have just studied.

However, in fact, we are not interested in the solution of the Schrödinger equation in the whole region $V > 0$. All that is important to us is the maximum value of $|\psi|^2$ on the surface $V = 0$. To find this value, we first consider the values of $|\psi(\mathbf{q})|^2$ on the caustic surface. As before, these are given by the expression (11.21). Moreover, the "Euclidean action function"

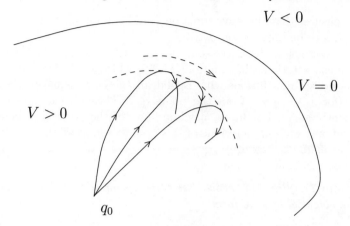

Figure 11.3.

$S(\mathbf{q})$ varies along the caustic surface, and

$$\frac{\partial S(\mathbf{q})}{\partial \mathbf{q}} = M\frac{d\mathbf{q}}{d\tau}, \tag{11.24}$$

where the left-hand side is the gradient of the Euclidean action at the point $\mathbf{q}$ of the caustic surface, and the right-hand side is the Euclidean momentum along that trajectory, which is tangential to the caustic surface at this point. This momentum is tangential to the caustic surface, so that the Euclidean action actually varies along the caustic (in Figure 11.3, the action increases in the direction shown by the arrow).

Let us now consider the minimum of the Euclidean action $S(\mathbf{q})$ *on the caustic surface.* Since the gradient of $S(\mathbf{q})$ is tangential to the caustic surface, at this minimum, the following is satisfied:

$$\frac{\partial S(\mathbf{q})}{\partial \mathbf{q}} = 0.$$

From (11.24), it is clear that at this minimum, the velocity of a particle is equal to zero, and, hence, the minimum is located on the boundary of the forbidden region, i.e. on the surface $V(\mathbf{q}) = 0$. This situation is illustrated in Figure 11.4, where the caustic surface is shown by a dashed curve, and the surface $V = 0$ by a continuous curve; trajectories with zero Euclidean energy are also shown by continuous curves; $\mathbf{q}_1$ is the point where the action on the caustic is minimal.

Since on the caustic, $S(\mathbf{q}) > S(\mathbf{q}_1)$, and the wave function continues to decrease between the caustic and the surface $V = 0$ (this is still the forbidden region of the system in question), the wave function, considered on the surface $V(\mathbf{q}) = 0$, has a maximum at the point $\mathbf{q}_1$. In other words, on the boundary of the classically forbidden region, the wave function is maximal at the point $\mathbf{q}_1$, where its value is given by expression (11.21) and $S(\mathbf{q}_1) = \frac{1}{2}S_B$, where S_B is determined by formula (11.19). This completes the justification of the prescription formulated at the beginning of this section.

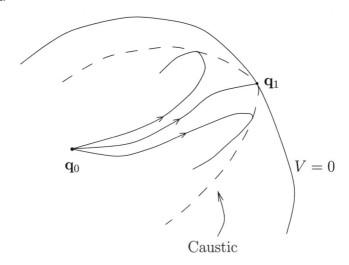

Figure 11.4.

Problem 8. *Consider the quantum mechanics of two variables q and y with the potential*

$$W(q, y) = \frac{1}{2}\omega^2 y^2 + V(q),$$

where $V(q)$ has the form shown in Figure 11.1. Find the bounce solution. Find the form of the curve $W = 0$, separating the classically allowed and forbidden regions, near the turning point of the bounce solution. Find classical solutions of the type considered in this section, which are close to the bounce. Find the caustic near the bounce turning point, draw everything in the plane (q, y) by analogy with Figure 11.4. Using the fact that the variables in the potential $W(q, y)$ separate, find the semiclassical wave function in the region between the caustic and the boundary of the forbidden region, and show that on the curve $W(q, y) = 0$, it is indeed maximal at the turning point of the bounce solution.

Thus, the problem of finding the semiclassical tunneling exponent of the decay of a metastable ground state reduces to finding a bounce solution $\mathbf{q}_B(\tau)$, i.e. a solution of the Euclidean classical equations of motion with zero Euclidean energy, which starts, for $\tau \to -\infty$, from the point of the local minimum of the potential $V(\mathbf{q})$ (i.e. a stable state of the classical system), comes to a stop at some point $\mathbf{q}_1$ of the surface $V = 0$ (the surface separating the classically forbidden and allowed regions) and returns to the local minimum for $\tau \to +\infty$. The wave function on the surface $V = 0$ is maximal at the point $\mathbf{q}_1$ and equal to $e^{-S_B/2}$ (modulo a pre-exponential factor), where S_B is the Euclidean action of the bounce solution, the leading exponential for the tunneling probability is e^{-S_B}.

Let us now consider the motion of a particle *after* it emerges from under the barrier, i.e. into the classically allowed region. At the point $\mathbf{q}_1$, the point of the maximum of the wave function on the surface $V = 0$, the gradient (of the leading exponential part) of the wave function is equal to zero. In other words, with greatest probability, a particle emerges from under the barrier at the point $\mathbf{q}_1$ with zero momentum (and, accordingly, with zero velocity). The subsequent motion of this particle can be found by solving the classical equations of Newtonian mechanics in conventional time, with the initial conditions

$$
\begin{aligned}
\mathbf{q}(t = 0) &= \mathbf{q}_1 \\
\frac{d\mathbf{q}}{dt}(t = 0) &= 0.
\end{aligned}
$$

The classical energy of this solution is equal to zero. Thus, in the classically allowed region, as well as under the barrier, just one trajectory is relevant. This situation is shown in Figure 11.5. If the bounce trajectory $\mathbf{q}_B(\tau)$ is known explicitly, then in order to find the classical trajectory after emergence from under the barrier, $\mathbf{q}(t)$, the following consideration can be employed. Let us analytically continue the bounce $\mathbf{q}_B(\tau)$ in the complex plane τ. For a purely imaginary $\tau = it$ (t real), the function

$$
\mathbf{q}_B(it) = \mathbf{q}(t) \tag{11.25}
$$

satisfies the conventional Newtonian equations

$$
M\frac{d^2\mathbf{q}}{dt^2} = -\frac{\partial V}{\partial \mathbf{q}}
$$

(this follows, after a change of notation, from the Euclidean Newton's equations (11.17), which the bounce satisfies). Furthermore, for $\tau = it = 0$, we have $\mathbf{q} = \mathbf{q}_B = \mathbf{q}_1$ and $\frac{d\mathbf{q}}{dt} = \frac{d\mathbf{q}_B}{d\tau} = 0$. Thus, analytic continuation (11.25) of the bounce solution to the region of imaginary τ is the desired classical path.

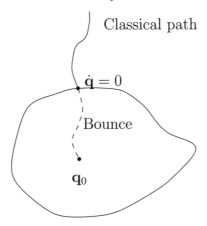

Figure 11.5.

Problem 9.   *Consider the theory with one variable q and potential*

$$V(q) = \frac{\mu^2}{2} q^2 - \frac{\lambda}{4} q^4.$$

Find the bounce solution $q_B(\tau)$, and its analytic continuation to the region of imaginary τ, and show that $q(t) = q_B(it)$ is indeed a classical solution in conventional time, describing the motion of a particle after emergence from under the barrier.

As in the theory with one variable, the question arises as to the minimum height of the barrier separating a local minimum of the potential (point $\mathbf{q}_0$) and the classically allowed region. This is again associated with an unstable solution $\mathbf{q}_S$ of the *static* equations of motion (sphaleron),

$$\frac{\partial V}{\partial \mathbf{q}}(\mathbf{q} = \mathbf{q}_S) = 0.$$

In the case of many variables, $\mathbf{q}_S$ is a saddle point of the potential: as the particle moves from $\mathbf{q}_S$ in one direction, the potential decreases (a particle moves down either toward a local minimum or toward the allowed region), while in all other perpendicular directions, it increases.[1] We note that the bounce solution $\mathbf{q}_B(\tau)$ (most probable path of escape from under the barrier), generally speaking, does not pass through the saddle point $\mathbf{q}_S$ (sphaleron).

---

[1] If a system has continuous symmetry, then sphalerons, generally speaking, form a continuous family, and have neutral directions (zero modes), along which the potential does not change. These zero modes correspond to symmetry transformations over one of the sphalerons.

The results studied in this section can be directly generalized to systems which are more complicated than (11.13), for example, to systems with action of type

$$S = \int dt \left[\frac{1}{2} \frac{dq_i}{dt} M_{ij}(\mathbf{q}) \frac{dq_j}{dt} - V(\mathbf{q}) \right]$$

or to systems where the coordinates satisfy some constraints. In the following chapters, we shall consider the generalization of these results to the case of field theory.

Problem 10. *Consider a system of two variables with action*

$$S = \int dt \left[\frac{1}{2} m_1 \left(\frac{dq_1}{dt} \right)^2 + \frac{1}{2} m_2 \left(\frac{dq_2}{dt} \right)^2 - V(q_1, q_2) \right],$$

where

$$V(q_1, q_2) = \frac{1}{2} \mu^2 (q_1^2 + q_2^2) - \frac{1}{4} \lambda (q_1^2 + q_2^2)^2 + \varepsilon q_2.$$

a) Suppose $m_2 \gg m_1$ and that ε is small. Find the bounce solution and the sphaleron. Does the bounce solution pass through the sphaleron? b) The same for $m_1 = m_2$. c) The same for $m_1 = m_2$ and $\varepsilon = 0$.

11.3 Tunneling in potentials with classical degeneracy

The problem of calculating the semiclassical tunneling exponent also arises for systems with degenerate minima of the classical potential. The simplest such system is the quantum mechanics of a single variable q with a symmetric potential $V(q) = -\frac{\mu^2}{2} q^2 + \frac{\lambda}{4} q^4$, shown in Figure 11.6. If the barrier separating the minima $q_0^{(+)}$ and $q_0^{(-)}$ were impenetrable, the system would have two degenerate ground states $\psi_0^{(+)}$ and $\psi_0^{(-)}$, whose wave functions would be confined near $q = q_0^{(+)}$ and $q = q_0^{(-)}$, respectively. When tunneling is taken into account degeneracy is lifted, which can be seen as follows. Let us define a reflection operator P, such that $(P\psi)(q) = \psi(-q)$. It commutes with the Hamiltonian and its eigenvalues are equal to $+1$ (symmetric wave functions) and -1 (antisymmetric wave functions). The lowest states of the Hamiltonian with $P = +1$ and -1 have the form $\psi_S = \frac{1}{\sqrt{2}} [\psi_0^{(+)} + \psi_0^{(-)}]$ and $\psi_A = \frac{1}{\sqrt{2}} [\psi_0^{(+)} - \psi_0^{(-)}]$, respectively, where we suppose that $\psi_0^{(+)}$ and $\psi_0^{(-)}$ are positive everywhere, so that

$P\psi^{(\pm)} = \pm\psi^{(\pm)}$. The function ψ_S has no nodes, and it is in fact the true ground state; ψ_A has one node at the point $q = 0$, so that it is the first excited state.

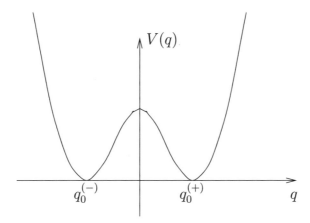

Figure 11.6.

The splitting is determined, in the leading semiclassical approximation, by the value of the semiclassical exponent for tunneling from the point $q_0^{(-)}$ to the point $q_0^{(+)}$,

$$S = \int_{q_0^{(-)}}^{q_0^{(+)}} \sqrt{2MV(q)}\,dq, \qquad (11.26)$$

where M is the particle mass.

To find the difference between the energies E_A and E_S of the states ψ_A and ψ_S, we write down the stationary Schrödinger equations, satisfied by these states:

$$-\frac{1}{2M}\psi_A'' + (V - E_A)\psi_A = 0$$

$$-\frac{1}{2M}\psi_S'' + (V - E_S)\psi_S = 0.$$

Multiplying the first of these by ψ_S and the second by ψ_A, and integrating over dq from $-\infty$ to 0, we obtain

$$-\frac{1}{2M}(\psi_A'\psi_S - \psi_S'\psi_A)(q=0) = (E_A - E_S)\int_{-\infty}^{0} \psi_A\psi_S\,dq. \qquad (11.27)$$

Taking into account that $\psi_S'(0) = \psi_A(0) = 0$, $\psi_S(0) = \sqrt{2}\psi_0^{(+)}(0)$, $\psi_A'(0) = \sqrt{2}\psi_0^{(+)\prime}(0)$, and also that the integral on the right-hand side of (11.27) is saturated in the region $q \approx q_0^{(-)}$ and is equal to $[-\frac{1}{2}]$, we obtain

$$E_A - E_S = \frac{2}{M} \psi_0^{(+)}(0) \frac{d\psi_0^{(+)}}{dq}(0). \tag{11.28}$$

This leads to (11.26), since

$$\psi_0^{(+)} \sim \frac{d\psi_0^{(+)}}{dq} \sim \exp\left(-\int_0^{q_0^{(+)}} \sqrt{2MV(q)}dq\right).$$

Equation (11.28) can also be brought to a form convenient for generalization to the quantum mechanics of several variables. For this, we take into account that $\psi_0^{(+)}(0) = \psi_0^{(-)}(0)$, $\psi_0^{(+)\prime}(0) = -\psi_0^{(-)\prime}(0)$. Further, the expression on the right-hand side of equation (11.28) is exponentially small, thus, in it, we may suppose that $\psi_0^{(\pm)}$ satisfies the Schrödinger equation with unperturbed energy E_0,

$$-\frac{1}{2M}\psi_0^{(\pm)\prime\prime} + (V - E_0)\psi_0^{(\pm)} = 0.$$

It follows from this equation that

$$j(q) = \left(\psi_0^{(-)}\frac{d\psi_0^{(+)}}{dq} - \psi_0^{(+)}\frac{d\psi_0^{(-)}}{dq}\right)$$

does not depend on q. Near $q = q_0^{(+)}$, we have $\psi_0^{(+)}$, $\psi_0^{(+)\prime} \sim 1$, and $\psi_0^{(-)\prime}(q_0^{(+)}) \sim \psi_0^{(-)}(q_0^{(+)})$. Thus,

$$E_A - E_S = \frac{1}{M}j(q=0) = \frac{1}{M}j(q \sim q_0^{(+)}) \sim \psi_0^{(-)}(q_0^{(+)}). \tag{11.29}$$

Consequently, the semiclassical exponential of the wave function $\psi_0^{(-)}$ at the right minimum $q_0^{(+)}$ is precisely the quantity that determines $(E_A - E_S)$, in the leading semiclassical approximation. Of course, hence, we again obtain from this consideration that

$$(E_A - E_S) = A\exp\left[-\int_{q_0^{(-)}}^{q_0^{(+)}} \sqrt{2MV}dq\right].$$

Problem 11. *Generalize the result (11.29) to the case of the quantum mechanics of many variables $q^1, q^2, \ldots, q^n$, with potential $V(q^1, q^2, \ldots, q^n)$, which is symmetric with respect to the transformation $q^1 \rightarrow -q^1$ and has minima at $q^1 = \pm q_0$, $q^2 = q^3 = \cdots = q^n = 0$.*

Repeating the discussion of Section 11.1, we can see straightforwardly that the expression (11.26) is related to a certain solution of the Euclidean equation of Newtonian mechanics:

$$M \frac{d^2 q}{d\tau^2} = \frac{\partial V}{\partial q}.$$

This solution has zero Euclidean energy,

$$\frac{M}{2} \left(\frac{dq}{d\tau} \right) - V(q) = 0.$$

It describes a particle in the potential $[-V(q)]$ (see Figure 11.7), which, for $\tau \to -\infty$, starts from the left hump of this potential (from the point $q_0^{(-)}$) and for $\tau \to +\infty$, ends at the right hump (at the point $q_0^{(+)}$). Unlike the bounce solution, considered in Section 11.1, the solution does not come back to the point $q_0^{(-)}$. Such solutions (and their generalizations to quantum mechanics with many degrees of freedom and to field theory) are known as *instantons*. Expression (11.26) represents the Euclidean action

$$S_E = \int_{-\infty}^{+\infty} d\tau \left[\frac{M}{2} \left(\frac{dq_I}{d\tau} \right)^2 + V(q_I) \right],$$

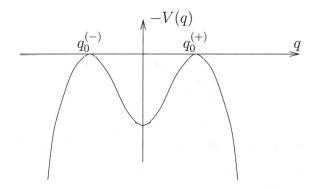

Figure 11.7.

calculated on the instanton $q_I(\tau)$. Explanation of the equality between the Euclidean action of the instanton and the tunneling exponent once again, as in Section 11.1, is provided by a semiclassical solution of the Schrödinger equation for $\psi_0^{(-)}$ in the region below the barrier, which reduces to the Euclidean Hamilton–Jacobi equation. The corresponding analysis is completely analogous to that undertaken in Section 11.1 and we shall not present it here.

We further note that calculation of the semiclassical exponential for the splitting between levels in the quantum mechanics of many variables (see Problem 11) also reduces to finding an instanton with precisely the properties described, and calculating its Euclidean action. The substantiation of this assertion is based on the results of Problem 11 and essentially repeats that given in Section 11.2.

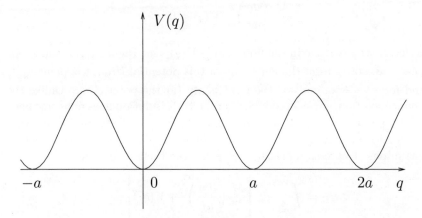

Figure 11.8.

Another example, where instantons of precisely the type described play a role, is that of a particle in a one-dimensional periodic potential $V(q)$, as shown in Figure 11.8. The instanton in this system is a classical solution in Euclidean time, which starts for $\tau \to -\infty$ from the point $q = 0$ and reaches the point $q = a$ for $\tau \to +\infty$. (The analogous solution, starting at $q = 0$ and ending at the point $q = -a$, is called an anti-instanton). The Euclidean action of the instanton gives the semiclassical exponent for tunneling between adjacent minima which, in turn, determines the energy of the Bloch wave with quasi-momentum θ. In order to explain this statement, we first note that in the absence of tunneling in the system there would be an infinite number of degenerate ground states, whose wave functions $\psi_n(q) = \psi_0(q - na)$ would be confined near the minima $q = na$, where n is an arbitrary integer. Here, ψ_0 is the ground state in the well near $q = 0$. When tunneling is taken into account, the wave functions $\psi_n(q)$ are no longer eigenfunctions of the Hamiltonian. In order to construct approximate wave functions, we define the operator T_a of translation by a,

$$T_a\psi(q) = \psi(q - a). \tag{11.30}$$

This operator is unitary, $T_a^\dagger T_a = 1$, and commutes with the Hamiltonian.

Problem 12. *Determine the action of the operator $T_a^\dagger$ on an arbitrary wave function and prove that the operator T_a is unitary.*

We note that

$$\psi_n(q) = (T_a)^n \psi_0(q). \tag{11.31}$$

Furthermore, the Hamiltonian and T_a are simultaneously diagonalizable, where T_a has eigenvalues with unit modulus,

$$T_a|\theta\rangle = e^{i\theta}|\theta\rangle, \tag{11.32}$$

where $\theta \in [0, 2\pi)$. An approximate eigenstate of the Hamiltonian and the operator T_a with energy close to zero should be constructed from states $|n\rangle$ with wave functions $\psi_n(q)$. From (11.32), we obtain

$$|\theta\rangle = \sum_{n=-\infty}^{+\infty} e^{-in\theta}|n\rangle.$$

Indeed, from (11.31) we have $T_a|n\rangle = |n+1\rangle$, and

$$T_a|\theta\rangle = \sum_{n=-\infty}^{+\infty} e^{-in\theta}|n+1\rangle = e^{i\theta} \sum_{n=-\infty}^{+\infty} e^{-i(n+1)\theta}|n+1\rangle,$$

whence (11.32) follows. The state $|\theta\rangle$ is called a Bloch wave, and the parameter θ is the quasi-momentum.

When tunneling is taken into account, the energy of the state $|\theta\rangle$ becomes dependent on θ. The leading semiclassical exponential in the energy is determined by the Euclidean action of the instanton S_I,

$$E(\theta) = \text{constant} - Ae^{-S_I}\cos\theta, \tag{11.33}$$

where A is the sub-leading pre-exponential factor.

Problem 13. *Find the energy of the state $|\theta\rangle$, i.e. show that equation (11.33) holds.*

We note that the quantum mechanics of a particle in a periodic potential considered here is used as a model of the behavior of electrons in an ideal crystal.

As a final quantum-mechanical example of tunneling, let us consider a physical pendulum of mass M and length l. Its dynamical coordinate is the angle which, as before, we shall denote by the letter q; here, the periodicity condition

$$\psi(q + 2\pi) = \psi(q) \tag{11.34}$$

is imposed upon the wave function. The potential for this pendulum is equal to

$$V(q) = Mgl(1 - \cos q),$$

where we have chosen the zero of the energy such that $V(0) = 0$. Formally, this system is the same as a particle in a periodic potential, shown in Figure 11.8, for $a = 2\pi$; the only difference is the extra condition (11.34). The under-barrier process for a pendulum is a full rotation, where q varies from 0 to 2π. By virtue of (11.34), the parameter θ, which would characterize different states of the pendulum, does not appear; in the language of the operator $T_{2\pi}$ introduced in (11.30), the property (11.34) implies that

$$T_{2\pi}\psi = \psi,$$

i.e. in terms of the model of Figure 11.8, we have to restrict ourselves to the sector with $\theta = 0$.

A parameter of the θ type can be introduced by considering *another* system, namely a physical pendulum in an Aharonov–Bohm potential. Let us suppose that the pendulum has electric charge e, and rotates around a thin solenoid with magnetic flux Φ (Figure 11.9, the magnetic field $\mathbf{H}$ is directed perpendicularly to the figure). In the plane of the pendulum there is a vector potential $\mathbf{A}$, whose circulation along the circle on which the pendulum moves is proportional to the magnetic flux:

$$\int_0^{2\pi} lA\,dq = \Phi.$$

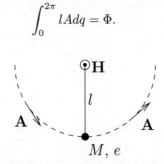

Figure 11.9.

Without loss of generality, we may suppose that $\mathbf{A}$ is directed tangentially to the pendulum circle.

The Hamiltonian for this pendulum is obtained, in the usual way, by introducing the covariant derivative,

$$H = \frac{1}{2M}\left[-i(\nabla - ie\mathbf{A})\right]^2 + V.$$

In terms of the variable q, it is equal to

$$H = -\frac{1}{2Ml^2}\left(\frac{d}{dq} - ielA\right)^2 + V(q) \tag{11.35}$$

where, as before, the condition (11.34) is imposed upon the wave function. Let us introduce a new wave function $\varphi(q)$, whereby

$$\psi(q) = \exp\left[ie\int_0^q lAdq\right]\varphi(q).$$

The new wave function will obey the Schrödinger equation of the standard form with Hamiltonian

$$H_\varphi = -\frac{1}{2M}\frac{d^2}{dq^2} + V(q),$$

i.e. with the Hamiltonian for the pendulum in the absence of a vector potential. However, the function $\varphi(q)$ will not be periodic; for $\varphi(q)$, instead of (11.34), we have

$$\varphi(q + 2\pi) = e^{-ie\Phi}\varphi(q).$$

This is precisely the condition $T_{2\pi}\varphi = e^{i\theta}\varphi$ with $\theta = e\Phi$, so that the system is equivalent to the θ-sector of the model of Figure 11.8. Formula (11.33) then describes the dependence of the energy of the pendulum ground state on the magnetic flux.

It is clear from the example considered that the interpretation of θ-states depends on the model; in the system with periodic potential of Figure 11.8, the θ-states are different states of *the same* system, while for the pendulum, these are states of different systems (pendula in the presence of different Aharonov–Bohm potentials).

Problem 14. *Show that the Hamiltonian (11.35), where for simplicity we assume that A does not depend on q, $A = \Phi/2\pi i$, corresponds to the classical Lagrangian*

$$\mathcal{L} = \frac{Ml^2}{2}\left(\frac{dq}{dt}\right)^2 + \frac{e\Phi}{2\pi}\frac{dq}{dt} - V(q). \tag{11.36}$$

We note that the second term makes no contribution to the classical equation of motion.
a) Making the formal substitution $t = -i\tau$, find the Euclidean action functional.
b) Find the instanton and its Euclidean action. Give an interpretation of the imaginary term in the Euclidean action.

To conclude this section, we note that in all the examples considered, it is also sensible to consider sphalerons, unstable static solutions of the equations of motion $\frac{\partial V}{\partial q} = 0$, which determine the height of the barrier separating the classical minima. For the potential of Figure 11.6, the sphaleron is the point $q = q_S = 0$; for that of Figure 11.8, it is the point of the maximum of the potential; for the physical pendulum, it is the point $q = \pi$, where the pendulum is upside down.

As in Section 11.2, in systems with many variables and degenerate minima of the potential, the sphaleron is a saddle-point solution of the static equations $\frac{\partial V}{\partial q} = 0$ with one negative mode. It can also be defined as follows. Let us consider, as an example, a system of many variables, with potential which is periodic in one of these variables q_1 and has minima (with $V = 0$) at points $q_1 = na$, $q_2 = q_3 = \cdots = 0$ (such a choice of coordinates is possible without loss of generality). Let us consider all possible paths C connecting adjacent minima of the potential. On each of the paths we introduce a parameter $\lambda \in [0,1]$, such that $\mathbf{q}(\lambda = 0) = (0, 0, \ldots, 0)$; $\mathbf{q}(\lambda = 1) = (a, 0, \ldots, 0)$. Let us consider the static energy as a function of the points on the path

$$V(\lambda) = V(\mathbf{q}(\lambda)).$$

Clearly, $V(0) = V(1) = 0$, and there exists a maximum value of $V(\lambda)$ at some point on the path, $\max_\lambda V(\lambda)$. This value is the height of the barrier along the given path C. To determine the minimum barrier height, we minimize $\max_\lambda V(\lambda)$ over all paths connecting adjacent minima:

$$V_S = \min_C \max_\lambda V(\lambda). \tag{11.37}$$

This will be the minimum barrier height, and the point where the minimax (11.37) is attained is the sphaleron $\mathbf{q}_S$.

In models like the physical pendulum, this discussion requires slight modification. In this case, there is just one minimum of the potential $V(\mathbf{q})$, however, there exist non-contractible paths, closed in the configuration space (the space of possible values of $\{\mathbf{q}\}$), starting and ending at the minimum of the potential. The tunneling process in these models involves motion under the barrier along such a non-contractible path. In the previous discussion for C, we now have to take all possible non-contractible closed paths, starting and ending in the classical ground state. The rest of the discussion is unchanged.

Chapter 12

Decay of a False Vacuum in Scalar Field Theory

12.1 Preliminary considerations

In this chapter we consider a model of a real scalar field in d-dimensional space–time $(d \geq 3)$ with the action

$$S = \int d^d x \left[\frac{1}{2} (\partial_\mu \varphi)^2 - V(\varphi) \right], \tag{12.1}$$

where the scalar potential $V(\varphi)$ has the form shown in Figure 12.1. We stress that $V(\varphi)$ represents the energy *density* of a homogeneous and static scalar field, with value equal to φ (everywhere in space).

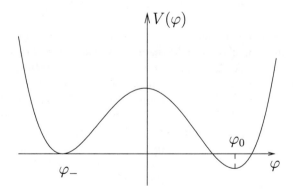

Figure 12.1.

249

At the classical level in the model there are two stable static states: the state $\varphi = \varphi_-$ (false vacuum), whose energy density is chosen to be zero, without loss of generality, and the state $\varphi = \varphi_0$ (true vacuum), where the energy density is negative, and the total energy is proportional to the spatial volume (and equal to $-\infty$ in the limit of infinite volume).

Let us suppose that at the initial moment in time, the system is in the false vacuum $\varphi = \varphi_-$. In quantum theory, generally speaking, a tunneling process is possible, ultimately leading the system to the true vacuum $\varphi = \varphi_0$. Assuming this process is of a semiclassical nature its probability will be exponentially small. The problem is to calculate the semiclassical exponential for this probability.

First and foremost, we must explain what is meant by the quantum theory of a *field* with the action (12.1). For our purposes, the following considerations will suffice. Let us suppose that $(d-1)$-dimensional space is discrete, i.e. it consists of individual points $\mathbf{x}_i$. The simplest variant is the cubic lattice; for this the index $\mathbf{i}$ in fact represents $(d-1)$ integer indices $\mathbf{i} = (i_1, \ldots, i_{d-1})$. Each site of this lattice has coordinates $(x_1, \ldots, x_{d-1})_\mathbf{i} = (ai_1, \ldots, ai_{d-1})$, where a is the distance between the sites of the lattice. At each site of the lattice, we assign a variable $\varphi_\mathbf{i}(t)$. We shall further suppose that the spatial size of the system is finite, but large; then the indices $i_1, \ldots, i_{d-1}$, will range over a finite number of values. Next, we can consider the system with the action

$$ S = \int dt \sum_{i_1, \ldots, i_{d-1}} a^{d-1} \left[\frac{1}{2} \left(\frac{d\varphi_i}{dt} \right)^2 - \frac{1}{2} (\nabla \varphi)_i^2 - V(\varphi) \right], \qquad (12.2) $$

where the gradient is discretized, for example, along the first coordinate it is equal to: $(\nabla_1 \varphi)_i = \frac{1}{a} (\varphi_{i_1+1, i_2, \ldots, i_{d-1}} - \varphi_{i_1, i_2, \ldots, i_{d-1}})$. Thus, we have a system with a finite, but very large, number of variables $\varphi_\mathbf{i}$. At the quantum level it becomes the quantum mechanics of a finite number of variables. It is physically clear that for sufficiently small a the system with action (12.2) is not distinguishable from a field theory with the action (12.1), at least at the classical level. At the quantum level, there are subtleties, associated with ultraviolet divergences and renormalization; however, for us, these subtleties will be unimportant. Indeed, we shall consider only leading semiclassical exponentials, whose calculation requires only a knowledge of classical theory (in conventional or Euclidean time). Thus, for our purposes, it suffices to view the theory (12.1) as a model with a finite, albeit very large number of variables, and to interpret the spatial coordinates $\mathbf{x}$ as enumerating the variables $\varphi(\mathbf{x}, t)$ (here, we consider time as a continuous variable, as usual in quantum mechanics).

Following this general remark, let us turn to the decay of the false vacuum. The false vacuum and true vacuum states are spatially homogeneous, however, transition between them cannot occur in a spatially homogeneous manner. Indeed, the tunneling exponent for spatially homogeneous tunneling is proportional to the spatial volume of the system, thus, spatially homogeneous transition occurs with zero probability in the limit of infinite volume (compare with the discussion at the end of Section 5.1). At the same time, an inhomogeneous tunneling process is possible, whose final result is transition to the true vacuum state. Namely, for the tunneling process to occur, it is sufficient that the system emerges from under barrier with zero static energy. In our case, by the static energy, we understand the quantity

$$E_S = \int d^{d-1}\mathbf{x} \left[\frac{1}{2}(\partial_i \varphi)^2 + V(\varphi)\right]. \tag{12.3}$$

We note that precisely this expression is the analogue of the potential in quantum mechanics, since it does not contain derivatives of φ with respect to time (this is best seen in the discrete version (12.2); the expression in the integrand in (12.2) is the quantum mechanical Lagrangian, the difference of a kinetic term, quadratic in $\dot{\varphi}_i$ and a potential term, independent of $\dot{\varphi}_i$; in the continuous limit the potential term reduces to (12.3)).

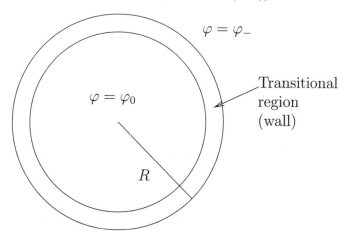

Figure 12.2.

One can achieve that $E_S = 0$, and that the field configuration has $\varphi \neq \varphi_-$ only locally in space, by considering a bubble of the true vacuum in the false one (Voloshin *et al.* 1974), see Figure 12.2. The static energy of such a bubble consists of the energy of the transitional region (wall), where the field changes from φ_0 to φ_- and the energy of the inside of the bubble. The energy of the wall is proportional to its area and positive, constant $\cdot R^{d-2}$, while the energy of the inside region is negative and proportional to the

volume of the bubble, constant $\cdot R^{d-1}$. Thus, the static energy is estimated
by the quantity

$$E_S = \mu R^{d-2} - c R^{d-1},$$

where $\mu > 0$ and $c \propto (-V(\varphi_0)) > 0$; its behavior as a function of R is
shown (for $d > 3$) in Figure 12.3. A bubble of size R_0 may be formed as
the result of the tunneling process. From Figure 12.3, it is clear that after
it has been formed, the bubble will expand, as a result of which the true
vacuum will occupy an increasingly large region of space. Ultimately, the
whole system will transit to a true vacuum state $\varphi = \varphi_0$.

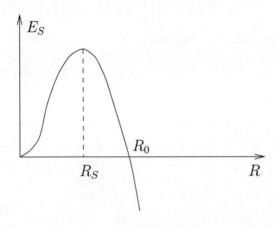

Figure 12.3.

We shall see that literally this picture arises only in the special case when
the energy of the true vacuum differs very little from the energy of the false
vacuum, i.e. when $|V(\varphi_0)|$ is small in comparison with all other parameters.
In the general case, at the moment of the spontaneous creation, the "wall"
has a size comparable with the size of the bubble itself, so that it does not
make sense to talk about the wall and the inside region at the moment
of creation. In the center of the bubble at the moment of its creation,
generally speaking, the field may be very different from φ_0. Nevertheless,
the picture we have described captures the main features of the bubble
creation process; it works perfectly for a very long time *after* the bubble
formation.

The qualitative considerations of this section also enable us to establish
the approximate structure of a sphaleron (critical bubble). A sphaleron in
our model is a configuration in $(d - 1)$-dimensional *space*, which (after a
slight "push") can evolve into either the true or the false vacuum. At the
qualitative level, a sphaleron is a bubble of size R_S (see Figure 12.3); its

instability is associated with the fact that slightly smaller bubbles collapse (so that the system returns to the false vacuum), while slightly larger bubbles expand unboundedly, and the true vacuum fills the whole space.

12.2   Decay probability: Euclidean bubble (bounce)

For a quantitative description of the under-barrier process of the creation of bubbles in the leading semiclassical approximation (Coleman 1977), we make use of the prescription formulated in Section 11.2. We write the action in the form

$$S = \int d^{d-1}\mathbf{x}dt \left[\frac{1}{2}\left(\frac{\partial\varphi}{\partial t}\right)^2 - \frac{1}{2}\left(\frac{\partial\varphi}{\partial \mathbf{x}}\right)^2 - V(\varphi) \right] \tag{12.4}$$

and make in it the formal substitution $t = -i\tau$. Up to a factor i, the action (12.4) becomes the Euclidean action

$$S_E = \int d^{d-1}\mathbf{x}dt \left[\frac{1}{2}\left(\frac{\partial\varphi}{\partial \tau}\right)^2 + \frac{1}{2}\left(\frac{\partial\varphi}{\partial \mathbf{x}}\right)^2 + V(\varphi) \right] \tag{12.5}$$

$$= \int d^d \left[\frac{1}{2}(\partial_\mu\varphi\partial^\mu\varphi) + V(\varphi) \right], \tag{12.6}$$

where we have again combined the coordinates $x^\mu = (\tau, \mathbf{x})$, and the summation over $\mu = 0, \ldots, d-1$ is performed with the Euclidean metric diag $(1, 1, \ldots, 1)$. As mentioned earlier in Section 11.1, the latter property is the reason for calling τ Euclidean time.

The field equations, following from the action (12.5) have the form

$$-\partial_\mu\partial^\mu\varphi + \frac{\partial V}{\partial\varphi} = 0. \tag{12.7}$$

Our problem is to find the bounce solution of these equations, which would tend to the false vacuum φ_- for $\tau \to \pm\infty$, and would have a "turning point." In field theory, a "turning point" means a "moment in time" (without loss of generality, we may suppose that it is $\tau = 0$) at which

$$\frac{\partial\varphi}{\partial\tau} = 0 \text{ for all } \mathbf{x}. \tag{12.8}$$

Indeed, we interpret $\mathbf{x}$ as a continuous index, enumerating variables; thus, (12.8) is a continuous analogue of the equation (11.16).

The requirement

$$\varphi(\tau, \mathbf{x}) \to \varphi_- \text{ as } \tau \to \pm\infty \tag{12.9}$$

and the requirement (12.8) can be simultaneously satisfied if we consider smooth spherically symmetric fields $\varphi(r)$, where $r = \sqrt{\tau^2 + \mathbf{x}^2} = \sqrt{x_\mu x^\mu}$, with asymptotics

$$\varphi(r \to \infty) = \varphi_-. \tag{12.10}$$

Indeed, condition (12.10) ensures that (12.9) is satisfied and

$$\partial_\tau \varphi(x^\mu) = \frac{\tau}{r} \frac{d\varphi}{dr} \tag{12.11}$$

is equal to zero everywhere at $\tau = 0$; the point $r = 0$ is not special in (12.11), since for smoothness of the field near $r = 0$, we require

$$\frac{d\varphi}{dr}(r = 0) = 0. \tag{12.12}$$

For spherically symmetric fields, equation (12.7) is brought to the form

$$\varphi'' + \frac{d-1}{r} \varphi' = \frac{\partial V}{\partial \varphi}, \tag{12.13}$$

where the prime denotes the derivative with respect to r.

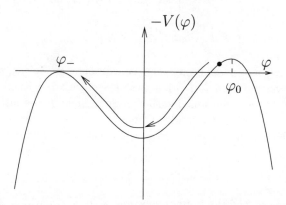

Figure 12.4.

We show that a solution of equation (12.13) with the boundary conditions (12.10), (12.12) indeed exists. Equation (12.13) is formally equivalent to the equation of Newtonian mechanics for a particle with coordinate φ, moving in "time" r in the potential $[-V(\varphi)]$ in the presence of a friction force, proportional to the particle speed, where the friction coefficient is inversely proportional to "time." Conditions (12.12) and (12.10) imply that at the initial time $r = 0$ the particle is at rest, while at the end of the motion ($r \to \infty$) it comes to rest precisely on the hump of the potential ($-V$) (Figure 12.4). The existence of such a solution follows from the following consideration. Let us consider a particle trajectory starting at $r = 0$ with velocity zero, as a function of the initial point. If the initial point is chosen such that

$$-V(\varphi(r = 0)) < 0,$$

then the particle certainly *does not roll up* to the top of the hump, the point φ_- (its "energy" is negative). If the initial point is chosen to be very close to the second hump $\varphi = \varphi_0$, then the particle will stay near this hump for a long time, moving with a very small acceleration and velocity. During this time, it will lose little energy (the velocity is small). Further movement will take place at large r, when the friction coefficient is small. Consequently, for such a choice of initial point, the particle will lose very little energy throughout the whole of its motion and it will *roll over* the hump φ_-. Thus, when the position of the initial point changes, the regime of not rolling up to φ_- is replaced by the regime of rolling over it, and, by continuity considerations, there exists an initial position at which the particle comes to rest precisely on the hump as $r \to \infty$. The solution in which we are interested actually exists.

The bounce configuration in d-dimensional Euclidean space–time is very simple. It represents a spherically symmetric Euclidean bubble, outside which the field tends rapidly to the false vacuum φ_- and within which it differs from φ_- substantially. In the center of the bubble, the field has value less than φ_0, but may be quite close to the true vacuum φ_0. The leading semiclassical exponent for the decay probability for the false vacuum is determined by the Euclidean action of this solution:

$$\Gamma \propto e^{-S_B}.$$

Problem 1. *Consider the theory with scalar potential*

$$V(\varphi) = -a\varphi^2 - b\varphi^3 + \lambda\varphi^4 + \text{constant},$$

where a, b, λ are positive parameters.
a) Find the region of values of the parameters a, b, λ, in which the potential has the form shown in Figure 12.1.
b) Using the considerations studied at the beginning of Section 7.1, find the region of values of the parameters a, b, λ, in which property (a) is satisfied and a weak coupling regime is realized in the true and false vacua.
c) Find the parametric dependence of the size of the Euclidean bubble and of its Euclidean action on combinations of the parameters a, b, λ. Show that in the weak coupling regime, the Euclidean action is large (which implies that the semiclassical approximation is applicable).

Let us now discuss the behavior of the bubble after its creation as a result of tunneling. As discussed in Section 11.2 the most likely values of the variables at the moment of emergence from under the barrier are their values at the turning point of the Euclidean solution. In field theory, a "turning point" is a $(d-1)$-dimensional surface $\tau = 0$ of d-dimensional Euclidean space. Thus, at the moment of emergence from under the barrier (the time at which the bubble materializes) the field configuration coincides with $\varphi(\mathbf{x}, \tau = 0)$. This is again a localized spherically symmetric (but now in $(d-1)$-dimensional space) configuration with $\varphi = \varphi_-$ outside the bubble and φ close to φ_0 in the center of the bubble. At the moment of materialization, the velocities are equal to zero (see Section 11.2); in our case, this means that

$$\frac{\partial \varphi}{\partial t}(\mathbf{x}, t = 0) = 0 \text{ for all } \mathbf{x},$$

where $t = 0$ is the moment the bubble materializes.

If the Euclidean (bounce) solution $\varphi_B(r)$ is known, then the behavior of the bubble outside a light cone in Minkowski space–time (after its materialization) can be found, using the fact that its field $\varphi(\mathbf{x}, t)$ is the analytic continuation of the solution $\varphi_B(\mathbf{x}, \tau)$ into the purely imaginary region $\tau = it$ (see Section 11.2). Outside the light cone it is easy to find this analytic continuation: for this, it suffices to replace $r = \sqrt{\mathbf{x}^2 + \tau^2}$ by $\rho = \sqrt{\mathbf{x}^2 - t^2}$ (the expression under the square root is positive outside the cone). Thus, the field of the bubble in conventional space–time has the form

$$\varphi(\mathbf{x}, t) = \varphi_B(\sqrt{\mathbf{x}^2 - t^2}).$$

The surfaces of constant φ are the hyperboloids $\mathbf{x}^2 - t^2 = \text{constant}$; they are shown in Figure 12.5. It is evident that the field deep inside the cone reaches its true vacuum value, which is also shown in Figure 12.5. From the point of view of a stationary observer, the size of the bubble wall (region where the field changes from φ_- to φ_0) decreases with time, and the velocity of the wall approaches the speed of light.

Problem 2. *Let us consider a bubble in $(3+1)$-dimensional space–time, which has expanded to a large size. In this case, the thickness of the bubble wall is small in comparison with the bubble size; the curvature of the wall is also small. Assuming that the shape of the wall does not change in its own reference frame, and neglecting the curvature of the wall, the wall can be characterized by a single variable $x^1(t)$ (if the wall is perpendicular to the first axis).*

a) By calculating the energy–momentum tensor, find the pressure on the

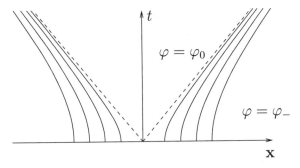

Figure 12.5.

wall and its acceleration in the inertial reference frame, instantaneously coinciding with the wall.

b) Move to a fixed inertial reference frame and find an equation for $x^1(t)$. Solve this equation. Draw the world line in the plane (x^1, t) (the section of the world surface of the wall by the plane $x^2 = x^3 = 0$).

To end this section, let us consider the sphaleron (critical bubble) in this model. It is an unstable solution of the static field equations and determines the height of the barrier separating the false and the true vacua. As mentioned in Section 12.1, the sphaleron must be a bubble in the false vacuum in $(d-1)$-dimensional space.

The static field equation in d-dimensional space–time has the form

$$-\partial_i\partial_i\varphi + \frac{\partial V}{\partial \varphi} = 0,$$

where the index i is spatial and takes values $i = 1, \ldots, d-1$. This equation coincides with equation (12.7), except that the Euclidean space is now $(d-1)$-dimensional rather than d-dimensional. Its solution again has spherical symmetry, $\varphi = \varphi(\sqrt{x_i x_i})$ and the equation for φ coincides with (12.13) with d replaced by $d-1$. In other words, the Euclidean bubble (bounce) for $(d-1)$-dimensional space–time coincides with the critical bubble (sphaleron) for d-dimensional space–time. The structure of this solution was discussed above.

Problem 3. *Show that the critical bubble is unstable and that among all the perturbations about it there exists a negative mode. Hint: use the existence of zero modes around the critical bubble and their properties under spatial rotations.*

We stress that we found the Euclidean bubble and the sphaleron, assuming a spherical symmetry of the solutions. This assumption, in fact, can be rigorously justified (Coleman *et al.* 1978).

Thus, in order to find the probability of tunneling (at temperature zero) we need to find the bounce solution (Euclidean bubble), and the probability of formation of a bubble of the true vacuum in the false vacuum will be equal to

$$\Gamma = Ae^{-S_B},$$

where S_B is the Euclidean action of the bounce, and A is the sub-leading pre-exponential factor. A bubble may be formed at any point of space, thus Γ is in fact the probability of bubble creation *per unit spatial volume per unit time*. In a large volume of space, bubbles will be created at different points at different times, they will expand, their walls will then collapse, the false vacuum will "boil." Ultimately, the energy of the false vacuum (more precisely, the difference between the energies of the false and true vacua) will turn into heat.

In complete analogy with Section 11.1, for sufficiently high temperatures (which are still small in comparison with the energy of a critical bubble in a false vacuum) bubble creation does not take place by tunneling, but by thermal jumps, with the formation of a critical bubble. The probability of such processes per unit volume per unit time is proportional to

$$\Gamma \propto e^{-F_S/T},$$

where F_S is the free energy of a critical bubble (sphaleron). In a number of cases, the free energy of a critical bubble coincides with the energy of a critical bubble in the theory (12.1), but with the scalar potential changed due to temperature corrections. The methods studied in this section are thus applicable to the study of critical bubbles at a finite temperature.

Potentials of the type shown in Figure 12.1 are characteristic for systems with first-order phase transition. Indeed, the theory of phase transitions (such as those which occurred in the early Universe) makes use of the solutions discussed in this section.

Problem 4. *In the model of Problem 1, find the parametric dependence of the size of a critical bubble and its energy on a, b and λ in the weak coupling regime. Give numerical estimates of the order of magnitude, in the case of parameters typical for the Standard Model (vacuum expectation value of the scalar field of the order of 250 GeV), assuming, for example, dimensionless coupling constants of the order of 0.3 (express the size of a bubble in centimeters). We note that the energy of a critical bubble gives an idea of the temperature of the phase transition.*

12.3   Thin-wall approximation

It is not possible to find an explicit form of the solution for a Euclidean bubble or a critical bubble in the general case. However, this can done in the special case when the difference between the energies of the false and true vacua is small in comparison with all other parameters of the model (Coleman 1977). Namely, let us consider a scalar potential of the special form

$$V(\varphi) = V_0(\varphi) - \varepsilon V_1(\varphi), \qquad (12.14)$$

where $V_0(\varphi)$ is symmetric under the transformation $\varphi \to -\varphi$ and has degenerate minima at $\varphi = -\varphi_0$ and $\varphi = +\varphi_0$, and $V_1(\varphi)$ is not invariant with respect to this transformation. We shall assume that $V_0(\pm\varphi_0) = 0$, $V_1(-\varphi_0) = 0$ and $V_1(+\varphi_0) = 1$. Let us further suppose that the parameter ε is small. Then, the homogeneous field $\varphi_- = -\varphi_0 + O(\varepsilon)$ will be a false vacuum with zero energy, and $\varphi = +\varphi_0 + O(\varepsilon)$ will be a true vacuum with energy density $(-\varepsilon)$. To be specific, we shall consider four-dimensional space–time, $d = 4$.

We shall seek a solution of equation (12.13) for a Euclidean bubble using a mechanical analogy (Figure 12.4). Since ε is small, the particle must lose a very small amount of energy when moving from the neighborhood of the hump $\varphi = \varphi_0$ to the neighborhood of the hump $\varphi = \varphi_-$. Far from the humps, the particle velocity is in any case finite for arbitrarily small ε, thus the main movement of the particle must take place at large r, when the friction force is small. During this motion, the friction force, as well as the part of the potential proportional to ε, can be neglected, so that equation (12.13) reduces to

$$\varphi'' = \frac{\partial V}{\partial \varphi}.$$

We are interested in the solution of this equation, which rolls down from the point $+\varphi_0$ and rolls up to the point $-\varphi_0$. We know this solution: it is the antikink in the symmetric potential V_0. The antikink is characterized by a single parameter R, its position. Thus, in the region of values of r, where the bounce $\varphi_B(r)$ is substantially different from $+\varphi_0$ and $-\varphi_0$, we have

$$\varphi_B(r) = \varphi_k(R - r).$$

This solution is shown in Figure 12.6. We note that $\varphi_B(r)$, for r near R (wall region), does not depend upon ε. For the force of friction to be indeed small, R must be large for small ε. The structure of a solution in Euclidean space corresponds exactly to Figure 12.2.

The parameter R is the only parameter of the solution, which has yet to be determined. To find it, we note that the Euclidean action must

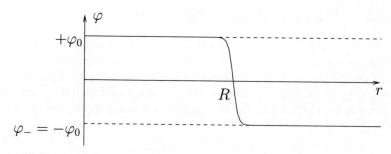

Figure 12.6.

be extremal in the class of configurations satisfying (12.10), (12.12). In particular, it must be extremal for variations of the parameter R. Let us calculate the Euclidean action (12.5) as a function of R. For spherically symmetric configurations

$$S_E = 2\pi^2 \int_0^\infty r^3 dr \left[\frac{1}{2} (\varphi')^2 + V_0(\varphi) - \varepsilon V_1(\varphi) \right] \qquad (12.15)$$

($2\pi^2$ is the area of a unit three-dimensional sphere). Far from the region of the kink (wall), the field φ_B does not depend on r; moreover, $V_0(\varphi) = 0$. The external region of the bubble ($r > R$, outside the wall) does not contribute to S_E, while the internal region yields the contribution

$$S_E^{\text{in}} = -\frac{\pi^2}{2} R^4 \varepsilon$$

modulo corrections of the order of εR^3. The contribution of the wall is proportional to R^3 and to leading order does not depend on ε; to calculate it, we set $r = R$ in the measure in (12.15) and we neglect εV_1. We obtain

$$S_E^{\text{wall}} = 2\pi^2 R^3 \mu,$$

where

$$\mu = \int_{-\infty}^{+\infty} dr \left[\frac{1}{2} (\varphi_k'(r))^2 + V_0(\varphi_k(r)) \right].$$

We note that the form of μ coincides with the energy of a kink in one-dimensional space; μ does not depend on ε. Thus, to leading order in ε and R the Euclidean action of the bounce of size R is equal to

$$S_B(R) = 2\pi^2 \mu R^3 - \frac{\pi^2}{2} R^4 \varepsilon.$$

The extremum of this expression is attained at

$$R = R_B = \frac{3\mu}{\varepsilon}. \qquad (12.16)$$

For the Euclidean action at the extremum, we finally obtain

$$S_B = \frac{27}{2}\pi^2\frac{\mu^4}{\varepsilon^3}. \qquad (12.17)$$

We note that the size of a Euclidean bubble (12.16) is indeed large for small ε. The action (12.17) is also large, so that the tunneling process for the creation of bubbles is strongly suppressed.

We recall, finally, that according to the discussions of Section 12.2, the bubble wall, after materialization of the bubble, moves along a hyperboloid $\mathbf{x}^2 - t^2 = R^2$.

Problem 5. *Find the size of a Euclidean bubble R_B directly from the equation (12.13), by considering the energy balance for the mechanical analogy (Figure 12.4).*

Problem 6. *Find the form of the wall (i.e. $\varphi_B(r)$ for r near R) and the value of the parameter μ for the potential*

$$V_0(\varphi) = \frac{\lambda}{4}(\varphi^2 - v^2)^2.$$

Show that the size of a Euclidean bubble and its action are large, not only in terms of ε^{-1}, but also in terms of λ^{-1} for small λ. Find the limit of applicability of the thin-wall approximation (hint: require the size R_B to be much greater than the width of the kink itself).

Problem 7. *Find the form, size and static energy of the critical bubble in the model (12.14) in four-dimensional space–time for small ε.*

Problem 8. *Show by direct calculation to leading order in ε, that the static energy of the bounce configuration at $\tau = 0$, i.e. the configuration $\varphi_B(\mathbf{x}, \tau = 0)$, is equal to zero. Use this calculation to give an alternative derivation of equation (12.16).*

Chapter 13

Instantons and Sphalerons in Gauge Theories

In Chapters 11 and 12 we saw that solutions of Euclidean equations of motion are useful for describing tunneling processes in quantum mechanics and scalar field theory. In this chapter, we consider Euclidean solutions in gauge theories and give interpretations of these in terms of tunneling processes. We also discuss unstable static solutions of field equations in gauge theories (sphalerons), which enable us to find the height of corresponding barriers.

13.1 Euclidean gauge theories

First and foremost, we need to understand how to make the transition to Euclidean time in gauge theories. The most naive transition, involving the substitution $t = -i\tau$, without any modification of the field A_μ is not appropriate, for the following reason. If the fields A_μ are not modified, then the components F_{0i}^a of the field strength become complex,

$$F_{0i}^a = \partial_0 A_i^a - \partial_i A_0^a + g f^{abc} A_0^b A_i^c \to i\partial_\tau A_i^a - \partial_i A_0^a + g f^{abc} A_0^b A_i^c. \quad (13.1)$$

The covariant derivatives $D_0\varphi$ of the scalar fields also take on complex values; the Euclidean action and the field equations will also be complex. This is clearly unlike what happens in quantum mechanics or scalar field theory.

Formula (13.1) suggests that at the same time as the substitution $t = -i\tau$, it is useful to make the substitution $A_0^a \to iA_0^a$. Then F_{0i}^a will be purely imaginary and the Euclidean action of the gauge field will be real:

$$S = \int d^{d-1}\mathbf{x}dt \left[\frac{1}{2}F^a_{0i}F^a_{0i} - \frac{1}{4}(F^a_{ij})^2 \right] \to iS_E,$$

where

$$S_E = \int d^d x \frac{1}{4}(F^a_{\mu\nu})^2,$$ (13.2)

where the Euclidean field strengths are given by the usual expressions, $F^a_{\mu\nu} = \partial_\mu A^a_\nu - \partial_\nu A^a_\mu + gf^{abc}A^b_\mu A^c_\nu$, where ∂_0 denotes ∂/∂_τ, and the summation over Greek subscripts is carried out with the Euclidean metric. Analogously, after the substitutions

$$t \to -i\tau, \quad A^a_0 \to iA^a_0, \quad A^a_i \to A^a_i, \quad \varphi \to \varphi,$$ (13.3)

the covariant derivative of the scalar field $D_0\varphi$ becomes purely imaginary, and the action of the scalar field becomes real and non-negative,

$$S_E = \int d^d x \left[(D_\mu\varphi)^\dagger D_\mu\varphi + V(\varphi^\dagger, \varphi) \right],$$ (13.4)

where, once again, $D_\mu\varphi$ is given by the usual expression,

$$D_\mu\varphi = (\partial_\mu - igT^a A^a_\mu)\varphi.$$

To elucidate the prescription (13.3), let us again consider the theory in Minkowski space–time and choose the gauge

$$A^a_0 = 0.$$ (13.5)

In this gauge, the action of the gauge field has the canonical form:

$$S = \int d^d x \left[\frac{1}{2}(\partial_0 A^0_i)^2 - \frac{1}{4}(F^a_{ij})^2 \right],$$ (13.6)

where derivatives with respect to time appear only in the first term in the integrand. We can then move to Euclidean time via the usual rule, $t \to -i\tau$, without modification of the variables A^a_i, and obtain

$$S_E = \int d^d x \left[\frac{1}{2}(\partial_\tau A^a_i)^2 + \frac{1}{4}(F^a_{ij})^2 \right].$$ (13.7)

The action of the scalar fields in the gauge (13.5) also has a canonical form and admits the standard transition to Euclidean time. In principle, we could stop at this point and search for Euclidean solutions in the gauge (13.5). However, in practice, this is not usually convenient. Thus, it is useful to restore the gauge invariance of the Euclidean action by introducing the Euclidean variable A_0^a in (13.7). Thus, we come to the Euclidean action (13.2), and analogously, (13.4) for scalar fields. We can search for Euclidean solutions in an arbitrary gauge and then we shall interpret them by transferring to the gauge (13.5).

Here, we have ignored one thing. The field equations, obtained from the action (13.6) by varying A_i^a, do not exhaust all the Yang–Mills equations $D_\mu F_{\mu\nu}^a = 0$, written in the gauge (13.5). The missing equation is Gauss's constraint

$$D_i F_{0i}^a = D_i(\partial_0 A_i)^a = 0 \qquad (13.8)$$

(and the analogous condition in theories involving scalar fields). As mentioned in Section 4.5, in classical theory Gauss's constraint can be viewed as an additional condition on the initial data. In quantum theory it is imposed upon the permissible states $|\psi\rangle$,

$$(D_i F_{0i}^a)|\psi\rangle = 0. \qquad (13.9)$$

At the semiclassical level, condition (13.9) implies that configurations of fields from which tunneling takes place (i.e. configurations of fields as $\tau \to -\infty$) must satisfy Gauss's constraint. For the solutions under consideration in this chapter, this property will be automatically satisfied since we shall solve the full system of Euclidean equations $D_\mu F_{\mu\nu} = 0$, following from the action (13.2).

Problem 1. *Find the field equations following from variation of the action (13.6) with respect to the variable A_i. Show that these equations, supplemented by Gauss's constraint (13.8), are equivalent to a full system of Yang–Mills equations, written in the gauge (13.5).*

Problem 2. *The same in the theory with scalar fields.*

13.2   Instantons in Yang–Mills theory

In this section we consider classical solutions with a finite Euclidean action in non-Abelian gauge theories without scalar fields in four-dimensional Euclidean space, namely instantons and anti-instantons (Belavin *et al.* 1975). An interpretation of these solutions will be given in the next section. The scale arguments of section 7.2 do not rule out the existence of these solutions. At the same time, the scale transformation $A_\mu^a(x) \to \lambda A_\mu^a(\lambda x)$ leaves the Euclidean action (13.2) invariant, i.e. it transforms one solution to another. Thus, the solutions must have a free parameter, the size of the instanton.

We shall find it convenient to use a matrix formulation of the gauge field $A_\mu = -igt^a A_\mu^a$ where the t^a are generators of the Lie algebra, normalized as usual, by the condition $\text{Tr}\, t^a t^b = \frac{1}{2}\delta^{ab}$. We shall assume that the gauge group G is simple. We recall that, in terms of matrix fields, the Euclidean action has the form

$$S_E = -\frac{1}{2g^2} \int \text{Tr}(F_{\mu\nu}^2)d^4x$$

(here and in what follows, the Euclidean metric is understood).

Let us first consider all possible configurations of fields with a finite action. For these, the field strength $F_{\mu\nu}$ must decrease sufficiently rapidly as $|x| \to \infty$, where $|x| \equiv r = \sqrt{x_\mu x_\mu}$. This means that the vector potentials A_μ must tend to a pure gauge as $|x| \to \infty$,

$$A_\mu = \omega\partial_\mu\omega^{-1} \quad \text{as } |x| \to \infty, \tag{13.10}$$

where $\omega(x) \in G$ is some gauge function. Let us consider a remote sphere S^3 in four-dimensional space. According to (13.10), any configuration with finite action defines a mapping of this sphere to the group G:

$$\omega : S^3 \to G.$$

Moreover, we know that $\pi_3(G) = Z$, for any simple Lie group G (see Chapter 8). Consequently, gauge functions divide into disjoint homotopy classes, enumerated by the integer Q. Configurations of gauge fields with finite Euclidean action also divide into disjoint homotopy classes according to their asymptotics (13.10). Here, assignment to a particular class does not depend on the choice of remote sphere S^3 in Euclidean space, and gauge-equivalent configurations belong to the same homotopy class.

Problem 3.   *Consider two remote spheres S_1 and S_2 with radii R_1 and R_2. Suppose that between these spheres the field A_μ has the form (13.10). Show that gauge functions $\omega(R_1, n_\mu)$ and $\omega(R_2, n_\mu)$ on these spheres belong to the same homotopy class (n_μ is a unit radius vector in four-dimensional space).*

Problem 4.   *Suppose $A_\mu(x)$ has the behavior (13.10), $A_\mu^\Omega(x)$ is a gauge-transformed field ($\Omega(x)$ is a gauge function which is smooth throughout the four-dimensional space). Show that $A_\mu(x)$ and $A_\mu^\Omega(x)$ belong to the same homotopy class.*

Without loss of generality, we may suppose that $\omega(x)$ in (13.10) depends only on angles. Indeed, if $\omega(r, n_\mu)$ depends on $r \equiv |x|$ (n_μ is a unit radius vector in four-dimensional space), then one can construct a gauge function $\Omega(r, n_\mu) = \omega(R, n_\mu)\omega^{-1}(r, n_\mu)$, where R is the radius of some fixed remote sphere S^3. For all sufficiently large r, the gauge function Ω belongs to the trivial (zero) homotopy class, which contains the unit element of the group G (indeed, $\Omega(R, n_\mu) = 1$, and as r varies, the function $\Omega(r, n_\mu)$ varies continuously). Thus, $\Omega(r, n_\mu)$ can be smoothly extended to the whole space, including the interior of a ball of radius R. Having performed the gauge transformation $\Omega(x)$ on the original vector potential $A_\mu(x)$ in the whole space, we obtain that the transformed vector potential belongs to the same homotopy class as $A_\mu(x)$ and has asymptotics with gauge function $(\Omega \cdot \omega)(r, n_\mu) = \omega(R, n_\mu)$, independent of r.

Thus, we arrive at a topological classification of Euclidean configurations of a gauge field with a finite Euclidean action. The minimum of the action over fields with a fixed topological charge Q (i.e. in each topological sector), if it exists, is a solution of the Yang–Mills equations. Our next task is to find an explicit form of the solution in the sector with $Q = 1$, the instanton (and for $Q = -1$, the anti-instanton), in the theory with gauge group $SU(2)$.

It is useful to write down an explicit expression for Q:

$$Q = \frac{1}{24\pi^2} \int d\sigma_\mu \varepsilon^{\mu\nu\lambda\rho} \mathrm{Tr}\left(\omega\partial_\nu\omega^{-1} \cdot \omega\partial_\lambda\omega^{-1} \cdot \omega\partial_\rho\omega^{-1} \right), \qquad (13.11)$$

where the integration is carried out over a remote sphere. We recall that the Greek indices take values 0,1,2,3; the antisymmetric tensor $\varepsilon^{\mu\nu\lambda\rho}$ is defined such that $\varepsilon^{0123} = 1$. To prove the representation (13.11) in the theory with gauge group $SU(2)$, we recall that any gauge function $\omega(n_\mu) \in SU(2)$ can be represented in the form

$$\omega(n) = v_\alpha(n)\sigma_\alpha, \qquad (13.12)$$

where $\alpha = 0, 1, 2, 3$; $\sigma_0 = 1$, $\sigma_i = -i\tau_i$ and the real functions $v_\alpha(n)$ satisfy the relation

$$v_\alpha v_\alpha = 1.$$

The relation (13.12) is a one-to-one correspondence between the manifold of the group $SU(2)$ and the sphere $S^3_{SU(2)}$; the topological number Q is in fact nothing other than the degree of the mapping of the remote sphere S^3 to the group $SU(2)$. One can check directly that (13.11) leads to an expression for the degree of the mapping of the sphere to the sphere, considered in Chapter 8.

Problem 5. *Write down the expression (13.11) in terms of $v_\alpha(\mathbf{n})$. Show that the expression obtained is indeed the degree of the mapping of the remote sphere S^3 to the sphere $S^3_{SU(2)}$.*

In the general case of an arbitrary simple gauge group, the proof of formula (13.11) is more complicated, and we shall not dwell on it here. The fact that the integral in (13.11) is a topological characteristic of the mapping $S^3 \to G$ is evident from the result of the following problem.

Problem 6. *Show for an arbitrary gauge group that the value of the integral (13.11) does not change when the function $\omega(x)$ changes smoothly on the sphere S^3.*

The next step involves the transformation of the surface integral (13.11) to a volume integral. For this, we recall that (see Problem 8 of Section 4.2)

$$\mathrm{Tr}\,(F_{\mu\nu}\tilde{F}_{\mu\nu}) = \partial_\mu K_\mu,$$

where

$$\tilde{F}_{\mu\nu} = \frac{1}{2}\varepsilon_{\mu\nu\lambda\rho}F_{\lambda\rho}$$

is the dual field, and

$$K_\mu = \varepsilon_{\mu\nu\lambda\rho}\mathrm{Tr}\,\left(F_{\nu\lambda}A_\rho - \frac{2}{3}A_\nu A_\lambda A_\rho\right). \tag{13.13}$$

Consequently, the integral of $\mathrm{Tr}\,(F_{\mu\nu}\tilde{F}_{\mu\nu})$ over the four-dimensional volume, reduces to the surface integral of K_μ over the remote three-dimensional sphere. On this sphere, the $F_{\mu\nu}$ vanish more rapidly than r^{-2}, by virtue of the finiteness of the action; furthermore, $A_\mu \sim r^{-1}$, and thus the first term in (13.13) does not contribute to the surface integral. We obtain, therefore,

$$Q = -\frac{1}{16\pi^2}\int d^4x\,\mathrm{Tr}\,(F_{\mu\nu}\tilde{F}_{\mu\nu}). \tag{13.14}$$

Then we can use a trick which we have already employed a number of times (in Sections 7.4 and 9.4). Let us consider the evident relation

$$-\int d^4x\,\mathrm{Tr}\,(F_{\mu\nu} - \tilde{F}_{\mu\nu})^2 \geq 0, \tag{13.15}$$

where we have equality only when

$$F_{\mu\nu} = \tilde{F}_{\mu\nu}$$

everywhere in the space. Taking into account the fact that $\mathrm{Tr}\,(\tilde{F}_{\mu\nu}\tilde{F}_{\mu\nu}) = \mathrm{Tr}\,(F_{\mu\nu}F_{\mu\nu})$, the inequality (13.15) leads to a lower bound for the Euclidean action in the sector with fixed Q,

$$S \geq \frac{8\pi^2}{g^2}Q.$$

This inequality is useful for positive Q; for negative Q one has to use the fact that $\mathrm{Tr}\,(F_{\mu\nu} + \tilde{F}_{\mu nu})^2 \geq 0$, where we have equality if $F_{\mu\nu} = -\tilde{F}_{\mu\nu}$. Repeating the previous argument, for negative Q we have:

$$S \geq \frac{8\pi^2}{g^2}|Q|. \tag{13.16}$$

In both cases, the absolute minimum in the sector with given Q is reached if the following *first*-order equations are satisfied:

$$F_{\mu\nu} = \tilde{F}_{\mu\nu}, \qquad Q > 0 \tag{13.17}$$

$$F_{\mu\nu} = -\tilde{F}_{\mu\nu}, \qquad Q < 0. \tag{13.18}$$

Equation (13.17) is known as the self-duality equation, while (13.18) is the anti-self-duality equation. They are, of course, simpler than the second-order equation $D_\mu F_{\mu\nu} = 0$, the original Yang–Mills equation. From the Bianchi identity (see Problem 13 of Chapter 4),

$$\varepsilon_{\mu\nu\lambda\rho} D_\nu F_{\lambda\rho} = 0,$$

it follows that any solution of equations (13.17) or (13.18) satisfies the general Yang–Mills equation; the converse does not hold.

Let us now find the instanton in the $SU(2)$ theory, namely, the solution of equations (13.17) in the sector with $Q = 1$. For this we first construct the asymptotics as $r \to \infty$. This is equal to pure gauge (13.10), where $\omega(n^\mu)$ must be the first non-trivial mapping of the sphere S^3 to $SU(2)$. In terms of the functions v_α, in (13.12), this means that we must have a mapping $S^3 \to S^3_{SU(2)}$ of degree one. The simplest choice for this is

$$v_\alpha(n_\mu) = n_\alpha,$$

such that

$$\omega = n_\alpha \sigma_\alpha.$$

Hence, the gauge fields as $r \to \infty$ behave like

$$A_\mu(r \to \infty) = \omega \partial_\mu \omega^{-1} = \sigma_\alpha \sigma_\beta^\dagger n_\alpha \frac{\partial_\mu \beta - n_\mu n_\beta}{r}. \tag{13.19}$$

Further,

$$\sigma_\alpha \sigma_\beta^\dagger = \delta_{\alpha\beta} + i\eta_{\alpha\beta a} \tau^a, \tag{13.20}$$

where the values of the symbols $\eta_{\alpha\beta a}$ are obtained by direct substitution of the expressions $\sigma_\alpha = (1, -i\tau)$, $\sigma_\alpha^\dagger = (1, i\tau)$. The symbols $\eta_{\alpha\beta a}$ are called (in physics) 't Hooft symbols. They are antisymmetric in the indices α, β and their non-zero components are

$$\eta_{0ia} = -\eta_{i0a} = \delta_{ia}$$

$$\eta_{ija} = \varepsilon_{ija}.$$

Also, by direct substitution, one can establish that the 't Hooft symbols are self-dual in the first indices

$$\frac{1}{2} \varepsilon_{\alpha\beta\gamma\delta} \cdot \eta_{\gamma\delta a} = \eta_{\alpha\beta a}. \tag{13.21}$$

Problem 7.   *Prove the relations (13.20) and (13.21) by direct substitution.*

Substituting (13.20) in (13.19), we have for the asymptotics of the field

$$A_\mu = -i\eta_{\mu a a}\frac{n_\alpha}{r}\tau_a, \quad r \to \infty. \tag{13.22}$$

Of course, this field belongs to the algebra of the Lie group $SU(2)$.

The asymptotics (13.22) suggests the following Ansatz for the field A_μ in the whole space,

$$A_\mu = f(r)\omega\partial_\mu\omega^{-1} = -i\eta_{\mu a a}\frac{n_\alpha}{r}f(r)\tau_a. \tag{13.23}$$

The field strength for such vector potentials is equal to

$$F_{\mu\nu} = 2i\eta_{\mu\nu a}\frac{f(1-f)}{r^2}\tau_a + i\left(2\frac{f(1-f)}{r^2} - \frac{f'}{r}\right)(n_\mu\eta_{\nu\alpha a}n_\alpha - n_\nu\eta_{\mu\alpha a}n_\alpha)\tau_a. \tag{13.24}$$

Problem 8.   *Check the validity of (13.24) by direct calculation.*

The self-duality equation can then be solved in explicit form. Since the first term in (13.24) is self-dual, in order to satisfy the self-duality equation it is sufficient to satisfy the following equation:

$$f' = \frac{2}{r}f(1-f). \tag{13.25}$$

Here, we have to require that the function $f(r)$ tends to unity as $r \to \infty$ (then the field (13.23) will have the asymptotics (13.22)) and to zero as $r \to 0$ (sufficiently rapidly that the field (13.23) is regular at the origin). A solution of equation (13.25) with these properties has the form

$$f = \frac{r^2}{r^2 + \rho^2}, \tag{13.26}$$

where ρ is an arbitrary constant of integration. We note that ρ is the size of the instanton, i.e. the size of the region where A_μ differs significantly from its asymptotics (13.22). As one might expect, the size of the instanton is an arbitrary parameter.

Thus, the instanton solution has the form

$$A_\mu^{\text{inst}} = -i\eta_{\mu\nu a}x_\nu\tau_a\frac{1}{r^2 + \rho^2}, \tag{13.27}$$

where we have taken into account that $n_\nu = x_\nu/r$. The field strength is given by the first term in (13.24) and is equal to

$$F_{\mu\nu}^{\text{inst}} = 2i\eta_{\mu\nu a}\tau_a\frac{\rho^2}{(r^2 + \rho^2)^2}.$$

Recalling the connection between matrix fields and their real components, $A_\mu = -ig\frac{\tau^a}{2}A_\mu^a$, we then write for the components

$$A_\mu^{a,\text{inst}} = \frac{2}{g}\frac{1}{r^2 + \rho^2}\eta_{\mu\nu a}x_\nu.$$

We note that the field $F_{\mu\nu}$ decreases as a power law, which should be the case for a massless field. The action of the instanton, according to (13.16), is equal to

$$S_I = \frac{8\pi^2}{g^2}. \tag{13.28}$$

Problem 9. *Check formula (13.28) by direct calculation.*

Problem 10. *The Yang–Mills action is invariant under the group $SO(4)$ of spatial rotations and the global group $SU(2)$ of gauge transformations, which do not depend on the point of space. Under which subgroup of this group $SO(4) \times SU(2)$ is the instanton solution invariant?*
   To find the anti-instanton, we use the fact that the gauge function

$$\omega^{-1} = \sigma_\alpha^\dagger n_\alpha \tag{13.29}$$

has topological number $Q = -1$. Repeating the calculations of this section, we obtain for the anti-instanton

$$A_\mu = -i\bar{\eta}_{\mu\nu a}x_\nu\tau_a\frac{1}{r^2 + \rho^2}, \tag{13.30}$$

where $\bar{\eta}_{\mu\nu a}$ is the anti-self-dual 't Hooft symbol,

$$\begin{aligned}\bar{\eta}_{0ia} &= -\bar{\eta}_{i0a} = -\delta_{ia}\\ \bar{\eta}_{ija} &= \varepsilon_{ija}.\end{aligned}$$

The action for the anti-instanton is also equal to $8\pi^2/g^2$.

Problem 11. *1) Show that*

$$\sigma_\alpha^\dagger\sigma_\beta = \delta_{\alpha\beta} + i\bar{\eta}_{\alpha\beta a}\tau^a. \tag{13.31}$$

2) Show that

$$\frac{1}{2}\varepsilon_{\alpha\beta\gamma\delta}\cdot\bar{\eta}_{\gamma\delta a} = -\bar{\eta}_{\alpha\beta a}.$$

3) Show that

$$\omega^{-1}\partial_\mu\omega = -i\bar{\eta}_{\mu\alpha a}\frac{n_\alpha}{r}\tau_a.$$

4) Show that the topological number of the gauge function (13.29) is equal to (−1).

5) Show that the field (13.30) satisfies the anti-self-duality equation (13.18).

13.3   Classical vacua and θ-vacua

Based on the considerations studied in Chapter 11, we can expect that the instantons in Yang–Mills theory will describe a certain tunneling process from the ground state. The possibility of a decay interpretation (of the type met in Section 11.2 and Chapter 12) must be immediately excluded, since the classical vacuum in theory (the configuration $\mathbf{A} = 0$) has the least possible classical energy, equal to zero. Thus, the only possible interpretation is tunneling between degenerate classical vacua (Gribov 1976, Jackiw and Rebbi 1976a, Callan *et al.* 1976). The instanton solution itself should suggest to us precisely between which states tunneling takes place: we saw in Section 11.3 that the initial and final states are determined by the asymptotics of a Euclidean solution as $\tau \to \infty$ and $\tau \to +\infty$, respectively, where τ is Euclidean time. Thus, we are interested in the asymptotics of the instanton as $\tau \to \pm\infty$.

As emphasized in Section 13.1, for the interpretation of instantons, we have to work in the gauge

$$A_0 = 0. \tag{13.32}$$

The solution (13.27) does not satisfy this condition, thus, we have to perform a gauge transformation over it, i.e. to consider

$$A_\mu = \Omega A_\mu^{\text{inst}}\Omega^{-1} + \Omega\partial_\mu\Omega^{-1} = \Omega(A_\mu^{\text{inst}} - \Omega^{-1}\partial_\mu\Omega)\Omega^{-1}. \tag{13.33}$$

Here and later in this section, A_μ without the superscript "inst" denotes the instanton field in the gauge (13.32). From the requirement $A_0 = 0$ we obtain

$$\Omega^{-1}\partial_0\Omega = A_0^{\text{inst}} = -ix_a\tau_a\frac{1}{\mathbf{x}^2 + x_0^2 + \rho^2}. \tag{13.34}$$

This equation determines the gauge function $\Omega(x^0, \mathbf{x})$ modulo a gauge function depending only on spatial coordinates. Taking into account the fact that for the instanton solution $F_{ij}^{\text{inst}} \to 0$ as $x^0 \equiv \tau \to \pm\infty$, we can make use of the residual gauge freedom to set

$$A_i(\mathbf{x}, \tau) \to 0, \quad \tau \to -\infty. \tag{13.35}$$

Thus, we shall consider tunneling from the trivial vacuum $\mathbf{A} = 0$.

We further take into account the fact that the spatial components of the original solution (13.27) have the form:

$$A_i^{\text{inst}} = \left[i\delta_{ia}\frac{x_0}{\mathbf{x}^2 + x_0^2 + \rho^2} - i\varepsilon_{ija}\frac{x_j}{\mathbf{x}^2 + x_0^2 + \rho^2} \right] T_a, \tag{13.36}$$

and decrease as $x^0 \equiv \tau \to -\infty$. This means that the requirement (13.35) reduces to the condition

$$\partial_i \Omega(\mathbf{x}, \tau \to -\infty) = 0$$

so that, without loss of generality, we can set

$$\Omega(\mathbf{x}, \tau \to -\infty) = 1. \tag{13.37}$$

The solution of equation (13.34) with condition (13.37) can then be found in explicit form:

$$\Omega(\mathbf{x}, \tau) = e^{-i\tau^a \hat{x}^a F(|\mathbf{x}|, \tau)}, \tag{13.38}$$

where $\hat{x}^a = x^a/|\mathbf{x}|$ is a unit radius vector in *three-dimensional* space, and

$$F(|\mathbf{x}|, \tau) = \frac{|\mathbf{x}|}{\sqrt{|\mathbf{x}|^2 + \rho^2}} \left(\arctan \frac{\tau}{\sqrt{|\mathbf{x}|^2 + \rho^2}} + \frac{\pi}{2} \right).$$

To find the asymptotics of the instanton solution as $\tau \to \infty$ in the gauge $A_0 = 0$, we use the fact that the original field (13.36) decreases as $\tau \to +\infty$. It then follows from (13.33) that

$$A_i(\mathbf{x}, \tau \to +\infty) = \Omega_1 \partial_i \Omega_1^{-1}, \tag{13.39}$$

where Ω_1 depends only on $\mathbf{x}$ and is equal to

$$\Omega_1(\mathbf{x}) = \Omega(\mathbf{x}, \tau \to +\infty).$$

From (13.38) we obtain in explicit form:

$$\Omega_1(\mathbf{x}) = e^{-i\tau^a \hat{x}^a F_1(|\mathbf{x}|)}, \tag{13.40}$$

where

$$F_1(|\mathbf{x}|) = \pi \frac{|\mathbf{x}|}{\sqrt{|\mathbf{x}|^2 + \rho^2}}. \tag{13.41}$$

We conclude that the instanton describes tunneling between the classical vacuum $\mathbf{A} = 0$ and the classical vacuum (13.39), which is a pure gauge and has zero energy.

In order to comprehend this result, we consider all possible classical vacua of Yang–Mills theory, i.e. pure gauge fields

$$A_i = \tilde{\Omega} \partial_i \tilde{\Omega}^{-1}, \tag{13.42}$$

where, as before, we work in the gauge $A_0 = 0$, and $\tilde{\Omega}(\mathbf{x})$ depends only on the spatial coordinates. We are interested in transitions between different pairs of vacua of the type (13.42). We note first that such transitions are only possible if the field A_i does not vary at spatial infinity, otherwise a transition would require an infinite kinetic energy (the integral $\int d^3x(\partial_0 A_i^0)^2$ would diverge) and an infinite action. In other words, transitions are only possible between those vacua whose gauge functions $\tilde{\Omega}(\mathbf{x})$ are the same at spatial infinity. Without loss of generality, we can set

$$\tilde{\Omega}(|\mathbf{x}| \to \infty) = 1 \qquad\qquad (13.43)$$

for all vacua.[1]

Some of the classical vacua are not separated by a potential barrier at all. By potential energy in this case, we mean the static energy of a gauge field (the term in the energy functional not containing derivatives of the field A_i with respect to time),

$$E_{\text{stat}} = -\frac{1}{2g^2} \int d^3x \text{Tr}\,(F_{ij}F_{ij})$$

(see formula (13.6) and the discussion in Sections 12.1 and 13.1). A barrier does not exist between a pair of classical vacua if and only if the gauge function $\tilde{\Omega}$ can be continuously deformed from its value in the first vacuum of the pair to the value in the second vacuum (in such a way that condition (13.43) is always satisfied). In this case, there exists a path *in the space of classical vacua* (i.e. a path with zero potential energy) connecting the two vacua of the pair.

We further note that since we are considering only functions $\tilde{\Omega}$ satisfying condition (13.43), we are concerned with mappings of the space R^3 with infinity identified to the gauge group G. Since R^3 with infinity identified is homotopic to the sphere S^3, these mappings, and the respective classical vacua, divide into homotopy classes characterized by elements of $\pi_3(G) = Z$. From what has been said earlier, it is clear that there is no barrier between classical vacua if and only if the corresponding gauge functions $\tilde{\Omega}(\mathbf{x})$ belong to the same homotopy class; these vacua are said to be topologically equivalent.[2]

On the other hand, there exists a barrier between topologically inequivalent vacua, and transitions between these take place by tunneling. The instanton describes the transition between the vacuum $\mathbf{A} = 0$ with

---

[1]Since $\pi_2(G) = 0$, this condition can always be satisfied by performing a gauge transformation which is smooth throughout space and does not depend on time.

[2]From the point of view of quantum theory, topologically equivalent vacua can be identified with one another. The operator D_iF_{ij} is an operator of infinitesimal gauge transformations, and the condition $D_iF_{ij}|\psi\rangle = 0$ implies that the wave function $|\psi\rangle$ is the same in all topologically equivalent vacua.

topological number zero and the vacuum (13.39) with topological number unity (the topological number of a vacuum and the topological number of a Euclidean configuration are related, but not identical concepts, and should not be confused). Indeed, for the topological number of the vacuum (13.42) one can write down an expression completely analogous to (13.11),

$$n(\tilde{\Omega}) = \frac{1}{24\pi^2} \int dx^3 \varepsilon^{ijk} \mathrm{Tr} \left(\tilde{\Omega}\partial_i\tilde{\Omega}^{-1} \cdot \tilde{\Omega}\partial_j\tilde{\Omega}^{-1} \cdot \tilde{\Omega}\partial_k\tilde{\Omega}^{-1} \right). \tag{13.44}$$

By direct substitution, one can show that $n(\Omega_1) = 1$, if $\Omega_1(\mathbf{x})$ is determined by formulae (13.40) and (13.41).

Problem 12. *Show by direct calculation that $n(\Omega_1) = 1$.*

In fact, explicit calculation of the gauge function $\Omega(\mathbf{x}, \tau)$, its asymptotics $\Omega_1(\mathbf{x})$ and topological number $n(\Omega_1)$ is not necessary. Indeed, let us show that any configuration with finite action and topological number Q, where Q is defined in a gauge-invariant way according to (13.14), interpolates in the gauge $A_0 = 0$ between vacua with topological numbers n, differing by Q. The fact that the action is finite means that $F_{\mu\nu} \to 0$, as $(\tau^2 + \mathbf{x}^2) \to \infty$, i.e. that A_μ is a pure gauge at infinity in space–time. Let us consider the infinitely remote cylinder in four-dimensional Euclidean space–time, shown in Figure 13.1. In the gauge $A_0 = 0$, the field A_i does not vary on the lateral surface of the cylinder (since $F_{0i} = 0$ on that surface); without loss of generality, it may be assumed to be zero.

The integral (13.14) then reduces to the difference of the surface integrals over the bases of the cylinder for $\tau \to +\infty$ and $\tau \to -\infty$ (compare with (13.11)). On the bases

$$\begin{aligned} A_i(\mathbf{x}, \tau \to -\infty) &= \Omega_0 \partial_i \Omega_0^{-1} \\ A_i(\mathbf{x}, \tau \to +\infty) &= \Omega_1 \partial_i \Omega_1^{-1}, \end{aligned}$$

where $\Omega_0(\mathbf{x})$ and $\Omega_1(\mathbf{x})$ are gauge functions characterizing the initial and final classical vacua. These integrals over the bases of the cylinder represent topological numbers of the vacua Ω_0 and Ω_1, so that

$$Q = n(\Omega_1) - n(\Omega_0), \tag{13.45}$$

as required.

Thus, the situation in non-Abelian four-dimensional gauge theories is analogous to the quantum-mechanical model with a periodic potential, considered in Section 11.3. Up to topologically trivial gauge transformations, there exists a discrete set of classical ground states, enumerated by the integer n. The field A_i (in the gauge $A_0 = 0$) in each of these vacua is a pure gauge, $A_i(\mathbf{x}) = \Omega_n(\mathbf{x})\partial_i\Omega_n^{-1}(\mathbf{x})$, where the topological number

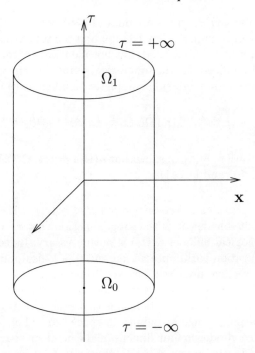

Figure 13.1.

of the vacuum Ω_n is equal to n. Tunneling between neighboring states is described by an instanton (if n is increased by one) and by an anti-instanton (if n is decreased).

This analogy can be developed further. Let us introduce into the quantum theory the gauge transformation operator with topological number 1, $T = T(\Omega_1)$, which acts such that $T^{-1}A_iT = \Omega_1 A_i\Omega_1^{-1} + \Omega_1\partial_i\Omega_1^{-1}$. This operator commutes with the Hamiltonian (we assume that the gauge invariance under time-independent transformations is preserved in the quantum theory in the gauge $A_0 = 0$; this assumption is valid in non-anomalous gauge theories). Thus, it is simultaneously diagonalizable with the Hamiltonian

$$T|\Psi_\theta\rangle = e^{i\theta}|\Psi_\theta\rangle. \tag{13.46}$$

In particular, from the states $|n\rangle$, with wave functions confined near the classical n-vacua, we construct states[3]

$$|\theta\rangle = \sum_{n=-\infty}^{+\infty} e^{-in\theta}|n\rangle \qquad (13.47)$$

which satisfy (13.46). Physical states obtained by the action of (localized) gauge-invariant operators on the θ-vacuum (13.47), have the same value of θ; if $|\Psi_\theta\rangle = \hat{O}|\theta\rangle$, where $\hat{O}$ is a gauge-invariant operator, then equation (13.46) is satisfied for $|\Psi_\theta\rangle$. Thus, the parameter θ is an integral of motion; it does not change in the process of evolution under the action of any gauge-invariant Hamiltonian. In other words, the parameter θ is yet another "universal constant," in addition to the coupling constant g.

We see that the presence of a complex vacuum structure and of instanton effects in four-dimensional non-Abelian gauge theories leads to the occurrence of an additional coupling constant θ. The dependence of physical quantities on θ is determined by the tunneling amplitude: matrix elements of gauge-invariant operators in the θ-vacuum contain contributions of the type $e^{i\theta}\langle n + 1|\hat{O}|n\rangle$, which are proportional to the overlap of the states $|n\rangle$ and $|n+1\rangle$, i.e. to the quantity

$$e^{-S_{\text{inst}}} = e^{-8\pi^2/g^2}$$

(at least when g is small). These considerations hold true not only in pure Yang–Mills theory, but also in more complex gauge theories[4] (with additional matter fields, the Higgs mechanism, etc.).

The emergence of the parameter θ is very significant in quantum chromodynamics, where CP is broken at $\theta \neq 0$ ('t Hooft 1976a,b). This contradicts experiments searching for the electric dipole moment of the neutron, if $\theta > 10^{-9}$. One solution of this "strong CP-problem" is to add new fields, and this leads to the prediction of a new particle, the axion (Peccei and Quinn 1977, Weinberg 1978, Wilczek 1978).

We note that the instanton can be given a slightly different interpretation (Manton 1983), which, in fact, is equivalent to the above. From the point of view of gauge-invariant quantities, topologically distinct classical vacua are equivalent, since they differ only by a gauge transformation. Let us identify these vacua. Then the situation becomes analogous to the quantum-mechanical model of the pendulum. Namely, in the gauge $A_0 = 0$, let us consider the configuration space, the set of all static configurations

---

[3]As already mentioned, physical states, including $|n\rangle$, are invariant under topologically trivial gauge transformations. Thus, the specific choice of $\Omega_1(\mathbf{x})$ is not significant.

[4]The dependence on θ disappears in gauge theories with massless fermions, and also in the Standard Model of electroweak interactions, because of the specific properties of fermions in fields of instanton-like configurations; see Appendix.

$A_i(\mathbf{x})$ with finite static energy, where gauge-equivalent configurations are identified. In this set there is a single ground state, the classical vacuum. Let us now consider possible paths in this set, $A_i(\mathbf{x}, \tau)$, beginning and ending at the classical vacuum. Then τ is a parameter along a path; we shall suppose that it varies from 0 to 1. For each of these paths, we can define the quantity

$$Q = -\frac{1}{8\pi^2} \int_0^1 d\tau \int d^3x \, \varepsilon^{ijk} \mathrm{Tr} \left(\partial_\tau A_i F_{ij} \right).$$

This quantity does not depend on the choice of parametrization of the path (i.e. it is invariant under the substitution $\tau \to \tau'(\tau)$) and coincides with (13.14), if formally we set $F_{0i} = \partial_0 A_i$. As in the previous section, we find that Q takes integer values for paths beginning and ending in the vacuum; paths with different Q cannot be continuously deformed into one another. Thus, in configuration space there exist non-contractible closed paths, analogous to a full rotation of a physical pendulum. The static energy along paths with $Q \neq 0$ takes non-zero values (since everywhere on the path $F_{ij} \neq 0$), i.e. evolution along such paths is possible at zero energy only via tunneling. This tunneling process is described by the instanton. The parameter θ is introduced analogously to in the model of a physical pendulum, by a change in the Lagrangian, namely by the addition to the original action (in Minkowski space) of a term

$$-\frac{\theta}{16\pi^2} \int \mathrm{Tr} \left(F_{\mu\nu} \tilde{F}_{\mu\nu} \right) d^4x$$

(compare with (11.36)). This term is a total derivative, and does not affect the classical field equations. At the same time, it is non-zero for instantons, on which it takes a purely imaginary value, and leads, as in the case of the pendulum, to a dependence of the physical quantities (for example, vacuum expectation values of gauge-invariant operators) on θ.

Finally, we stress that in four-dimensional space–time, everything that was said in the previous sections goes over (with minor modifications) to theories with an arbitrary *non-Abelian* simple gauge group. The generalization to *semi-simple* gauge groups is also not complicated: instantons exist independently for each simple subgroup. At the same time, in *Abelian* four-dimensional theories, there are no instantons, there is no vacuum structure associated with the latter and there is no additional parameter θ.

To end this section, we shall briefly discuss the topological classification of classical vacua in the Abelian Higgs model in two-dimensional space–time. Classical vacua in the gauge $A_0 = 0$ are static pure gauge configurations

$$A_1(x^1) \;=\; \frac{1}{e} \partial_1 \alpha(x^1) \tag{13.48}$$

$$\varphi(x^1) \ = \ e^{i\alpha(x^1)} \cdot v. \tag{13.49}$$

The analogue of equation (13.43) is the requirement

$$e^{i\alpha(x^1)} = 1 \quad x^1 \rightarrow \pm\infty. \tag{13.50}$$

Without loss of generality, we can set

$$\alpha(x^1 \rightarrow -\infty) = 0,$$

then the admissible values of the gauge function as $x^1 \rightarrow +\infty$ are equal to

$$\alpha(x^1 \rightarrow +\infty) = 2\pi n, \quad n = 0, \pm 1, \pm 2, \ldots. \tag{13.51}$$

The integer n characterizes different vacua; it is analogous to the vacuum topological number (13.44). Indeed, under condition (13.50), the space can be interpreted as a circle of large length. The gauge function $e^{i\alpha(x^1)}$ maps this circle S_R^1 to the group $U(1)$, which is also a circle $S_{U(1)}^1$. The degree of the mapping $S_R^1 \rightarrow S_{U(1)}^1$ is given by the integer n, which appears in (13.51). We note that for pure gauge fields, the topological number of the vacuum can be written in the form

$$n = \frac{i}{2\pi} \int \tilde{\Omega} \partial_1 \tilde{\Omega}^{-1} dx^{-1} = \frac{1}{2\pi} \int A_1 dx^1,$$

where $\tilde{\Omega} = e^{i\alpha(x^1)}$ is the gauge function of the vacuum configuration.

The gauge fields $A_1(x^0, x^1)$, interpolating between topologically distinct vacua, for $x^0 \rightarrow \pm\infty$ tend to vacuum configurations of the type of (13.48) with gauge functions with different topological numbers (as before, we work in the gauge $A_0 = 0$). Such configurations $A_1(x^0, x^1)$ can be characterized by the quantity

$$q = \frac{e}{4\pi} \int \varepsilon_{\mu\nu} F_{\mu\nu} d^2 x,$$

which is analogous to the topological number (13.14) of four-dimensional configurations. In the gauge $A_0 = 0$ we have, after integration over time,

$$q = \frac{e}{2\pi} \left[\int_{-\infty}^{+\infty} A_1 dx^1 \right]_{x^0=-\infty}^{x^0=+\infty} = n(\tilde{\Omega}_1) - n(\tilde{\Omega}_0),$$

where $\tilde{\Omega}_0$ and $\tilde{\Omega}_1$ are gauge functions of the initial and final vacua. This formula is completely analogous to (13.45).

To describe tunneling between neighboring classical vacua, we need to find a solution of the Euclidean field equations with $q = 1$. In fact, we already know this solution, it is the Abrikosov–Nielsen–Olesen vortex, which is interpreted as an instanton in Euclidean two-dimensional space–time.

Thus, the Abelian Higgs model in two-dimensional space–time has the same structure of the gauge vacua as four-dimensional non-Abelian gauge theories. In this model, the vortex solution serves as the instanton.

13.4 Sphalerons in four-dimensional models with the Higgs mechanism

Let us now consider the question of the height of the barrier separating classical topologically distinct vacua. As mentioned in Sections 11.2 and 11.3, a key role here is played by the sphaleron, a static unstable solution of the classical equations of motion. Namely, in systems with a finite number of variables, considered in Chapter 11, the barrier height is equal to the value of the potential at the saddle point (sphaleron). In field theory, we are interested in the saddle point of the static energy functional.

In four-dimensional Yang–Mills theory without the Higgs mechanism, the static energy has the form

$$E_{\text{stat}} = -\frac{1}{2g^2} \int d^3x \operatorname{Tr}(F_{ij}^2).$$

The scale argument of Section 7.2 shows that the functional E_{stat} does not have extremal configurations. Indeed, under the scale transformations $A_i(\mathbf{x}) \to \lambda A_i(\lambda \mathbf{x})$, we have $E_{\text{stat}} \to \lambda E_{\text{stat}}$, which implies that there are no extrema. In other words, configurations of large size may have arbitrarily small energy (if the $A_i(x)$ are concentrated in a region with size of the order of unity, then $A_i(\lambda x)$ are concentrated in a region with size of the order of $1/\lambda$). The energy along a path connecting topologically distinct vacua can be arbitrarily small.

To clarify why this property does not lead to a large probability of tunneling, let us consider the set of configurations of the form

$$A_i(\mathbf{x}, t) = \beta(t)\Omega_1 \partial_i \Omega_1^{-1}, \quad A_0 = 0, \tag{13.52}$$

where Ω_1 is a gauge function of the type of (13.40), (13.41). The important thing for us is that $\Omega_1(\mathbf{x})$ is non-trivial and that $\Omega_1 = \Omega_1(\mathbf{x}/\rho)$, where ρ is an arbitrary scale. For $\beta = 0$ and $\beta = 1$, configurations of the type (13.52) are classical vacua with $n = 0$ and $n = 1$, respectively. For configurations of the form (13.52) we have

$$F_{ij} = \beta(1 - \beta)[\Omega \partial_i \Omega^{-1}, \Omega \partial_j \Omega^{-1}],$$

and the static energy is equal to

$$E_{\text{stat}} = [\beta(1-\beta)]^2 \int -\frac{1}{2g^2} \text{Tr} \left([\Omega \partial_i \Omega^{-1}, \Omega \partial_j \Omega^{-1}] \right)^2 d^3 x.$$

Making the change of variables $\mathbf{x} = \rho \mathbf{y}$, we obtain

$$E_{\text{stat}} = \frac{1}{g^2 \rho} [\beta(1-\beta)]^2 C_1,$$

where C_1 does not depend on ρ. Analogously, the kinetic energy,

$$E_{\text{k}} = -\frac{1}{g^2} \int d^3 x \, \text{Tr} \, (\partial_0 A_i)^2$$

for configurations (13.52), has the form

$$E_{\text{k}} = \frac{\rho}{2g^2} C_2 \dot{\beta}^2,$$

where C_2 is also independent of ρ. Thus, if we restrict ourselves to fields of the form (13.52), then we obtain a mechanical system with a single variable $\beta(t)$ and action (in conventional time):

$$S_\beta = \int dt (E_{\text{k}} - E_{\text{stat}}) = \int dt \left\{ \frac{\rho}{2g^2} C_2 \dot{\beta}^2 - \frac{C_1}{g^2 \rho} [\beta(\beta-1)]^2 \right\}. \quad (13.53)$$

We note that the potential in this system is proportional to $1/\rho$, so that the barrier between states $\beta = 0$ and $\beta = 1$ is small for large ρ. At the same time, the effective mass (coefficient in front of $\dot{\beta}^2$ in (13.53)) is proportional to ρ. As a result, the tunneling exponent does not depend on ρ:

$$\int_0^1 \sqrt{2MV} d\beta = \int_0^1 \sqrt{\frac{\rho}{g^2} C_2 \cdot \frac{C_1}{g^2 \rho} [\beta(\beta-1)]^2} d\beta$$

$$= \frac{1}{g^2} \sqrt{C_1 C_2} \int_0^1 \beta(\beta-1) d\beta.$$

The example of configurations of the form (13.52) shows that the tunneling exponent is finite, despite the fact that the potential barrier is arbitrarily low, and that this property is associated with a growth of the kinetic term as the spatial size of the configuration increases.

The fact that the potential barrier between topologically distinct vacua is arbitrarily low in the four-dimensional Yang–Mills theories is directly associated with the masslessness of the vector fields. In theories with the Higgs mechanism, configurations of gauge fields with large size have large energy and the barrier is of finite height. In these models, the sphaleron must exist (Manton 1983, Klinkhamer and Manton 1984). Our next task is to determine the structure of the sphaleron and to estimate the barrier height in a theory with gauge group $SU(2)$ and a Higgs field doublet:

$$\phi = \begin{pmatrix} \phi_1 \\ \phi_2 \end{pmatrix}.$$

This model could be viewed as the limiting case of the Standard Model, when the gauge constant g' of the group $U(1)$ is equal to zero, i.e. $\sin\theta_W = 0$. This model was discussed in Section 6.2; we recall that modulo global transformations, the following are satisfied in the trivial vacuum:

$$
\begin{aligned}
A_\mu &= 0 && (13.54) \\
\phi^{\text{vac}} &= \frac{1}{\sqrt{2}} \begin{pmatrix} 0 \\ \phi_0 \end{pmatrix},
\end{aligned}
$$

where, in the notation of Section 6.2,

$$\phi_0 = \frac{\mu}{\sqrt{\lambda}}.$$

Small perturbations about this vacuum have masses

$$
\begin{aligned}
m_V &= \frac{g\phi_0}{2} \\
m_\chi &= \sqrt{2\lambda}\,\phi_0.
\end{aligned}
$$

It will be convenient for us to choose the origin for the energy such that in classical vacua $E = 0$; then the Euclidean action in the model is written in the form

$$S_E = \int d^4x \left[-\frac{1}{2g^2}\operatorname{Tr} F_{\mu\nu}^2 + (D_\mu\phi)^\dagger D_\mu\phi + \lambda\left(\phi^\dagger\phi - \frac{1}{2}\phi_0^2\right)^2 \right],$$

and the static energy is equal to

$$E_{\text{stat}} = \int d^3x \left[-\frac{1}{2g^2}\operatorname{Tr} F_{ij}^2 + (D_i\phi)^\dagger D_i\phi + \lambda\left(\phi^\dagger\phi - \frac{1}{2}\phi_0^2\right)^2 \right]. \quad (13.55)$$

In the gauge $A_0 = 0$ the classical vacua are gauge-transformed fields (13.54)

$$
\begin{aligned}
A_i(\mathbf{x}) &= \tilde\Omega\partial_i\tilde\Omega^{-1} \\
\phi(\mathbf{x}) &= \tilde\Omega\phi^{\text{vac}}.
\end{aligned}
$$

As in the previous section, they are determined by the gauge function $\tilde\Omega(\mathbf{x})$ and are characterized by the topological number n.

To guess the structure of the sphaleron solution, let us consider, first, Euclidean configurations with

$$Q = -\frac{1}{16\pi^2} \int d^4x \, \text{Tr} \, (F_{\mu\nu}\tilde{F}_{\mu\nu}) = 1.$$

In the gauge $A_0 = 0$, these interpolate between neighboring classical vacua with topological numbers $n(\tilde{\Omega})$ differing by one. It will be convenient for us to leave the gauge $A_0 = 0$ and again consider the fields $A_\mu(x)$ with asymptotics

$$A_\mu(x) \to \omega\partial_\mu\omega^{-1}, \quad x^2 \equiv r^2 \to \infty$$

in four-dimensional Euclidean space, where $w(x) = \sigma_\alpha n_\alpha$ (in the notation of Section 13.2). The configurations of the Higgs field of interest must have a finite Euclidean action, therefore asymptotically,

$$\phi(x) \to w(x)\phi^{\text{vac}}$$

(in this case only, the covariant derivative $D_\mu\phi$ and the scalar potential are equal to zero on the infinitely remote sphere S^3 in four-dimensional space). If we then consider configurations of the form

$$
\begin{aligned}
A_\mu(x) &= f(r)\omega\partial_\mu\omega^{-1} \\
\phi(x) &= h(r)\omega\phi^{\text{vac}},
\end{aligned}
\tag{13.56}
$$

where f and h depend only on $r = \sqrt{x^2} = \sqrt{\mathbf{x}^2 + \tau^2}$ and satisfy the evident conditions $f(\infty) = h(\infty) = 1$, $f(0) = h(0) = 0$, then one can show that they have the following properties:

- they have $SO(4)$ symmetry under rotations of four-dimensional space, supplemented by global transformations from the full global group of the theory;

- under inversion of Euclidean time $\tau \to -\tau$, configurations with $Q = 1$ transform into configurations with $Q = -1$, and the same functions f and g.

Problem 13. *Verify these two properties. Hint: use the result of Problem 10 and the additional global $SO(3)$ symmetry, considered in Problem 6 of Section 6.2.*

The second of the above properties implies that at time $\tau = 0$ the field configuration is half-way between neighboring vacua (after it is transformed to the gauge $A_0 = 0$). Thus, we can hope that modulo a gauge transformation, the sphaleron will be a configuration of type (13.56) taken at $\tau = 0$. Of course, only the spatial components of the vector potential $A_i(\mathbf{x}, \tau = 0)$ are important as far as we are concerned. Since the energy is invariant under a gauge transformation taking the configuration (13.56) to the gauge $A_0 = 0$, there is no need to perform this gauge transformation in order to find the barrier height. Thus, we may expect that modulo a gauge transformation, the sphaleron configuration will have the following structure:

$$A_i(\mathbf{x}) = f(|\mathbf{x}|)\omega_0 \partial_i \omega_0^{-1}$$
$$\phi(\mathbf{x}) = h(|\mathbf{x}|)\omega_0 \phi^{\mathrm{vac}},$$

where $\omega_0(\mathbf{x}) = \omega(\mathbf{x}, \tau = 0) = -i\tau_a \hat{x}_a$, and, as before, we have denoted $\hat{x}_a = x_a/|\mathbf{x}|$. In explicit form,

$$A_i(x) = -i\varepsilon_{ija}\hat{x}_j \tau_a \frac{f(|\mathbf{x}|)}{|\mathbf{x}|}$$
$$\phi(|\mathbf{x}|) = \phi_0 h(|\mathbf{x}|)(-i\tau_a \hat{x}_a)\begin{pmatrix} 0 \\ 1 \end{pmatrix}, \tag{13.57}$$

where f and g satisfy the conditions

$$f(\infty) = h(\infty) = 1$$
$$f(0) = h(0) = 0.$$

Since any configuration of this structure lies (modulo a gauge transformation) half-way between two neighboring vacua, in order to obtain the sphaleron solution, we need to find the minimum of the static energy among configurations of the form (13.57): at this minimum the barrier height will be minimal, whereas the negative mode must fall outside the class of configurations (13.57), i.e. have a different spatial and group structure.

Problem 14. *Show that the Ansatz (13.57) passes through the field equations, i.e. the static field equations reduce to two ordinary differential equations for the functions $f(|\mathbf{x}|)$ and $h(|\mathbf{x}|)$. Show that these equations can be obtained by finding the extrema of the energy functional (13.55) in the class of configurations of the form (13.57). We note that this property is a consequence of the $SO(4)$ symmetry of the field (13.56) (which, on the surface $\tau = 0$, leads to $SO(3)$ symmetry of the field (13.57)).*

These arguments suggesting the possibility of searching for the sphaleron in the form (13.57) are by no means rigorous. A solution of the field equations with the Ansatz (13.57) is relatively easily found numerically.[5]

---

[5]This solution was found by Dashen *et al.* (1974), and also independently by Soni (1980) and Boguta (1983). Its interpretation as a sphaleron was given by Manton (1983) and Klinkhamer and Manton (1984).

Analysis of the actual number of negative modes around this solution is a very much more complicated problem. It has been shown, using numerical calculation (Yaffe 1989), that among the fluctuations around a solution of the form (13.57) there is indeed exactly one negative mode for $m_\chi < 12m_V$; when the model parameters are related in this way, the sphaleron indeed has the form (13.57). Otherwise, the sphaleron does not have the form (13.57). We note that in the electroweak theory the region $m_\chi < 12m_V$ covers all the physically interesting region of the Higgs boson masses.

The dimensional estimation of the static energy of the sphaleron repeats the estimation of the mass of the magnetic monopole, given in Section 9.1, word for word. Traditionally, the sphaleron energy is written in the form

$$E_{\text{sph}} = \frac{2m_V}{\alpha_V} B\left(\frac{m_\chi}{m_V}\right),\tag{13.58}$$

where $\alpha_V = g^2/4\pi$, and the function B varies within the range 1.56–2.27 as m_χ ranges from small to large values (this is the result of numerical analysis (Klinkhamer and Manton 1984)).

Problem 15. *Estimate the sphaleron energy by the variational method, choosing for $f(|\mathbf{x}|)$ the function corresponding to the instanton configuration,*

$$f = \frac{\mathbf{x}^2}{\mathbf{x}^2 + \rho^2},$$

and taking for $h(|\mathbf{x}|)$ the function

$$h = \frac{|\mathbf{x}|}{\sqrt{\mathbf{x}^2 + \rho^2}}.$$

Take ρ^2 as a variable parameter.

As has already been stressed on a number of occasions, sphalerons are useful for describing processes of transition between topologically distinct vacua at finite temperatures. We shall discuss this point in Section 17.4, but here, we merely indicate that such processes are of considerable interest for cosmology.

To conclude this section, we note that the scale argument of Section 7.2 prohibits the existence of instantons in four-dimensional theories with the Higgs mechanism. This does not mean that there is no tunneling between topologically distinct vacua in these theories; however, its analysis requires special methods ('t Hooft 1976b, Affleck 1981) which are briefly discussed in the Appendix. In Yang–Mills–Higgs theories the lower bound (13.16) is a strict inequality, so that in any case in theories with a small coupling constant, the probability of tunneling processes of instanton type is exponentially small, $\Gamma \lesssim \exp(-16\pi^2/g^2)$.

Supplementary Problems for Part II

Problem 1. *Non-topological solitons.*[1]

Let us consider the theory of two real scalar fields φ^a in $(1 + 1)$-dimensional space–time. We choose the Lagrangian in the form

$$\mathcal{L} = \frac{1}{2} \sum_a \partial_\mu \varphi^a \partial_\mu \varphi^a - V_0(\varphi) - V_1(\varphi),$$

where

$$V_0(\varphi) = \frac{\lambda}{4} \left(\sum_a \varphi^a \varphi^a - v^2 \right)^2$$

$$V_1(\varphi) = -\varepsilon \varphi^1.$$

We note that for $\varepsilon = 0$ the Lagrangian is invariant under global $O(2)$ symmetry. The term $V_1(\varphi)$ in the potential breaks this symmetry explicitly.

1. Find the dimensions of the constants λ, v and ε.

2. Find the ground state and the spectrum of small perturbations about it for small ε.

3. Prove that for sufficiently small ε there exists a soliton, a static local minimum of the energy functional which is a solution of the field equations and has finite energy.

4. Show that this soliton can be deformed to the ground state, in such a way that all intermediate field configurations have finite energy. Thus, the soliton is not of a topological nature.

5. Estimate the height of the energy barrier between the soliton and the ground state for small ε.

---

[1] In four-dimensional space–time these solutions describe domain walls. Walls with this structure arise in certain models with axions and are of cosmological interest (Vilenkin and Everett 1982, Kibble *et al.* 1982).

Problem 2.   *Potential of the interaction of vortices.*

Consider an Abelian Higgs model in $(2+1)$-dimensional space–time and choose $m_H \gg m_V$ (m_H and m_V are the masses of the Higgs and the vector boson, respectively).

1. Find the interaction energy of two vortices separated by distance r in the following cases:

   (a) $r \gg m_V^{-1}$, large distances;

   (b) $m_V^{-1} \gg r \gg m_H^{-1}$, small distances; in this case we limit ourselves to logarithmic accuracy, i.e. we suppose that

$$|\ln(rm_V)| \gg 1, \quad |\ln(rm_H)| \gg 1.$$

2. The same for the vortex and the anti-vortex.

Problem 3.   *Multi-instanton solutions ('t Hooft 1976c, Jackiw et al. 1977).*

Consider the Yang–Mills theory in Euclidean four-dimensional space–time. Choose the following Ansatz,

$$A_\mu^a = -\frac{1}{g}\bar\eta_{\mu\nu a}\partial_\nu(\ln\phi),$$

where $\phi(x^\mu)$ is a real function.

1. Show that the self-duality equations $F_{\mu\nu} = \tilde F_{\mu\nu}$ reduce to the four-dimensional Laplace equation

$$\partial_\mu\partial_\mu\phi = 0$$

at all points where $\phi \neq 0$.

2. Show that singularities in the function ϕ of type $1/x^2$ lead to pure gauge singularities in the vector potential A_μ^a, i.e. to singularities which can be removed by a singular gauge transformation.

3. Find n-instanton solutions of the given structure. Show by explicit computation that for these,

$$\int F_{\mu\nu}\tilde F_{\mu\nu}d^4x \propto n.$$

4. Find in explicit form the gauge transformations taking the solution obtained in point 3 with $n = 1$ to the standard form, given in the main text.

Problem 4. *Sphaleron in the Abelian Higgs model (Bochkarev and Shaposhnikov 1987, Grigoriev and Rubakov 1988).*

Find the sphaleron in the Abelian Higgs model in $(1 + 1)$-dimensional space–time.

Problem 5. $SU(2)$ *sigma model in* $(3 + 1)$-*dimensional space–time (Skyrme 1961).*

Suppose the field $U(x)$ is a 2×2 matrix from the group $SU(2)$ at each point of space–time (equivalently, the field may be considered as taking values in the three-dimensional sphere of unit radius – we recall that the $SU(2)$ group manifold is S^3; in this sense the model is analogous to **n**-field models). We require the Lagrangian to be invariant under global transformations from the group $SU(2) \times SU(2)$; where the first $SU(2)$ will be denoted by $SU(2)_L$ and the second by $SU(2)_R$. The transformation rule is chosen in the form

left : $\qquad U(x) \to U'(x) = \omega_L U(x), \quad \omega_L \in SU(2)_L$

right : $\qquad U(x) \to U'(x) = U(x)\omega_R, \quad \omega_R \in SU(2)_R.$

1. Show that the monomials

$$L_2 = \operatorname{Tr}\left(\partial_\mu U^\dagger \partial_\mu U\right),$$
$$L_4 = \operatorname{Tr}\left([U^\dagger \partial_\mu U, U^\dagger \partial_\nu U][U^\dagger \partial_\mu U, U^\dagger \partial_\nu U]\right)$$

   are invariant under $SU(2)_L \times SU(2)_R$. Here, as usual, $[,]$ denotes the matrix commutator.

2. Consider the Lagrangian

$$\mathcal{L} = \frac{F^2}{2}L_2 + \frac{g^2}{16}L_4.$$

   Find the dimensions of the constants F and g. Show that $\mathcal{L}$ contains at most two derivatives of the field U with respect to time. Find the energy functional and show that the energy is non-negative.

3. Show that for the ground state, one can choose $U(x) = 1$. By considering small perturbations about the ground state, find their spectrum.

4. Consider static configurations $U(\mathbf{x})$ with finite energy. Show that they divide into topological sectors, characterized by an integer topological number n. Thus, in the model, the existence of topological solitons is possible.

5. Show that for $g^2 = 0$ there are no static solitons.

6. Let us consider configurations of the form

$$U(\mathbf{x}) = e^{i\tau^a n^a f(r)},$$
$$n^a = x^a/r.$$

In what sense are they spherically symmetric? Find values of $f(0)$ and $f(\infty)$ for which $U(\mathbf{x})$ are non-singular, and the static energy is finite. Find the connection between $f(0), f(\infty)$ and the topological number n.

7. Estimate the mass and size of a soliton for $g^2 > 0$.

Problem 6. *Scattering in the field of a vortex (Alford and Wilczek 1989).* Let us consider a vortex, a static soliton in an Abelian Higgs model in $(2 + 1)$-dimensional space–time. Let us denote the configuration of the vortex by $A_\mu^c(\mathbf{x})$, $\phi^c(\mathbf{x})$. We add another complex scalar field to the model, namely the scalar field ξ with charge q and Lagrangian

$$\mathcal{L} = (D_\mu \xi)^* (D_\mu \xi) - m_\xi^2 \xi^* \xi,$$

where

$$D_\mu \xi = \partial_\mu \xi - ieqA_\mu \xi.$$

We shall consider the fields A_μ^c, ϕ^c as background.

1. Write down the equation for the field ξ in the external field of the vortex. Find the decomposition of its solutions in terms of the eigenfunctions of the energy E and the angular momentum $L = -i\frac{\partial}{\partial \theta}$, where θ is the polar angle in the plane (x^1, x^2). Find the corresponding radial equations and their solutions far from the center of the vortex for low energies $E \ll m_V, m_H, m_\xi$.

2. In the unitary gauge, where $\phi^c(\mathbf{x})$ is real, the field $\xi(\mathbf{x}, t)$ far from the vortex can be interpreted (at least for low energies) as a wave function of some particle. Find the scattering cross section for this particle off the vortex at a fixed angle θ for low energies and for arbitrary q.

Problem 7. *Klein–Gordon equation in a monopole field.* Let $A_i^a(\mathbf{x})$, $\phi^a(\mathbf{x})$ be the classical monopole field in the $SU(2)$ model, considered in the main text. Let us introduce into the theory yet another scalar field $\xi(x)$, a doublet under the gauge group $SU(2)$, with Lagrangian

$$\mathcal{L}_\xi = (D_\mu \xi)^\dagger (D_\mu \xi) - m^2 \xi^\dagger \xi,$$

where $D_\mu \xi = (\partial_\mu - ig\frac{\tau^a}{2} A_\mu^a) \xi$, and g is the gauge coupling constant.

1. Considering the monopole field as background, write down the equation for the field ξ (schematically this equation can be written in the form $K\xi = 0$; determine the operator K). Using the fact that the monopole field is invariant under spatial rotations supplemented by gauge transformations, find the analogue of the angular momentum operator (i.e. a generalization of $\mathbf{L} = [\mathbf{r} \times \mathbf{p}]$, $\mathbf{p} = -i\frac{\partial}{\partial \mathbf{x}}$), which commutes with the operator K. Find an explicit form for the lowest "monopole harmonics," i.e. the eigenfunctions of the analogue of the angular momentum with the lowest eigenvalue.

2. By considering solutions for fields ξ with a fixed energy $\xi = e^{-iEx^0}\xi_E(\mathbf{x})$, write down a system of radial equations for the lowest monopole harmonics. Find the solution of this system for $E \ll m_V, m_H$ (m_V and m_H are the mass of the vector and the Higgs field, respectively) far from the monopole core, $r \gg m_V^{-1}, m_H^{-1}$.

Problem 8. *Euclidean bubble in φ^4 model (Fubini 1976, Lipatov 1977).* Let us consider a model with one scalar field in $(3+1)$-dimensional space–time with Lagrangian

$$\mathcal{L} = \frac{1}{2}\partial_\mu\varphi\partial_\mu\varphi + \frac{\lambda}{4}\varphi^4.$$

The scalar potential $V(\varphi) = -\frac{\lambda}{4}\varphi^4$ is unbounded from below, thus the model has no ground state. Loosely speaking, the ground state corresponds to the field $\varphi = \infty$. Nevertheless, one can ask the following questions.

1. Is the state $\varphi = 0$ stable under small perturbations with finite energy?

2. If it is, find in explicit form the Euclidean bubble corresponding to the decay of the state $\varphi = 0$ by tunneling. Assuming that $\lambda \ll 1$, find the semiclassical exponent for the decay probability.

Problem 9. *The solution in the **n**-field model considered in Section 7.4 can be viewed as an instanton in two-dimensional Euclidean space–time.*

1. Give an interpretation of this solution in terms of the tunneling process which takes place in **n**-field theory in two-dimensional Minkowski space–time, whose action is of the form

$$S = \int d^2x \frac{1}{2g^2}(\partial_\mu\varphi_a\partial_\mu\varphi_a),$$

$$\mu = 0,1; \quad a = 1,2,3; \quad \sum_{a=1}^{3}\varphi_a\varphi_a = 1.$$

2. Let us modify this model by adding to the action a term

$$\Delta S = -\int d^2 x \, \lambda (\varphi_3 - 1)^2,$$

where λ is a real parameter. Find the classical vacuum and sphaleron in the model with action $(S + \Delta S)$ (Mottola and Wipf 1989).

Part III

Chapter 14

Fermions in Background Fields

14.1 Free Dirac equation

Free particles with spin $1/2$, fermions, in four-dimensional space–time are described by wave functions with four components, $\psi_\alpha(x^0, \mathbf{x})$, $\alpha = 1, 2, 3, 4$. It is convenient to represent these in the form of columns with four components:

$$\psi = \begin{pmatrix} \psi_1 \\ \psi_2 \\ \psi_3 \\ \psi_4 \end{pmatrix}.$$

In the absence of background fields, the wave functions of fermions with mass m satisfy the Dirac equation

$$i\gamma^\mu \partial_\mu \psi - m\psi = 0, \tag{14.1}$$

where the γ^μ are 4×4 Dirac matrices. The relativistic relation $p^2 = m^2$ will hold if the Dirac matrices satisfy the equation

$$\{\gamma^\mu, \gamma^\nu\} = 2\eta^{\mu\nu}, \tag{14.2}$$

where the parentheses denote the anticommutator, and $\eta^{\mu\nu}$ is the Minkowski metric tensor. Indeed, acting on equation (14.1) by the operator $(-i\gamma^\mu \partial_\mu - m)$ we obtain, taking into account (14.2), that each component of the wave function satisfies the Klein–Gordon equation

$$(\partial^\mu \partial_\mu + m^2)\psi = 0,$$

or, in the momentum representation, $(p^2 - m^2)\tilde{\psi}(p) = 0$.

Equation (14.2) implies, in particular, that $(\gamma^0)^2 = 1$. Multiplying equation (14.1) by γ^0 and introducing the notation $\beta = \gamma^0$, $\alpha^i = \gamma^0\gamma^i$, we obtain the Dirac equation in a form analogous to the Schrödinger equation,

$$i\frac{\partial\psi}{\partial x^0} = \left[\alpha^i\left(-i\frac{\partial}{\partial x^i}\right) + m\beta\right]\psi. \qquad (14.3)$$

Precisely in this form, the Dirac equation is used, as a rule, in quantum mechanics.

Equation (14.2) is the defining equation for Dirac matrices. In particular, the dimension of the Dirac column (four) is determined by the fact that the minimum size of matrices satisfying this equation is 4×4. This assertion is proved in textbooks on quantum mechanics and quantum field theory and we shall not dwell on the proof here. The specific choice of the Dirac matrices is determined by considerations of convenience. We shall, as a rule, use the representation

$$\gamma^\mu = \begin{pmatrix} 0 & \sigma^\mu \\ \tilde{\sigma}^\mu & 0 \end{pmatrix}, \qquad (14.4)$$

where 0 is the 2×2 zero matrix, σ^μ are 2×2 matrices, where

$$\sigma^0 = \tilde{\sigma}^0 = 1,$$

the σ^i are the Pauli matrices, and

$$\tilde{\sigma}^i = -\sigma^i.$$

Any other representation of the Dirac matrices is obtained from this by a unitary transformation, $\gamma^\mu \to U\gamma^\mu U^{-1}$, where U is a unitary matrix of size $A \times A$.

Problem 1.   *Show that the matrices (14.4) indeed satisfy equation (14.2).*

In this representation of γ-matrices, the matrices α^i and β have the form

$$\beta \equiv \gamma^0 = \begin{pmatrix} 0 & 1 \\ 1 & 0 \end{pmatrix}, \qquad \alpha^i \equiv \gamma^0\gamma^i = \begin{pmatrix} -\sigma^i & 0 \\ 0 & \sigma^i \end{pmatrix}. \qquad (14.5)$$

It is convenient to introduce yet another matrix γ^5, which anticommutes with all γ^μ,

$$\gamma^5 = -i\gamma^0\gamma^1\gamma^2\gamma^3.$$

In the representation (14.4), it is equal to

$$\gamma^5 = \begin{pmatrix} 1 & 0 \\ 0 & -1 \end{pmatrix}. \qquad (14.6)$$

where the blocks are 2×2 matrices. It is this simple form of the matrix γ^5 that determined our choice (14.4). Finally, it is useful to define the matrices

$$\sigma^{\mu\nu} = \frac{i}{2}[\gamma^\mu, \gamma^\nu].$$

In the representation (14.4) they have the form

$$\sigma^{\mu\nu} = \begin{pmatrix} \tilde{\tau}^{\mu\nu} & 0 \\ 0 & \tau^{\mu\nu} \end{pmatrix}, \tag{14.7}$$

where

$$\begin{aligned} \tau^{0i} &= -\tau^{i0} = i\sigma^i \\ \tilde{\tau}^{0i} &= -\tilde{\tau}^{i0} = -i\sigma^i \\ \tau^{ij} &= \tilde{\tau}^{ij} = \varepsilon^{ijk}\sigma^k. \end{aligned}$$

The matrices

$$s^i = \frac{1}{4}\varepsilon^{ijk}\sigma^{jk}$$

are spin operators. Indeed, the full three-dimensional angular momentum operators

$$L_i = -i\varepsilon_{ijk}x_j\partial_k + s_i$$

commute with the Dirac Hamiltonian

$$H_D = -i\alpha^i\frac{\partial}{\partial x^i} + \beta m \tag{14.8}$$

entering (14.3). In the representation (14.4) we have

$$s^i = \begin{pmatrix} \frac{1}{2}\sigma^i & 0 \\ 0 & \frac{1}{2}\sigma^i \end{pmatrix}. \tag{14.9}$$

Problem 2. *Check formulae (14.6), (14.7) and (14.9).*

Problem 3. *Show that, in any representation of the Dirac matrices, the relations $[L_i, H_D] = 0$ and $[L_i, L_j] = i\varepsilon_{ijk}L_k$ hold, so that L_i indeed satisfy the commutation relations for conserved angular momentum.*

The Dirac Hamiltonian in equation (14.3) is invariant under spatial reflection, supplemented by unitary transformation of the wave function ψ. Indeed, if $\psi(\mathbf{x}, x^0)$ satisfies equation (14.3) (or, what amounts to the same thing, the Dirac equation (14.1)) then the function

$$\psi_P(\mathbf{x}, x^0) = P\psi(-\mathbf{x}, x^0) \tag{14.10}$$

will satisfy the same equation, if

$$P^{-1}\beta P = \beta$$
$$P^{-1}\alpha^i P = -\alpha^i.$$

Here, P is a unitary matrix. From the definition of the matrices β and α^i and the anticommutation relation (14.2), it is clear that for P, one can take the matrix

$$P = \gamma^0 \tag{14.11}$$

(we note that γ^0 is unitary). Thus, if we know one solution of the Dirac equation, then out of it we can construct another solution using formula (14.10).

Let us now discuss yet another property of the Dirac equation, namely its invariance under C-conjugation. If ψ satisfies the Dirac equation, then the complex conjugate column satisfies the equation

$$(-i\gamma^{\mu*}\partial_\mu - m)\psi^* = 0.$$

Let us introduce the unitary matrix C such that

$$C\gamma^{\mu*}C^{-1} = -\gamma^\mu. \tag{14.12}$$

Then the column

$$\psi^C = C\psi^* \tag{14.13}$$

will again satisfy the Dirac equation (14.1). In the representation of γ-matrices (14.4) the matrix C can be chosen in the form

$$C = \begin{pmatrix} \mathbf{0} & -\varepsilon \\ \varepsilon & \mathbf{0} \end{pmatrix}, \tag{14.14}$$

where ε is a 2×2 matrix with matrix elements $\varepsilon_{\alpha\beta}$ $(\alpha, \beta = 1, 2)$, the components of an antisymmetric tensor of rank 2. In other words, $\varepsilon = i\sigma_2$.

Problem 4. *Show that (14.12) is valid for the matrix C of the form (14.14) and the γ-matrices chosen in the form (14.4). Show that the matrix (14.14) is unitary.*

Sometimes it is useful to consider systems in space–time of dimension $d = 2$ or $d = 3$. Let us briefly look at the properties of the Dirac equation in these dimensions.

For $d = 2$, there are just two matrices γ^μ $(\mu = 0, 1)$, and their minimum size is 2×2. Correspondingly, the wave function ψ is a two-component column, and the γ-matrices can be chosen in the form

$$\gamma^0 = \tau_1, \quad \gamma^1 = i\tau_2, \tag{14.15}$$

where the τ_i are Pauli matrices. In this case, the matrix[1] γ^5 is conveniently defined by the formula

$$\gamma^5 = -\gamma^0\gamma^1.$$

In the representation (14.15) we have $\gamma^5 = \tau^3$. As before, the P-transformation matrix is given by formula (14.11). The C-conjugation matrix in this representation is equal to

$$C = \tau^3. \tag{14.16}$$

There is no analogue of angular momentum in $(1+1)$ dimensions, since there are no rotations in one-dimensional space.

In three-dimensional space–time, the minimum size of the γ-matrices is 2×2 and the wave functions ψ also comprise two-component columns. The γ-matrices can be chosen in the form

$$\gamma^0 = \tau^3, \quad \gamma^1 = -i\tau^1, \quad \gamma^2 = -i\tau^2. \tag{14.17}$$

We stress that there is no analogue of γ^5 in three-dimensional space–time.

Problem 5. *Show that there does not exist a 2×2 matrix which is anticommutative with all three matrices of (14.17).*

The C-conjugation matrix can be defined as before; in the representation (14.17) it is equal to
$$C = \tau^1. \tag{14.18}$$

Problem 6. *Show that in two-dimensional space–time the γ-matrices (14.15) satisfy equation (14.2), and the matrix (14.16) satisfies equation (14.12).*

Problem 7. *The same in three-dimensional space–time for the matrices (14.17) and (14.18).*

Let us now discuss the special case of *massless* fermions. Let us consider the situation in four-dimensional space–time. For $m = 0$ the Dirac Hamiltonian (14.8) commutes with the matrix γ^5 (in other words, the operator $i\gamma^\mu\partial_\mu$ is anticommutative with γ^5). Thus, the solutions of the Dirac equation divide into left-handed solutions, for which

$$\gamma^5\psi_L = \psi_L,$$

i.e.

$$\frac{1+\gamma^5}{2}\psi_L = \psi_L, \quad \frac{1-\gamma^5}{2}\psi_L = 0, \tag{14.19}$$

---

[1] As before, for the matrix which anticommutes with all γ^μ, we use the notation γ^5, although it is the third (not the fifth) in the set of two-dimensional Dirac matrices.

and right-handed ones, satisfying the equations

$$\gamma^5\psi_R = -\psi_R, \quad \frac{1+\gamma^5}{2}\psi_R = 0, \quad \frac{1-\gamma^5}{2}\psi_R = \psi_R. \tag{14.20}$$

In our chosen representation of γ-matrices we have

$$\psi_L = \begin{pmatrix} \chi \\ 0 \end{pmatrix}; \quad \psi_R = \begin{pmatrix} 0 \\ \eta \end{pmatrix}, \tag{14.21}$$

where χ and η are two-component columns. We note that often for an arbitrary wave function, one introduces its left- and right-handed components with the structure (14.21), in other words, one defines

$$\psi_L = \frac{1+\gamma^5}{2}\psi; \quad \psi_R = \frac{1-\gamma^5}{2}\psi.$$

The operators $\frac{1+\gamma^5}{2}$ and $\frac{1-\gamma^5}{2}$ are projections onto the left- and right-handed components of the wave function of general form. The right- and left-handed components of the wave function transform independently under Lorentz transformations.

For $m = 0$, no contradiction arises in the case when one assumes that physically only left-handed solutions are realized, while right-handed solutions are ignored (or conversely). In other words, one can consider only two-component spinors χ and write down the following equation for them:

$$i\tilde{\sigma}^\mu\partial_\mu\chi = 0. \tag{14.22}$$

The two-component spinors χ are said to be Weyl spinors, and equation (14.22) is the Weyl equation. We shall examine the interpretation of the Weyl equation and its solutions in the next section. Here, we recall only that the Weyl equation, *instead of* the Dirac equation provides an adequate description of massless particles with left helicity.

Weyl's equation is not invariant under C-conjugation. In the language of four-component fermions, C-conjugation takes a left-handed wave function to a right-handed one: if $\psi = \psi_L$ satisfies equation (14.19), the $C\psi^*$ satisfies equation (14.20). In the representation of γ-matrices used, this is evident from formula (14.14) for the matrix C. Weyl's equation is invariant, however, under CP-transformation, which is the subject of the next problem.

Problem 8. *Suppose the left-handed two-component spinor $\chi(x^0, \mathbf{x})$ satisfies the Weyl equation (14.22). Find a 2×2 unitary matrix U_{CP} such that the two-component spinor $U_{CP}\chi^*(x^0, -\mathbf{x})$ also satisfies the Weyl equation (14.22). This transformation is clearly a combination of a C-conjugation and spatial reflection P.*

An analogue of the Weyl equation also exists in the theory of massless fermions in two-dimensional space–time.

Problem 9. *Find the analogue of the Weyl equation in two-dimensional space–time.*

Instead of the left-handed massless fermions, it would be possible to consider right-handed fermions, satisfying the right-handed Weyl equation

$$i\sigma^\mu \partial_\mu \eta = 0.$$

Whether any particular massless fermion is left- or right-handed is an experimental question. It is known that neutrinos (if they are indeed massless) are left-handed fermions.

To end this section, we note that the Dirac equation and the Weyl equation have the property of probability conservation. For example, for the Dirac fermion, the integral

$$\int d^3x \psi^\dagger(x^0, \mathbf{x}) \psi(x^0, \mathbf{x}) \tag{14.23}$$

is conserved. This is evident from the Hermiticity of the Dirac Hamiltonian (14.8). This property can be formulated in a different manner. Let us introduce the Dirac-conjugate spinor (row of four elements)

$$\bar{\psi} = \psi^\dagger \gamma^0. \tag{14.24}$$

Taking into account that γ^0 is Hermitian $(\gamma^0)^2 = 1$, and γ^i are anti-Hermitian, we obtain from the Dirac equation that $\bar{\psi}$ satisfies the equation

$$\bar{\psi}(i\gamma^\mu \overleftarrow{\partial_\mu} + m)\psi = 0,$$

where the arrow over ∂_μ denotes that the derivative acts on $\bar{\psi}$. From this equation and the Dirac equation it follows that the current

$$j^\mu = \bar{\psi}\gamma^\mu \psi$$

is conserved:

$$\partial_\mu j^\mu = 0. \tag{14.25}$$

Taking into account that

$$\int d^3x \psi^\dagger \psi = \int d^3x \bar{\psi}\gamma^0 \psi = \int d^3x j^0,$$

we obtain that the conservation of the probability (14.23) is directly related to equation (14.25). The same argument applies to left-handed (or right-handed) fermions, obeying the Weyl equation.

14.2   Solutions of the free Dirac equation. Dirac sea

Let us discuss solutions of the Dirac equation (14.1), first for $m \neq 0$. In other words, we are interested in the spectrum of the Dirac Hamiltonian (14.8) and its eigenfunctions. We shall seek eigenfunctions in the form

$$\psi_{\mathbf{p}}(x) = e^{-i\omega x^0 + i\mathbf{p}\mathbf{x}} u_{\mathbf{p}}, \tag{14.26}$$

where ω and $\mathbf{p}$ are interpreted, as usual in quantum mechanics, as the particle energy and momentum. For $u_{\mathbf{p}}$ we have the equation

$$(\alpha^i p_i + \beta m) u_{\mathbf{p}} = \omega u_{\mathbf{p}}. \tag{14.27}$$

In the representation of γ-matrices used, where the matrices β and α^i have the form (14.5), it is convenient to rewrite equation (14.27), by introducing left- and right-handed components of the spinor $u_{\mathbf{p}}$,

$$u_{\mathbf{p}} = \begin{pmatrix} u_{\mathbf{p},L} \\ u_{\mathbf{p},R} \end{pmatrix}. \tag{14.28}$$

$u_{\mathbf{p},L}$ and $u_{\mathbf{p},R}$ are two-component spinors. From (14.27) it follows that they satisfy the system of equations (we shall omit the subscript $\mathbf{p}$ in the wave functions)

$$\begin{aligned} (-\boldsymbol{\sigma}\mathbf{p} - \omega)u_L + m u_R &= 0 \\ (\boldsymbol{\sigma}\mathbf{p} - \omega)u_R + m u_L &= 0. \end{aligned} \tag{14.29}$$

The condition for consistency of this system, in other words, the requirement $\det(\alpha\mathbf{p} + \beta m - \omega) = 0$ has the form

$$p^2 \equiv \omega^2 - \mathbf{p}^2 = m^2 \tag{14.30}$$

(this equation, however, is evident from the considerations of the beginning of Section 14.1). When this equation is satisfied we have that u_L is arbitrary and u_R is expressed in terms of u_L,

$$u_R = \frac{\boldsymbol{\sigma}\mathbf{p} + \omega}{m} u_L. \tag{14.31}$$

Since there are two linearly independent two-component columns u_L, there exist two independent solutions of the Dirac equation with fixed momentum $\mathbf{p}$ and energy ω, subject to relation (14.30). The physical meaning of this fact is most apparent when one chooses the two linearly independent columns u_L as the eigenvectors of the operator $\frac{\boldsymbol{\sigma}\mathbf{p}}{|\mathbf{p}|}$. This operator has eigenvalues $+1$ and -1 (since its square is equal to 1), and two linearly independent columns $u_{L,\pm}$ can be chosen such that

$$\frac{\boldsymbol{\sigma}\mathbf{p}}{|\mathbf{p}|}u_{L,\pm} = \pm u_{L,\pm}.$$

Taking into account (14.31), we obtain for the whole of the four-component column (14.28)

$$\frac{\mathbf{s}\mathbf{p}}{|\mathbf{p}|}u_{\pm} = \pm\frac{1}{2}u_{\pm},$$

where $\mathbf{s}$ is the spin matrix (14.9). Thus, the two linearly independent solutions of the Dirac equation with fixed momentum correspond (for $\omega > 0$) to states with positive and negative projections of the spin on the direction of motion of the particle, the helicities.

The energy values ω may be either positive $\omega = +\sqrt{\mathbf{p}^2 + m^2}$, or negative $\omega = -\sqrt{\mathbf{p}^2 + m^2}$. In other words, the spectrum of the Dirac Hamiltonian is unbounded from below. Dirac proposed an elegant method for handling this difficulty, which made the theory consistent and led to the concept of the antifermion. Let us consider a system in a box of large but finite size L. On the boundary of this box we impose periodicity conditions, i.e. we require

$$\psi\left(x^1 = -\frac{L}{2}, x^2, x^3, x^0\right) = \psi\left(x^1 = +\frac{L}{2}, x^2, x^3, x^0\right), \tag{14.32}$$

and analogously for other spatial coordinates. Then, the spectrum of the Dirac Hamiltonian will be discrete; the wave functions (14.26) satisfy the condition (14.32) only if

$$\mathbf{p} = \mathbf{p}_n = \frac{2\pi}{L}\mathbf{n},$$

$$\mathbf{n} = (n_1, n_2, n_3), \quad n_{1,2,3} = 0, \pm 1, \pm 2, \ldots.$$

Correspondingly, the energy takes on a discrete set of values

$$\omega_{\mathbf{n}} = \pm\sqrt{\mathbf{p}_n^2 + m^2}.$$

Between levels with positive energy and levels with negative energy, there is a gap of width $2m$. The spectrum of the Dirac Hamiltonian in a finite volume is shown schematically in Figure 14.1.

Let us next raise the question of the lowest state of the system (the vacuum), without, however, imposing any constraints on the possible number of particles in the system. Clearly, it is favorable to have the largest

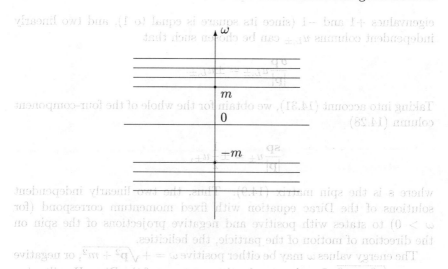

Figure 14.1. Schematic plot of the spectrum of the Dirac Hamiltonian. Density and degeneracy of the levels are not accounted for.

number of particles at negative levels. At the same time, the particles are fermions, thus at most one particle can be placed in any given state (Pauli principle). Consequently, the lowest energy state of the system will be such that all negative levels are occupied (i.e. in each state with negative energy there is exactly one fermion), and all positive levels will be empty. This ground state is shown schematically in Figure 14.2. The system of occupied negative energy levels is descriptively called a Dirac sea.

We must further assume that the Dirac vacuum has zero energy and zero momentum (and also zero electric and other charges), i.e. reference these quantities from their vacuum values.

Let us now discuss elementary excitations over the Dirac vacuum. When a fermion is added to the system, it can occupy a positive energy level only (it is not possible to place the fermion at any level with negative energy because of the Pauli principle). The wave function of this fermion will be described by the solution of the Dirac equation for positive ω. Such an excitation is shown in Figure 14.3.

Another type of excitation occurs when one *removes* a particle from the Dirac sea. In this case, a level with negative energy becomes empty, so that the total energy again increases: if one removes a fermion from an energy level characterized by momentum $\mathbf{p}$ and energy $-\omega_{\mathbf{p}} = -\sqrt{\mathbf{p}^2 + m^2}$, the energy of the state, referenced from the vacuum energy, will be equal to $+\omega_{\mathbf{p}}$, and the momentum will be equal to $(-\mathbf{p})$. Such an excitation in solid-state theory is called a hole, and in particle physics it corresponds to an

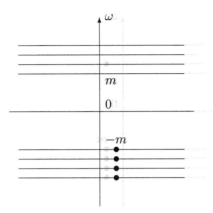

Figure 14.2. Schematic plot of the ground state. Degeneracy of the levels is not accounted for.

antifermion. If the fermion carries an electric charge q, then the antifermion has charge $(-q)$, since its state is obtained by removing a fermion from the Dirac sea. Fermions and antifermions have the same mass and dispersion law $\omega_{\mathbf{p}} = \sqrt{\mathbf{p}^2 + m^2}$. A state with an antifermion is shown in Figure 14.4.

If the system contains a fermion and a hole, as in Figure 14.5, then the fermion may move from a positive level to a negative one, emitting energy (for example, producing photons). This corresponds to the fermion–anti-fermion annihilation. The inverse process is possible with the addition of energy, and corresponds to the creation of a fermion–antifermion pair. In both these processes the fermion number, defined as the *difference* between the number of fermions and the number of antifermions, is conserved.

Of course, for each type of fermion, there is a corresponding type of antifermion: for the electron, it is the positron; for the proton, the antiproton, etc.

The notion of particles and holes is completely adequate in solid-state physics. In particle physics it is formal and not really adequate. A more symmetric description of fermions and antifermions is achieved in quantum field theory where a unique field operator $\Psi(x)$ describes particles and antiparticles at the same time. The fermion field does not have a classical limit (because of the Pauli principle); for precisely this reason we do not discuss it in the main text of this book. We stress that the description of free fermions and antifermions (and also of fermions and antifermions in background fields) in terms of the Dirac sea is equivalent to their description

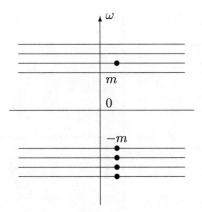

Figure 14.3. State with one fermion over the Dirac vacuum.

in quantum field theory.

One further remark relates to bosons. If the Klein–Gordon equation is interpreted as an equation for a wave function of a particle (boson), then there is also a difficulty with negative levels and the unboundedness of the energy spectrum from below. Since bosons are not subject to the Pauli principle, this difficulty cannot be resolved using a construction similar to the Dirac sea. The difficulty with the negative energies of bosons is only resolved in quantum field theory.

Let us now consider the solutions of the Weyl equation (14.22). In the momentum representation this equation has the form (compare with (14.29))

$$(\omega + \sigma \mathbf{p}) u_L = 0.$$

This system of equations for the two components of the spinor ψ_L is self-consistent and has a unique solution if

$$\omega = \pm |\mathbf{p}|.$$

For $\omega > 0$ this solution satisfies the relation

$$\frac{\mathbf{s}\mathbf{p}}{|\mathbf{p}|} u_L = -\frac{1}{2} u_L,$$

where $\mathbf{s} = \frac{1}{2}\boldsymbol{\sigma}$ is the spin operator. Thus, the Weyl equation describes massless fermions with negative helicity (projection of the spin onto the

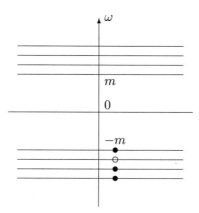

Figure 14.4. State with antifermion over the Dirac vacuum. The open circle denotes an unoccupied level in the Dirac sea.

direction of motion). This explains the term "left-handed fermions" (by analogy with photons or electromagnetic waves with a left-circular polarization); the term "fermions with a left chirality" is also used.

Neither is it difficult to find the solutions of the free Dirac equation in space–time of dimension two or three. In all cases, there exist solutions with both positive and negative energies, where the spectra of the positive and negative energies coincide up to sign. This last statement is evident from the property of C-symmetry: if $\psi(x^0, \mathbf{x})$ satisfies the Dirac equation with positive frequency ω, then

$$\psi^C(x^0, \mathbf{x}) = C\psi^*(x^0, \mathbf{x})$$

is a solution of the same equation with frequency $(-\omega)$.

Problem 10. *Find all the eigenfunctions of the Dirac Hamiltonian in space–time with $d = 2$ and $d = 3$.*

Problem 11. *Find all solutions of the "Weyl equation" (see Problem 9) in two-dimensional space–time. In which directions do the particles described by these wave functions and the antiparticles (holes in the Dirac sea) move?*

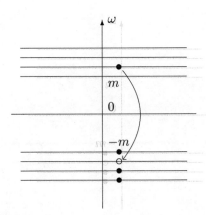

Figure 14.5. Annihilation of a fermion with an antifermion.

14.3 Fermions in background bosonic fields

In the following chapters we shall consider a number of interesting effects which arise when fermions interact with background bosonic fields of the soliton and instanton type. Thus, here, we shall consider what the Dirac equation looks like in background bosonic fields.

First, we recall that fermions in an external electromagnetic field are described by a Dirac equation analogous to (14.1) (or (14.3)), but with the conventional derivative replaced by the covariant derivative[2]

$$i\gamma^{\mu} D_{\mu}\psi - m\psi = 0, \qquad (14.33)$$

where, as usual,

$$D_{\mu} = \partial_{\mu} - ieA_{\mu},$$

is a solution of the same equation with frequency $(-\omega)$.

and e is the fermion electric charge. We note that the left-hand side of this equation transforms covariantly under gauge transformations of the electromagnetic group $U(1)$, i.e. under transformations

$$\psi(x) \to \psi'(x) = e^{i\alpha(x)}\psi(x), \qquad (14.34)$$

$$A_{\mu}(x) \to A'_{\mu}(x) = A_{\mu}(x) + \frac{1}{e}\partial_{\mu}\alpha(x). \qquad (14.35)$$

---

[2]Generally speaking, this equation may also contain terms associated with an anomalous fermion magnetic moment, a fermion electric dipole moment, etc.

Indeed, under these transformations we have

$$(i\gamma^\mu D_\mu \psi - m\psi) \rightarrow e^{i\alpha}(i\gamma^\mu D_\mu \psi - m\psi).$$

If ψ satisfies the Dirac equation in the external field A_μ, then ψ' satisfies the Dirac equation in the external field A'_μ. In other words, under a gauge transformation of the external field, the set of solutions of the Dirac equation remains intact, modulo a phase transformation (14.34). We shall say that the Dirac equation (14.33) is invariant under the gauge transformations (14.34), (14.35).

Non-Abelian global or gauge symmetries appear in the Dirac equation if fermions can be assigned internal quantum numbers. Suppose, for example, that the system (still without external fields) has two types of fermions. Then a state of the system in which there is just one fermion of any type can be described by a wave function in the form of a column

$$\psi = \begin{pmatrix} \psi_1(x) \\ \psi_2(x) \end{pmatrix}, \tag{14.36}$$

each of whose components, ψ_1 or ψ_2, is a Dirac spinor (for massless fermions these may be Weyl spinors). Thus, a fermion of the first type is described by a column

$$\begin{pmatrix} \psi_1(x) \\ 0 \end{pmatrix}, \tag{14.37}$$

and a fermion of the second type is described by a column

$$\begin{pmatrix} 0 \\ \psi_2(x) \end{pmatrix}.$$

In the general case of a column (14.36), the quantity $(\psi_1^\dagger \psi_1)(\mathbf{x})$ represents the probability that the system contains a fermion of the first type at the point $\mathbf{x}$, and, analogously for $(\psi_2^\dagger \psi_2)(\mathbf{x})$.

If the masses of the fermions are the same, then in the absence of external fields, each component of the column (14.36) satisfies the Dirac equation (14.1). These two equations can be combined by writing for the column (14.36) the equation

$$(i\gamma^\mu \partial_\mu - m) \cdot \mathbf{1} \cdot \psi = 0, \tag{14.38}$$

where $\mathbf{1}$ is the unit 2×2 matrix acting on the space of two-component columns (14.36) (of course, we shall not usually write it down explicitly, when that does not lead to confusion). Equation (14.38) is invariant under global (independent of the coordinates) transformations from the group $SU(2)$,

$$\psi(x) \rightarrow \omega\psi(x), \quad \omega \in SU(2) \tag{14.39}$$

(it is also invariant under the group $U(1)$ of global phase rotations $\psi \to e^{i\alpha}\psi$, but that property is not important to us here). We stress that the transformation (14.39) does not act on the Lorentz index, but on the "internal" indices, enumerating the fermion types. By (formal) analogy with the conventional spin, transformations (14.39) are called "internal spin" (or isospin) transformations. Here, one could say that the column (14.36) describes a fermion which can be found in different isospin states: for example, a fermion of the first type described by the column (14.37) can be said to be in a state with the third isospin projection $(+1/2)$, while a fermion of the second type is in a state with the third isospin projection $(-1/2)$.

Historically, the first example of internal symmetry was the isotopic symmetry of strong interactions. If one neglects the small difference in mass between the proton and the neutron, and also the electromagnetic and weak interactions, then the proton and neutron have the same properties. The wave function of the nucleon (proton or neutron) can then be written in the form (14.36), where $\psi_1(x)$ and $\psi_2(x)$ are the wave functions of the proton and neutron components of the nucleon; the free Dirac equation for the nucleon has exactly the form (14.38). The term "isospin" (isotopic spin) has precisely that origin; subsequently it was used for other symmetries described by the group $SU(2)$ (for example, we speak of weak isospin, which is related to the weak interaction, see Sections 6.3 and 14.4).

The construction studied can be generalized to the case when fermions are found in three or more states (in other words, there are three or more fermion types with the same mass). Thus, quarks may occur in three states (they are called colors); accordingly, a quark is described by a wave function

$$\begin{pmatrix} \psi_1 \\ \psi_2 \\ \psi_3 \end{pmatrix} \qquad (14.40)$$

and the Dirac equation for the quark wave function is invariant under the action of the internal (color) group $SU(3)_C$. Further generalization issues from the remark that the columns (14.36) or (14.40) can be interpreted as vectors in the space of the fundamental representation of the group $SU(2)$ or $SU(3)_C$, respectively. There is no need to limit oneself to the groups $SU(N)$ and the fundamental representations. Namely, one can choose an arbitrary compact group G of internal symmetries and an arbitrary representation of it $T(G)$, and require that the fermionic wave function belongs to the space of the representation $T(G)$, and that the Dirac equation be invariant under transformations

$$\psi \to T(g)\psi,$$

for all $g \in G$. We again stress that the matrix $T(g)$ acts on the internal and not the Lorentz indices of the fermion wave function.

Problem 12. *Let G be a simple global symmetry group, and $T(G)$ its complex n-dimensional irreducible representation. Write down the most general form of the free Dirac equation for fermions in the representation $T(G)$, which is invariant under the action of the group G. Show that it is actually invariant under the action of the group $SU(n)$. (This property holds, generally speaking, only for free fermions).*

The next step is to construct the interaction of fermions with non-Abelian gauge fields. In complete analogy with Chapter 4, this interaction must ensure invariance of the Dirac equation under gauge transformations with a non-Abelian gauge (internal) symmetry group. In the example with the internal group $SU(2)$ and a fermion doublet (14.36), the Dirac equation is required to be invariant under conventional gauge transformations of the gauge field of the group $SU(2)$,

$$A_\mu \to A'_\mu = \omega A_\mu \omega^{-1} + \omega \partial_\mu \omega^{-1}, \quad \omega(x) \in SU(2),$$

(where $A_\mu = -ig\frac{\tau^a}{2}A^a_\mu$), performed simultaneously with the transformations

$$\psi(x) \to \psi'(x) = \omega(x)\psi(x).$$

We already know the prescription for constructing the covariant quantities: we need to replace the conventional derivative by the covariant derivative,

$$D_\mu = \partial_\mu + A_\mu.$$

Thus, we come to the Dirac equation, which is invariant under gauge transformations from the group $SU(2)$,

$$(i\gamma^\mu D_\mu - m)\psi = 0, \tag{14.41}$$

which is completely analogous to the Dirac equation (14.33) in the presence of an electromagnetic field. We note that in equation (14.41), and in what follows, we use very compressed notation; if we write down this equation, preserving all the indices, we have

$$\left\{ i\gamma^\mu_{\alpha\beta} \left[\delta_{ij}\partial_\mu - i\frac{g}{2}(\tau^a)_{ij}A^a_\mu \right] - m\delta_{\alpha\beta}\delta_{ij} \right\} \psi_{\beta j} = 0, \tag{14.42}$$

where $\alpha, \beta = 1, \ldots, 4$ are the Lorentz indices of the spinor; $i, j = 1, 2$ are the isotopic indices; $a = 1, 2, 3$; and summation over repeated indices is understood.

In the general case of an arbitrary gauge group G and representation $T(G)$, under which fermions are transformed, the Dirac equation in the external gauge field has the previous form (14.41), where the covariant derivative is equal to

$$D_\mu = \partial_\mu + T(A_\mu). \tag{14.43}$$

Here, we have a complete analogy with Chapter 4.

For $m = 0$, as the original equation we could take not the Dirac equation but the Weyl equation. Repeating the above steps, we would then obtain the Weyl equation in the external gauge field. For a left-handed fermion this has the form

$$i\bar{\sigma}^\mu D_\mu \chi = 0,$$

i.e. it differs from the free Weyl equation (14.22) by the replacement of the conventional derivative by the covariant derivative (14.43). The Lorentz index of the spinor χ, as before, takes values $1, 2$.

Let us now turn to a discussion of the interaction of fermions with scalar fields. Let us begin with the simplest case of a single real scalar field $\varphi(x)$ and a single fermion. The main requirement on the Dirac equation for the fermion wave function in the external classical field φ is the Lorentz covariance and the Hermiticity of the Dirac Hamiltonian. Under Lorentz transformations, the scalar field at a fixed point of space–time is not transformed, therefore it can be included in the Dirac equation by analogy with the fermion mass term. Thus, we arrive at the equation

$$i\gamma^\mu \partial_\mu \psi - m\psi - h\varphi\psi = 0, \tag{14.44}$$

where h is a real coupling constant. This simplest coupling of the fermion with the scalar field is called Yukawa coupling. We stress that literally in the form (14.44), Yukawa coupling can be introduced for Dirac fermions but not for Weyl fermions.

Problem 13. *Find the dimension of the constant h in four-dimensional, and also in d-dimensional space–time.*

Problem 14. *Show that the Dirac Hamiltonian, corresponding to equation (14.44), in a real background field $\varphi(x)$, is Hermitian if and only if the constant h is real.*

It is convenient to construct a further generalization using the concept of action for fermions. To introduce this, we begin by discussing the Dirac equation (14.44). This can be viewed as the extremality condition for a functional,

$$S_F = \int d^4 x \left[\bar{\psi}(x) i\gamma^\mu \partial_\mu \psi(x) - m\bar{\psi}(x)\psi(x) - h\varphi(x)\bar{\psi}\psi(x) \right], \tag{14.45}$$

with respect to arbitrary variations of the variable $\bar{\psi}(x)$, which is a four-component row. Such a functional is called an action for fermions. The functional S_F is real, if $\bar{\psi}$ is taken to be the row[3] $\psi^\dagger \gamma^0$ (see also (14.24)). Indeed, for S_F^* we have

$$S_F^* = \int d^4x \left(-i\partial_\mu \psi^\dagger \gamma^{\mu\dagger}\gamma^0 \psi - m\psi^\dagger \gamma^0 \psi - h\varphi\psi^\dagger \gamma^0 \psi \right).$$

The equality $S_F^* = S_F$ is obtained after integration by parts in the first term and using the identity $\gamma^{\mu\dagger}\gamma^0 = \gamma^0 \gamma^\mu$. One can further show that S_F is a Lorentz scalar. The requirements that the action should be real and Lorentz invariant are equivalent to the requirements that the Dirac Hamiltonian should be Hermitian and that the Dirac equation should be Lorentz covariant.

The fermion action is a natural and necessary object in quantum field theory, where $\bar{\psi}(x)$ and $\psi(x)$ are treated as fermion field operators. There, S_F is placed on an equal footing with the boson field action. We have already mentioned that, unlike boson field operators, the operators $\psi(x)$ and $\bar{\psi}(x)$ do not have a classical limit. For our purposes, S_F will serve as an auxiliary object, which will enable us to obtain Dirac equations having the properties of Lorentz and gauge covariance, and for which the quantum-mechanical Dirac Hamiltonians are Hermitian.

Analogously, the Dirac equation (14.41), in the presence of gauge fields, can be obtained from variation of the action

$$S_F = \int d^4x \left[\bar{\psi}(x)i\gamma^\mu D_\mu \psi(x) - m\bar{\psi}(x)\psi(x) \right] \tag{14.46}$$

with respect to $\bar{\psi}$. Here, if ψ transforms according to a unitary representation $T(G)$ of the gauge group G, then $\bar{\psi} = \psi^\dagger \gamma^0$ transforms according to the conjugate representation: for gauge transformations $\omega(x) \in G$ we have

$$\psi(x) \rightarrow \psi'(x) = T(\omega(x))\psi(x)$$
$$\bar{\psi}(x) \rightarrow \bar{\psi}'(x) = \bar{\psi}(x)T^\dagger(\omega(x)),$$

or, in components,

$$\psi_i \rightarrow \psi_i' = [T(\omega)]_{ij}\psi_j$$
$$\bar{\psi}_i \rightarrow \bar{\psi}_i' = \bar{\psi}_j[T(\omega)]_{ij}^*, \tag{14.47}$$

where i, j are indices, corresponding to the internal symmetry. The functional S_F is clearly invariant under gauge transformations.

---

[3] The fact that, when taking variations of S_F, we need to consider $\bar{\psi}$ as an independent variable is analogous to the prescription for the variation of the action for a complex scalar field, where φ and φ^* are considered as independent variables; see Section 2.4.

One generalization of expressions (14.45), (14.46) for the fermion action to the case in which the scalar field transforms non-trivially under gauge (or global) transformations, is quite evident. Suppose ψ_i transforms under the representation $T(G)$ of the gauge group G, i.e. the transformation law for ψ and $\bar\psi_i$ has the form (14.47). Suppose the scalar field transforms under the real representation $T_S(G)$, i.e. the fields φ_a are real, and under gauge transformations we have

$$\varphi_a \to \varphi'_a = [T_S(\omega)]^b_a \varphi_b.$$

Suppose further that from $\bar\psi, \psi$ and φ one can form a real gauge-invariant combination of the form

$$\bar\psi_i(\Lambda^a)_{ij}\psi_j\varphi_a = \bar\psi\Lambda^a\psi\varphi_a, \qquad (14.48)$$

where $(\Lambda^a)_{ij}$ are some coefficients. Then a fermion action of the form

$$S_F = \int d^4x \left[\bar\psi i\gamma^\mu D_\mu\psi - m\bar\psi\psi - h\bar\psi\Lambda^a\psi\varphi_a \right] \qquad (14.49)$$

will be Lorentz invariant, gauge invariant and real for a real coupling constant h (in the case of global symmetry, instead of the covariant derivative D_μ, (14.49) will contain the conventional derivative ∂_μ). The requirement for invariance of the expression (14.48) is equivalent to the requirement that for all $\omega \in G$, the following holds:

$$\Lambda^a[T_S(\omega)]^b_a T(\omega) = T(\omega)\Lambda^b. \qquad (14.50)$$

Here, we have omitted the indices i, j relating to the fermion representation $T(G)$ and we consider Λ^a and $T(\omega)$ as matrices with these indices. Equation (14.50) determines the possible form of the matrices Λ^a. In other words, $\varphi_a\psi_j$ belongs to the representation $T_S \times T$ of the Lie group G; we require that $T_S \times T$ contains the representation T and that Λ operates as the projection $T_S \times T \to T$.

In terms of representations of the *Lie algebra* of the group G, equation (14.50) has the form

$$\Lambda^a(T_S^q)^b_a = [T^q, \Lambda^b], \qquad (14.51)$$

where T^a and T_S^q are generators of the Lie algebra in the representations T and T_S, and $q = 1, \ldots, \dim G$.

By varying the action (14.49) with respect to $\bar\psi$, we obtain the Dirac equation

$$i\gamma^\mu D_\mu\psi - m\psi - h\Lambda^a\varphi_a\psi = 0. \qquad (14.52)$$

Its left-hand side transforms covariantly (according to the representation $T(G)$) under gauge transformations.

Problem 15.   *Show that equation (14.50) is necessary and sufficient for the left-hand side of equation (14.52) to transform according to the representation $T(G)$.*

Problem 16.   *Consider infinitesimal transformations*

$$\omega = 1 + \varepsilon^q t^q, \qquad (14.53)$$

where t^q are generators of the Lie algebra of the group G and ε^q are infinitesimal real parameters. Show that for such transformations equation (14.50) is equivalent to (14.51), taking into account the definitions $T(t^q) = T^q$, $T_S(t^q) = T_S^q$.

Problem 17.   *Write down the Dirac Hamiltonian corresponding to equation (14.52). Show, under the conditions formulated earlier, that it is Hermitian.*

In fact, essentially the same construction was used (in the special case of the fundamental and adjoint representations of the group $SU(2)$) in the second example of Section 4.1 to construct the Lagrangian of a cubic interaction of scalar fields.

As an important example for what is to come, let us consider the group $SU(N)$ as a symmetry group, let us choose the fermions in the fundamental representation of this group and the scalars in the adjoint representation (this means, in particular, that φ_a are real fields, $a = 1, \ldots, N^2 - 1$). Then $T^q = it^q$, where t^q are Hermitian matrices of the generators of the group $SU(N)$ (unlike the anti-Hermitian generators entering (14.53)), and $(T_S^q)_a^b = f^{qab}$ are structure constants. For the matrix Λ^a, we can choose t^a. Indeed, the left-hand side of equation (14.51) in this case has the form $t^a f^{qab}$, while the right-hand side is equal to

$$[it^q, t^b] = -f^{qbc} t^c,$$

so that equation (14.51) is satisfied, taking into account the antisymmetry of the structure constants. Thus, the covariant Dirac equation in this example has the form

$$(i\gamma^\mu D_\mu - m - h t^a \varphi_a)\psi = 0. \qquad (14.54)$$

Let us briefly discuss how fermions behave in background fields corresponding to the classical *ground state* of the scalar and gauge fields. To be specific, let us consider the case of the group $SU(2)$ (global or gauge), of the scalar field in the adjoint (real triplet) representation and of fermions in the fundamental (complex doublet) representation. We recall that a scalar field in the ground state is constant in space–time, and in

gauge theory, at the same time, $A_\mu = 0$. In this background, the Dirac
equation with the Yukawa coupling has the form (14.54), where $t^a = \tau^a/2$
are Hermitian generators of the group $SU(2)$. The cases of unbroken and
broken symmetry need to be considered separately. If the scalar potential
$V(\varphi)$ is such that in the ground state $\varphi_a = 0$ and the symmetry is not
broken, the equation (14.54) reduces to two Dirac equations with mass m
for the two components ψ_i, where $i = 1, 2$ is the group index. Thus, the
system has two types of fermions with the same mass, as we observed at
the beginning of this section. If in the ground state $\varphi_a \neq 0$, i.e. the $SU(2)$
symmetry is broken, the situation becomes more interesting. Without loss
of generality, we can choose the ground state of the scalar field such that
only the third component is non-zero, $\varphi_a = v\delta_{a3}$, where the real parameter
v is determined by the form of the scalar potential. Equation (14.47) in
this field has the form

$$\left(i\gamma^\mu \partial_\mu - m - \frac{hv}{2}\tau^3 \right) \psi = 0, \qquad\qquad (14.55)$$

where, as before,

$$\psi = \begin{pmatrix} \psi_1 \\ \psi_2 \end{pmatrix}$$

is an isospinor in the internal space. Equation (14.55) is equivalent to the
two equations

$$\left(i\gamma^\mu \partial_\mu - m - \frac{hv}{2} \right) \psi_1 = 0$$

$$\left(i\gamma^\mu \partial_\mu - m + \frac{hv}{2} \right) \psi_2 = 0.$$

Hence it is clear that the components ψ_1 and ψ_2 describe fermions with
masses

$$m_1 = \left| m + \frac{hv}{2} \right|$$

and

$$m_2 = \left| m - \frac{hv}{2} \right|,$$

respectively. As a result of the spontaneous symmetry breaking, the
fermions have acquired *different* masses; the fermion spectrum has ceased
to be $SU(2)$ invariant.

Problem 18. *Consider the model with gauge group $SU(3)$, fermions in the fundamental representation and a scalar field in the adjoint (real octet) representation. Suppose that the scalar potential is such that in the ground state only the eighth component of the scalar field is non-zero, $\varphi_a = v\delta_{a8}$ (modulo $SU(3)$ transformations). Find the unbroken subgroup of $SU(3)$. Find the mass spectrum of fermions satisfying the Dirac equation (14.54), explain its structure.*

A less trivial generalization of the fermion action (14.45) and of the corresponding Dirac equation can be obtained if one writes the action in terms of the left- and right-handed components (which are two-component columns) χ and η, such that

$$\psi = \begin{pmatrix} \chi \\ \eta \end{pmatrix} \tag{14.56}$$

(see (14.21)). Note that

$$\bar{\psi} = \psi^\dagger \gamma^0 = (\eta^\dagger, \chi^\dagger).$$

In the representation of γ-matrices (14.4), we shall have for the action (14.45)

$$S_F = \int d^4x \left[\chi^\dagger i\tilde{\sigma}^\mu \partial_\mu \chi + \eta^\dagger i\sigma^\mu \partial_\mu \eta - m(\chi^\dagger\eta + \eta^\dagger\chi) - h(\varphi\chi^\dagger\eta + \varphi\eta^\dagger\chi) \right].$$
$$\tag{14.57}$$

All the terms in this expression are individually Lorentz invariant (we do not require invariance under spatial reflections). The expression (14.57) can be generalized, by assuming that the left- and right-hand components χ and η transform *differently* under gauge (or global) transformations. Here, as before, we need to require that S_F be real and gauge invariant.

As a first example, let us consider the theory with gauge group $U(1)$ and complex scalar field with charge e. The Yukawa interaction of type $\varphi\eta^\dagger\chi$ will be gauge invariant, if the sum of the charges of χ and $\eta^\dagger$ is equal to $(-e)$. For example, the charge of the left-handed component χ can be chosen equal to $(-\frac{1}{2}e)$, and the charge of the right-handed component η, equal to $(+\frac{1}{2}e)$. Here, the mass term of type $m\eta^\dagger\chi$ will not be gauge invariant and it is not possible to include it in the fermion action. Thus, we arrive at the action

$$S_F = \int d^4x \left[\chi^\dagger i\tilde{\sigma}^\mu D_\mu^{(-)}\chi + \eta^\dagger i\sigma^\mu D_\mu^{(+)}\eta - h(\varphi^*\chi^\dagger\eta + \varphi\eta^\dagger\chi) \right], \tag{14.58}$$

where $D_\mu^{(\pm)} = \partial_\mu \mp i\frac{e}{2}A_\mu$. When the constant h is real it will have all the necessary properties. We obtain the equations for the two-component columns of χ and η, in the usual way, by varying the action of S_F with respect to $\chi^\dagger$ and $\eta^\dagger$, assuming them to be independent of χ and η,

$$
\begin{aligned}
i\bar{\sigma}^\mu D_\mu^{(-)}\chi - h\varphi^*\eta &= 0 \\
i\sigma^\mu D_\mu^{(+)}\eta - h\varphi\chi &= 0.
\end{aligned}
\tag{14.59}
$$

This is a rather non-trivial generalization of the Dirac equation for the fermion wave function ψ. The left-hand sides of equation (14.59) transform covariantly under gauge transformations

$$\varphi \to e^{i\alpha}\varphi, \quad \varphi^* \to e^{-i\alpha}\varphi^*$$

$$\chi \to e^{-i\alpha/2}\chi, \quad \eta \to e^{i\alpha/2}\eta,$$

$$A_\mu \to A_\mu + \frac{1}{e}\partial_\mu\alpha.$$

The gauge covariance property of equation (14.59) follows from the gauge invariance of the action (14.58).

Problem 19. *Construct the quantum-mechanical Hamiltonian corresponding to equations (14.59), i.e. write equations (14.59) in the form*

$$i\frac{\partial\psi}{\partial x^0} = H_D\psi,$$

where ψ is the wave function (14.56). Show that the Hamiltonian H_D is Hermitian for real h.

We note that if the values of the bosonic fields in the ground state are equal to $A_\mu = 0$, $\varphi = v$ (so that the model incorporates the Higgs mechanism), equations (14.59) become the free Dirac equation for the wave function (14.56), $(i\gamma^\mu\partial_\mu - m)\psi = 0$, with fermion mass $m = hv$.

Generalizations of this example include the fermionic sector of the Standard Model, which we shall describe next.

14.4 Fermionic sector of the Standard Model

In Section 6.3, we considered the bosonic sector of the standard electroweak theory. In this section, we shall describe its fermionic sector. We shall use the notation introduced in Section 6.3.

Let us begin with a discussion of leptons, fermions which do not participate in strong interactions. In nature, there are three types of charged lepton, e^-, μ^- and τ^-, and three types of neutrino ν_e, ν_μ and ν_τ, together with the corresponding antiparticles. The masses of all neutrinos, if they have a mass, are very small and in what follows, we shall assume that neutrinos are massless. The neutrino is a fermion with a left chirality. The pairs (ν_e, e^-), (ν_μ, μ^-) and (ν_τ, τ^-) behave in the same way under all interactions (e–μ–τ universality). Thus, it is sufficient to consider one pair of leptons, for example, (ν_e, e^-).

It is a key experimental fact that only *left-handed* components of the electron and the (left-handed) neutrino interact with $W^\pm$-boson fields, while the right-handed component of the electron does not interact with $W^\pm$-bosons. The $W^\pm$-boson fields form part of the gauge fields of the subgroup $SU(2)$ of the gauge group of the electroweak interactions, $SU(2) \times U(1)$. Consequently, the right-handed component of the electron e_R^- must be a singlet under $SU(2)$, and the left-handed components ν_e and e_L^- are naturally combined into an $SU(2)$ doublet,

$$L_L = \begin{pmatrix} \nu_{e,L} \\ e_L^- \end{pmatrix}. \tag{14.60}$$

We find the weak hypercharges of the doublet L_L and the singlet e_R^- using formula (6.35), which relates the third component of the weak isospin, the electric charge and the weak hypercharge. For example, for ν_e we have $Q = 0$, $T^3 = \frac{1}{2}$, thus $Y_{\nu_e} = -\frac{1}{2}$. For e_L^- we have $Q = -1$, $T^3 = -\frac{1}{2}$ and, consequently $Y_{e_L} = -\frac{1}{2}$. As one might expect, the weak hypercharge of the components of the column (14.60) is the same; the column L_L transforms under gauge transformations from the subgroup $U(1)$ as a whole,

$$L_L \to e^{iY_L\alpha} L_L,$$

with

$$Y_L = -\frac{1}{2}.$$

For the right-handed component of the electron we have $T_3 = 0$ and from (6.35) we obtain

$$Y_{e_R^-} = -1.$$

Thus, the terms with covariant derivatives in the electron and neutrino action have the form

$$S_F^k = \int d^4x \left(L_L^\dagger i\bar\sigma^\mu D_\mu L_L + e_R^\dagger i\sigma^\mu D_\mu e_R \right), \tag{14.61}$$

where

$$D_\mu L_L = \left(\partial_\mu - ig\frac{T^a}{2}A_\mu^a + ig'\frac{1}{2}B_\mu \right) L_L, \tag{14.62}$$

$$D_\mu e_R = (\partial_\mu + ig'B_\mu)e_R. \tag{14.63}$$

We recall that e_R, e_L and ν_e are two-component columns, Lorentz (right- and left-handed) spinors.

The action (14.16) describes massless fermions. Thus, we need to introduce another term, responsible for the electron mass. It is not possible to introduce an explicit mass term of the type $m(e_L^\dagger e_R + e_R^\dagger e_L)$, since that is not invariant under gauge transformations. Thus, we introduce an interaction of the Yukawa type between fermions and the Higgs doublet φ and use the fact that in the ground state

$$\varphi = \varphi^{(v)} = \begin{pmatrix} 0 \\ \frac{v}{\sqrt{2}} \end{pmatrix}. \tag{14.64}$$

The Yukawa term in the fermion action must contain the right-handed fermion component (more precisely, the conjugate quantity), $e_R^\dagger$, as well as the left-handed component L_L and the Higgs field (see (14.57)). Taking into account the weak hypercharges of leptons and of the Higgs field (for which $Y_\varphi = \frac{1}{2}$) we can write down a unique gauge-invariant Yukawa term of the form

$$h_e[e_R^\dagger(\varphi^\dagger L_L) + (L_L^\dagger \varphi)e_R]. \tag{14.65}$$

Here, $(\varphi^\dagger L_L)$ denotes convolution with respect to the internal indices of the $SU(2)$ doublets; $(\varphi^\dagger L_L)$ is invariant under the group $SU(2)$ and is a two-component column from the Lorentz point of view: if $\alpha = 1, 2$ and $i = 1, 2$ are the Lorentz and the internal indices, respectively, then

$$e_R^\dagger(\varphi^\dagger L_L) = (e_R^\dagger)_\alpha \varphi_i^* L_{i\alpha} \tag{14.66}$$

(summation over i, α is understood) and analogously for the second term in (14.65). The invariance of (14.65) under the weak hypercharge subgroup $U(1)$ follows from the fact that

$$-Y_{e_R} - Y_\varphi + Y_{L_L} = 0.$$

Finally, the expression (14.65) is real for real h_e.

In the vacuum, the field φ takes the value (14.64) and the expression (14.65) becomes the electron mass term

$$h_e \frac{v}{\sqrt{2}}(e_R^\dagger e_L + e_L^\dagger e_R).$$

Thus, in the Higgs vacuum the electron has mass

$$m_e = h_e \frac{v}{\sqrt{2}}, \tag{14.67}$$

and the neutrino remains massless. We conclude that the action of the electron and the electron neutrino has the form

$$S_{e,\nu_e} = \int d^4x \left\{ L_L^\dagger i\tilde\sigma^\mu D_\mu L_L + e_R^\dagger i\sigma^\mu D_\mu e_R - h_e[e_R^\dagger(\varphi^\dagger L_L) + (L_L^\dagger \varphi)e_R] \right\}. \tag{14.68}$$

If we are interested only in electromagnetic interactions, i.e. we suppose that $W_\mu^\pm = Z_\mu = 0$ and $\varphi = \varphi^{(v)}$, then the non-zero components of the original gauge fields are (see (6.47))

$$A_\mu^3 = A_\mu \sin\theta_W$$
$$B_\mu = A_\mu \cos\theta_W,$$

where A_μ is the electromagnetic vector potential. In the presence of these fields the covariant derivatives (14.62), (14.63) take the form

$$D_\mu L_L = \left(\partial_\mu - ig\frac{\tau^3}{2} \sin\theta_W A_\mu + ig'\frac{1}{2} \cos\theta_W A_\mu \right) L_L$$
$$= (\partial_\mu - ieQ_L A_\mu)L_L$$
$$D_\mu e_R = (\partial_\mu + ig' \cos\theta_W A_\mu)e_R$$
$$= (\partial_\mu - ieQ_R A_\mu),$$

where

$$Q_L = \begin{pmatrix} 0 & 0 \\ 0 & -1 \end{pmatrix}$$
$$Q_R = -1,$$

and we have used the relationships between the constants g, g', e and the angle θ_W, described in Section 6.3. Thus, if there is only an electromagnetic field, then the action (14.68) becomes the action of electrodynamics

$$S_{e,\nu_e} = \int d^4x \left[e_L^\dagger i\tilde\sigma^\mu(\partial_\mu + ieA_\mu)e_L + e_R^\dagger i\sigma^\mu(\partial_\mu + ieA_\mu)e_R + \nu_L^\dagger i\tilde\sigma^\mu \partial_\mu \nu_L \right.$$
$$\left. - m_e(e_R^\dagger e_L + e_L^\dagger e_R) \right]$$
$$= \int d^4x \left[\bar{e}i\gamma^\mu(\partial_\mu + ieA_\mu)e + \nu_L^\dagger i\tilde\sigma^\mu \partial_\mu \nu_L - m_e\bar{e}e \right].$$

The equations which follow from this action include the Dirac equation for a massive electron with charge $(-e)$ in the electromagnetic field A_μ and the free Weyl equation for a massless electron neutrino (which is electrically neutral).

Leptons of two other generations are introduced into the model in complete analogy with the electron and the electron neutrino. They form left-handed doublets

$$\begin{pmatrix} \nu_\mu \\ \mu^- \end{pmatrix}_L, \quad \begin{pmatrix} \mu_\tau \\ \tau^- \end{pmatrix}_L, \quad Y = -\frac{1}{2},$$

and right-handed singlets

$$\mu_R^-, \quad \tau_R^-, \quad Y = -1.$$

The full leptonic action is the sum of the actions for the leptons of the three generations,

$$S_{\text{lept}} = S_{e,\nu_e} + S_{\mu,\nu_\mu} + S_{\tau,\nu_\tau}, \tag{14.69}$$

where each term has the structure (14.68), and only the Yukawa constants h_e, h_μ and h_τ are different. For example, the constant h_μ is related to the muon mass by an equation, analogous to (14.67),

$$m_\mu = h_\mu \frac{v}{\sqrt{2}}.$$

Let us now consider the strongly interacting particles, quarks. In nature, there exist six types of quark: u, d, s, c, b and t. Each of these quarks may be in three different states, so that, in fact, there are three types of u-quark, three types of d-quark, etc. (Bogolyubov *et al.* 1965, Han and Nambu 1965, Miyamoto 1965). These different states are called quark colors, and quarks are said to form a color triplet,

$$u = \begin{pmatrix} u_1 \\ u_2 \\ u_3 \end{pmatrix}, \quad d = \begin{pmatrix} d_1 \\ d_2 \\ d_3 \end{pmatrix}, \quad \text{etc.} \tag{14.70}$$

Strong interactions of quarks are described by a gauge theory with the unbroken color gauge group $SU(3)_C$ of quantum chromodynamics (Fritzsch *et al.* 1973, Gross and Wilczek 1973, Politzer 1973). The columns (14.70) transform according to the fundamental representation of $SU(3)_C$, where the left- and right-handed components of the quark wave functions transform in the same way. Free quarks, or gluons, gauge bosons for the group $SU(3)_C$, do not exist in nature; the strongly interacting particles, hadrons, are colorless objects, whose components are quarks and gluons. Loosely, the proton can be said to consist of two u-quarks and one d-quark, $p = (uud)$; the neutron consists of two d-quarks and one u-quark, $n = (ddu)$. According to this, the u-quark and the d-quark have electric charges

$$Q_u = \frac{2}{3}, \quad Q_d = -\frac{1}{3}. \tag{14.71}$$

In what follows, we shall not write down the color index of quarks, and summation over this index, where necessary, will be implicitly understood.

From the point of view of electroweak interactions, quarks, like leptons, divide into three pairs (called families or generations), namely $(u, d), (c, s)$ and (t, b). The electroweak interactions of quarks of each of the generations are the same,[4] thus, for the time being, we shall limit ourselves to quarks of the first generation, u and d. Their left-handed components form a doublet under the electroweak subgroup $SU(2)$,

$$Q_L = \begin{pmatrix} u_L \\ d_L \end{pmatrix},$$

while the right-handed components are singlets under $SU(2)$. In complete analogy with the leptons, we find from (14.71) and (6.35) the weak hypercharges

$$Y_{Q_L} = \frac{1}{6}, \quad Y_{u_R} = \frac{2}{3}, \quad Y_{d_R} = -\frac{1}{3}.$$

These properties uniquely define the form of the covariant derivatives for the doublet Q_L and the singlets u_R and d_R. Now, all quarks have non-zero mass. The Yukawa term leading to the mass of the d-quark is written, in complete analogy with (14.66), as

$$h_d \left[d_R^\dagger (\varphi^\dagger Q_L) + (Q_L^\dagger \varphi) d_R \right], \tag{14.72}$$

where the mass of the d-quark (in the vacuum (14.64)) is equal to

$$m_d = h_d \frac{v}{\sqrt{2}}.$$

We note that $-Y_{d_R} - Y_\varphi + Y_{Q_L} = 0$, so that the terms (14.72) are invariant under all of the $SU(2) \times U(1)$ group of electroweak interactions. In order to introduce the Yukawa term, providing the mass of the u-quark (in the vacuum (14.64)), we recall that the fundamental representation of the group $SU(2)$ is equivalent to its conjugate. Namely, if φ_i, $i = 1, 2$ transforms according to the fundamental representation of the group $SU(2)$, then $\tilde{\varphi}_i = \varepsilon_{ij} \varphi_j^*$ also transforms according to the fundamental representation of $SU(2)$. Thus, both $(\varphi^\dagger Q_L)$ and $(\tilde{\varphi}^\dagger Q_L)$ are singlets under $SU(2)$.

Problem 20. *Suppose that u_i and v_i are two-component columns transforming according to the fundamental representation of the group $SU(2)$, i.e. the transformation $\omega \in SU(2)$ acts as $u_i \to \omega_{ij} u_j$, $v_i \to \omega_{ij} v_j$.*

---

[4]Modulo Yukawa terms, see below.

Show that $(\tilde{u}^\dagger v) = \tilde{u}_i^* v_i = \varepsilon_{ij} u_j v_i$ *is invariant under the action of the group* $SU(2)$.

In addition to the Yukawa term (14.72), we can also write down another Yukawa term

$$h_u \left[u_R^\dagger (\tilde{\varphi}^\dagger Q_L) + (Q_L^\dagger \tilde{\varphi}) u_R \right]. \tag{14.73}$$

Since $\tilde{\varphi}$ is expressed in terms of φ^*, the weak hypercharge of $\tilde{\varphi}$ is equal to $Y_{\tilde{\varphi}} = -Y_\varphi = -\frac{1}{2}$. Because of this, $-Y_{u_R} - Y_{\tilde{\varphi}} + Y_{Q_L} = 0$ and the term (14.73) is invariant under $SU(2) \times U(1)$. In the vacuum (14.64) we have

$$\tilde{\varphi} = \begin{pmatrix} \frac{v}{\sqrt{2}} \\ 0 \end{pmatrix}$$

(we recall that v is real), and (14.73) becomes the mass term for the u-quark with

$$m_u = h_u \frac{v}{\sqrt{2}}.$$

Summarizing, we write the action for the u- and d-quarks in the form

$$S_{(u,d)} = \int d^4x \left\{ Q_L^\dagger i \tilde{\sigma}^\mu D_\mu Q_L + u_R^\dagger i \sigma^\mu D_\mu u_R + d_R^\dagger i \sigma^\mu D_\mu d_R \tag{14.74} \right.$$

$$\left. -h_d \left[d_R^\dagger (\varphi^\dagger Q_L) + (Q_L^\dagger \varphi) d_R \right] - h_u \left[u_R^\dagger (\tilde{\varphi}^\dagger Q_L) + (Q_L^\dagger \tilde{\varphi}) u_R \right] \right\},$$

where

$$D_\mu Q_L = \left(\partial_\mu - ig \frac{\tau^a}{2} A_\mu^a - ig' \frac{1}{6} B_\mu \right) Q_L$$

$$D_\mu u_R = \left(\partial_\mu - ig' \frac{2}{3} B_\mu \right) u_R$$

$$D_\mu d_R = \left(\partial_\mu + ig' \frac{1}{3} B_\mu \right) d_R.$$

We recall that each type of quark, i.e. each fermion Q_L, u_R, d_R, is a triplet transforming according to the fundamental representation of the color group $SU(3)_C$. Here $Q_L^\dagger, u_R^\dagger$ and $d_R^\dagger$ transform according to the representation of the group $SU(3)_C$ conjugate to the fundamental representation (i.e. they are antitriplets). Convolution with respect to the color index, which is implicit in (14.74), ensures the invariance under $SU(3)_C$ (there, the covariant derivatives include the gauge fields of the group $SU(3)_C$, which we have not written down).

One direct generalization of the action (14.74) to the case of three generations of quarks might be the expression

$$S_{\text{quark}} = S_{(u,d)} + S_{(c,s)} + S_{(t,b)},$$ (14.75)

where each of the terms has the structure (14.74), but with different Yukawa constants.

However, such a trivial generalization fails to take account of one important property, quark mixing (Cabbibo 1963, Kobayashi and Maskawa 1973). The complication relates to the Yukawa terms which are not of the most general form in (14.75). To construct Yukawa terms of general form we introduce a generation index $A = 1, 2, 3$ and combine quarks with the same quantum numbers into triplets $Q_L^A = (Q_L^{(u,d)}, Q_L^{(c,s)}, Q_L^{(t,b)})$, $U_R^A = (u_R, c_R, t_R)$, $D_R^A = (d_R, s_R, b_R)$. The most general form of real gauge-invariant Yukawa terms is the following:

$$\left[h_{AB}^{(\text{down})} D_R^{A\dagger} (\varphi^\dagger Q_L^B) + \text{c.c.} \right] + \left[h_{AB}^{(\text{up})} U_R^{A\dagger} (\tilde{\varphi}^\dagger Q_L^B) + \text{c.c.} \right],$$ (14.76)

where $h_{AB}^{(\text{down})}$ and $h_{AB}^{(\text{up})}$ are arbitrary complex 3×3 matrices. Some of this arbitrariness can be eliminated by change of variables, nevertheless, the model still has four additional (over and above the gauge constants and the quark masses) parameters, three mixing angles and a phase. This is important for particle physics. We note that in the case of massless neutrinos, there is no such complication in the leptonic sector.

Problem 21. *Find the Z-charge of each quark and lepton, i.e. the constant in front of the term of type $\psi_L^\dagger \bar{\sigma}^\mu \psi_L Z_\mu$ or $\psi_R^\dagger \sigma^\mu \psi_R Z_\mu$, in the fermion action.*

Let us now discuss an important property of the fermionic part of the action of the Standard Model,

$$S_F = S_{\text{lept}} + S_{\text{quark}},$$ (14.77)

and of the Dirac equations which follow from that. Let us first consider electrons, electron neutrinos and their antiparticles. The corresponding system of Dirac equations in the external boson fields follows from the action (14.69), (14.68) and has the form

$$\begin{aligned} i\bar{\sigma}^\mu D_\mu L_L - h_e \varphi e_R &= 0 \\ i\sigma^\mu D_\mu e_R - h_e (\varphi^\dagger L_L) &= 0. \end{aligned}$$ (14.78)

In the presence of the W-boson fields in equation (14.78) there exist terms mixing the electron and neutrino components of the wave function L_L.

These arise from the covariant derivative (14.78) and have the form

$$
i\tilde{\sigma}^\mu \left(-i\frac{g}{2}\right) (\tau^1 A_\mu^1 + \tau^2 A_\mu^2) \begin{pmatrix} \nu_{e,L} \\ e_L \end{pmatrix} = \frac{g}{2}\tilde{\sigma}^\mu \begin{pmatrix} 0 & A_\mu^1 - iA_\mu^2 \\ A_\mu^1 + iA_\mu^2 & 0 \end{pmatrix} \begin{pmatrix} \nu_{e,L} \\ e_L \end{pmatrix}
$$

$$
= \frac{g}{\sqrt{2}}\tilde{\sigma}^\mu \begin{pmatrix} 0 & W_\mu^+ \\ W_\mu^- & 0 \end{pmatrix} \begin{pmatrix} \nu_{e,L} \\ e_L \end{pmatrix}.
$$

Thus, in the external W-boson fields the electron number and the electron neutrino number are not individually conserved: e may transform into ν_e and conversely. Moreover, as we discussed in Section 14.2, particle–antiparticle pairs may be created. At the same time, the difference

$$
L_e = \begin{pmatrix} \text{electron number} \\ + \\ \text{electron neutrino} \\ \text{number} \end{pmatrix} - \begin{pmatrix} \text{positron number} \\ + \\ \text{electron antineutrino} \\ \text{number} \end{pmatrix} = N_e + N_{\nu_e} \quad (14.79)
$$

is conserved.[5] This quantity is called the electron lepton number: its conservation is clearly related to the fact that the action of the electrons and the neutrino $S_{(e,\nu_e)}$ is separated from the rest of the total action (14.77). The conservation of L_e means, for example, that boson fields of the Standard Model cannot lead to a transition of the electron to the muon (or conversely) without emission of a corresponding neutrino and antineutrino.

Analogously, the muon lepton number

$$
L_\mu = N_\mu + N_{\nu_\mu} \quad (14.80)
$$

and the third-generation lepton number

$$
L_\tau = N_\tau + N_{\nu_\tau} \quad (14.81)
$$

are conserved. In quantum theory the W-boson fields and the Z-boson fields can emerge virtually; precisely virtual W- and Z-bosons give rise to weak interaction of quarks and leptons for low energies. The above considerations about the conservation of L_e, L_μ and L_τ also apply for virtual boson fields. The conservation of L_e and L_μ is manifest, for example, in the fact that the following decay process is forbidden:

$$
\mu \to e + \gamma \quad (\Delta L_\mu = 1, \quad \Delta L_e = -1).
$$

---

[5] In the next sections, we shall discuss the possibility of non-conservation of L_e and other fermion numbers, associated with a highly non-trivial mechanism of fermion level crossing. The considerations studied here are valid for topologically trivial external fields.

Such a decay has not actually been experimentally observed and there are strong experimental bounds on its probability. The decay of the muon in nature occurs with conservation of L_e and L_μ, the dominant decay channel being

$$\mu \rightarrow e + \bar{\nu}_e + \nu_\mu.$$

Finally, in the quark sector, the baryon number

$$B = \frac{1}{3} \left[(\text{total number of quarks}) - (\text{total number of antiquarks}) \right]$$

is conserved. The factor $\frac{1}{3}$ was introduced here, so that the baryon numbers of the proton and the neutron are equal to 1.

The baryon numbers of individual generations are not conserved because of mixing (14.76). The total baryon number is conserved in nature with high accuracy: it is the baryon and lepton number conservation that ensures that the proton, the lightest particle with non-zero baryon number, is stable (it has been experimentally shown that the proton lifetime is greater than 10^{32} years!).

Thus, in the Standard Model, there are four conserved[6] global quantum numbers, L_e, L_μ, L_τ and B.

---

[6]See previous footnote.

Chapter 15

Fermions and Topological External Fields in Two-dimensional Models

In this chapter, we shall consider two effects arising due to the interactions of fermions with topologically non-trivial external bosonic fields, namely the fractionalization of the fermion number and the non-conservation of the fermion quantum numbers due to level crossing. These two phenomena occur in both two- and four-dimensional models. Since two-dimensional theories are easier to analyze, in this chapter we shall consider precisely models in two-dimensional space–time.

Let us make one general remark. In many cases the interaction of fermions with bosonic fields can be analyzed by assuming that the scalar and vector fields are external (and classical) and by studying the Dirac equation in these background fields. Of course, such a description is approximate: the presence of fermions, generally speaking, deforms the configuration of bosonic fields. However, in a number of cases, this effect represents a small correction, and we shall neglect it. We shall comment on this point below, when considering specific models.

15.1 Charge fractionalization

One phenomenon which arises when fermions interact with solitons is the fractionalization of the fermion number and the electric charge (Jackiw and Rebbi 1976b). This phenomenon was in fact experimentally observed in one-dimensional systems of condensed-matter physics. In this section, we shall discuss charge fractionalization, for the example of two-dimensional fermions, interacting with a kink.

Thus, let us consider fermions interacting à la Yukawa with a real scalar field φ in two-dimensional space–time. The Dirac equation in the external field φ has the form

$$(i\gamma_\mu \partial_\mu - h\varphi)\psi = 0, \tag{15.1}$$

where ψ is the two-component column,

$$\psi = \begin{pmatrix} \psi_1 \\ \psi_2 \end{pmatrix}, \tag{15.2}$$

and the two-dimensional Dirac matrices have the form (14.15). Let us consider the case when the external field φ is the field of the kink, described in Section 7.1. This field is static and has the asymptotics

$$\varphi_k(x^1 \to \pm\infty) = \pm v, \tag{15.3}$$

where v is the value of the field φ in one of the ground states, $v > 0$. We shall suppose that φ_k is antisymmetric in space,

$$\varphi_k(-x^1) = -\varphi_k(x^1).$$

We note that the field

$$\varphi_a(x^1) = \varphi_k(-x^1) = -\varphi_k(x^1) \tag{15.4}$$

is the field of the antikink and has the asymptotics

$$\varphi_a(x^1 \to \pm\infty) = \mp v.$$

Since the kink field is static, it makes sense to speak about fermion energy in that background field. In other words, solutions of the Dirac equation in the external static field $\varphi(x^1)$ can be sought in the form

$$\psi(x^0, x^1) = e^{i\omega x^0} \psi_\omega(x^1). \tag{15.5}$$

The wave function with fixed energy satisfies the equation

$$H_D \psi_\omega = \omega \psi_\omega(x^1), \tag{15.6}$$

where the Dirac Hamiltonian is equal to

$$H_D = -i\alpha \frac{\partial}{\partial x^1} + h\varphi(x^1)\beta, \tag{15.7}$$

and the matrices α and β have the form $\beta = \gamma^0 = \tau^1$, $\alpha = \gamma^0\gamma^1 = -\tau^3$.

If the field φ is in the ground state, $\varphi = v$, then the Hamiltonian describes a free fermion with mass $m = hv$. In the external field of the kink, a fortiori, there exist fermion states in a continuous spectrum with energies greater than m, and there may exist (and as we shall see now, there indeed do exist) discrete levels with energies less than m. In any case, if one considers fermions with energies of the order of m, their presence will have a small effect on the kink field, if the energy of the kink field itself is significantly greater than m. Thus, our approximation, in which the kink field is assumed to be external, and the effect of the fermions on the kink small, is valid for

$$M_k \gg hv, \qquad (15.8)$$

where M_k is the mass of the kink. In the model with scalar potential $V(\varphi) = \lambda(\varphi^2 - v^2)^2$, we have $M_k \sim \sqrt{\lambda}v^3$ (see Section 7.1) and equation (15.8) has the form

$$v^2 \gg \frac{h}{\sqrt{\lambda}}.$$

We shall consider precisely this case in what follows.

Let us turn to a discussion of the spectrum of the Dirac Hamiltonian (15.7) in the kink background field $\varphi = \varphi_k(x^1)$. The key point for what follows is the existence of a zero mode, a localized state with energy $\omega = 0$ (Dashen *et al.* 1974a). For $\omega = 0$, equation (15.6) in terms of the components of the column (15.2) has the form

$$i\partial_1\psi_1 + h\varphi_k\psi_2 = 0$$
$$-i\partial_1\psi_2 + h\varphi_k\psi_1 = 0$$

or, equivalently,

$$\partial_1(\psi_1 - i\psi_2) + h\varphi_k(\psi_1 - i\psi_2) = 0$$
$$\partial_1(\psi_1 + i\psi_2) - h\varphi_k(\psi_1 + i\psi_2) = 0.$$

It is clear from these equations that the combinations $(\psi_1 \pm i\psi_2)$ are equal to

$$(\psi_1 \pm i\psi_2) = A_\pm \exp\left[\pm \int_0^{x^1} h\varphi_k(x^1)dx^1\right],$$

where $A_\pm$ are, for the present, arbitrary constants. If we now recall the asymptotics of the kink field (15.3), we obtain that for large $|x^1|$,

$$\int_0^{x^1} h\varphi_k(x^1)dx^1 \sim hv|x^1|.$$

Thus, the combination $(\psi_1 + i\psi_2)$ increases exponentially as $|x^1| \to \infty$ (if $A_+ \neq 0$), and the combination $(\psi_1 - i\psi_2)$ decreases exponentially as

$x^1 \to +\infty$, and also as $x^1 \to -\infty$. Consequently, there is precisely one normalizable fermion mode, for which $(\psi_1 + i\psi_2) = 0$, i.e. $A_+ = 0$. In terms of two-component spinor the wave function of the zero mode is equal to

$$\psi_0^{(k)} = A \begin{pmatrix} 1 \\ i \end{pmatrix} \exp\left[-\int_0^{x^1} h\varphi_k(x^1)dx^1 \right], \tag{15.9}$$

where A is a normalization constant.

It is clear that the zero mode $\psi_0^{(k)}$ must be invariant under C-conjugation (up to a phase factor); if this were not the case, then the C-conjugate function

$$\psi^{(c)} = C\left[\psi_0^{(k)}\right]^*$$

would be *another* solution of the Dirac equation with zero energy.[1] The invariance of the zero mode under C-conjugation is not difficult to check explicitly.

Problem 1. *Show by direct computation that the wave function (15.9) is invariant under C-conjugation (up to a phase factor).*

The fermion zero mode in the antikink background field can be found using the property (15.4). It follows from this property that under spatial reflection (with matrix $P = \gamma^0$) the Dirac equation in the kink background field becomes the Dirac equation in the antikink background field. Thus, the spectrum of the eigenvalues of the Dirac operator in the antikink background field is the same as the spectrum of the Dirac operator in the kink background field, and the eigenfunctions are, correspondingly, interrelated by a P-transformation. In particular, the zero mode in the antikink background field is equal to

$$\psi_0^{(a)} = -iP\psi_0^{(k)}(-x^1)$$

(the factor $(-i)$ is introduced for convenience), or, in explicit form

$$\psi_0^{(a)} = A \begin{pmatrix} 1 \\ -i \end{pmatrix} \exp\left[\int_0^{x^1} h\varphi_a(x^1)dx^1 \right], \tag{15.10}$$

where we have used the explicit form of the matrix $P = \gamma^0$ and the property (15.4).

Problem 2. *Show by direct computation that the function (15.10) is the unique solution of the Dirac equation with zero energy in the antikink background field.*

_____

[1] The fact that the Dirac equation (15.1) is invariant under C-conjugation in the presence of a real external field $\varphi(x^1)$ is verified in complete analogy with Section 14.1.

Problem 3. *Show that, in the presence of topologically trivial static external fields $\varphi(x^1)$ (such that $\varphi(x^1 \to \pm\infty) = v$), there are no fermion zero modes.*

Let us now discuss the consequences of the existence of the fermion mode with zero energy in the kink background field. In the first place, there are two degenerate states of the system in the solitonic sector. In one of these the fermion level corresponding to the zero mode is populated, in the other it is not; since the energy of the fermion at the zero level is equal to zero, the total energies (masses) of these states are identical. The difference between the fermion numbers of these states is equal to one,

$$N_f^{(k)} - N_e^{(k)} = 1, \tag{15.11}$$

where $N_f^{(k)}$ denotes the fermion number of the state with the filled zero level in the kink background field and $N_e^{(k)}$ is the fermion number of the state with the empty zero level. Analogously, there are two degenerate states of the antikink, and the difference between their fermion numbers is also equal to one,

$$N_f^{(a)} - N_e^{(a)} = 1.$$

Let us show that one should assign *half-integer* fermion numbers to the corresponding states

$$
\begin{aligned}
N_f^{(k)} &= N_f^{(a)} = \frac{1}{2} \\
N_e^{(k)} &= N_e^{(a)} = -\frac{1}{2}.
\end{aligned}
\tag{15.12}
$$

Let us first note that because of the symmetry of the theory under spatial reflection the fermion numbers of states of the same type are equal for the kink and the antikink,

$$
\begin{aligned}
N_f^{(k)} &= N_f^{(a)} \\
N_e^{(k)} &= N_e^{(a)}.
\end{aligned}
\tag{15.13}
$$

Let us now consider a Gedanken experiment, in which a kink–antikink pair is created adiabatically slowly. The scalar field in this process changes with time, as shown in Figure 15.1. We are interested in the behavior of the system of fermion levels with time, in this process. In other words, at a fixed moment of time x^0, we need to find the eigenvalues of the Dirac Hamiltonian in the external field $\varphi(x^1, x^0)$, where x^0 is viewed as a parameter, i.e. to solve the eigenvalue problem

$$\left[-i\alpha \frac{\partial}{\partial x^1} + h\varphi(x^0, x^1)\beta \right] = \omega(x^0)\psi.$$

The set of eigenvalues $\omega(x^0)$ is the desired system of fermion levels; it depends on x^0 as a parameter, i.e. the levels move (adiabatically slowly) with time.

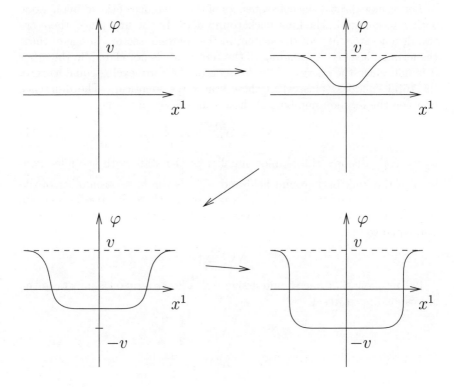

Figure 15.1.

By virtue of the C-symmetry, positive and negative levels move symmetrically: if at a given moment of time x^0 there is a positive level with energy ω, then there is also a negative level with energy $(-\omega)$. Initially, the system of levels corresponds to the free fermions with mass $m = hv$. At the end of the process, when the kink and the antikink move an infinite distance apart, the system has two levels with zero energy. Thus, the system of fermion levels changes with time, as shown in Figure 15.2.

Suppose that at the beginning of the process the system of fermions is in the Dirac vacuum: all levels with negative energy are filled, and all levels with positive energy are empty. In this case, the fermion number of the system is equal to zero. When the external field changes, fermions do not jump from level to level. Consequently, in the final state, all negative levels and *a single zero* level will be filled. This may be the level localized in the

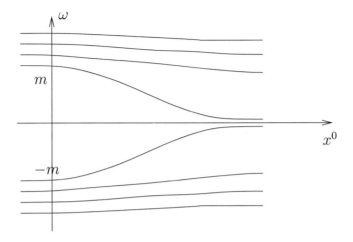

Figure 15.2.

kink or the level localized in the antikink (actually, these two possibilities are each realized with probability $\frac{1}{2}$, which is evident from P-invariance). At the same time, the total fermion number of the system does not change, i.e. at the end of the process it remains equal to zero, as before. If, for example, the filled zero level is localized in the kink, and the level in the antikink is empty, then the fact that the fermion number is equal to zero in the final state means that

$$N_f^{(k)} + N_e^{(a)} = 0.$$

Together with equations (15.11) and (15.13), this leads to half-integer fermion numbers (15.12).

If the fermions carry an electric charge e (and the scalar field is electrically neutral), then we must conclude that the electric charge of a kink with a filled zero fermion level is equal to $\frac{1}{2}e$ (and the charge of a kink with an empty zero level is equal to $(-\frac{1}{2}e)$)! At first glance, this conclusion is paradoxical, since the system contains no elementary excitations (particles) with half-integer electric charge. The resolution of this apparent paradox lies in the fact that the Dirac sea in the external field $\varphi(x^1)$ can carry an electric charge, which is not necessarily integer; the Dirac sea is said to be polarized. This statement is supported by direct computation of the electric charge (and the fermion number) of the Dirac sea in the kink background field; this calculation is carried out in the framework of quantum field theory, and we shall not present it here.

15.2   Level crossing and non-conservation of fermion quantum numbers

In this section we shall consider a very simple model, in which the phenomenon of fermion level crossing and the associated non-conservation of fermion quantum numbers occurs, namely, the theory of two-dimensional massless fermions, interacting with an Abelian (external) gauge field. The phenomenon of level crossing is closely related to the quantum anomaly (Adler 1969, Bell and Jackiw 1969) in the divergence of the corresponding fermionic current; therefore, the resulting non-conservation of fermion quantum numbers, being of a non-perturbational nature, is sometimes said to be anomalous.   The anomalous non-conservation of fermion numbers was discovered in the context of instanton processes by 't Hooft (1976a,b), who used a very different formalism from that studied in this and forthcoming sections (the Euclidean path integral formalism, in which the non-conservation of fermion quantum numbers arises as a result of the existence of Euclidean fermion zero modes). An interpretation of 't Hooft's mechanism in terms of fermion level crossing was given by Callan *et al.* (1978) and Kiskis (1978). Two-dimensional models were considered from this point of view by Nielsen and Ninomiya (1983) and Ambjorn *et al.* (1983). In the following chapters we shall see that anomalous non-conservation of fermion quantum numbers is inherent in a number of systems in four-dimensional space–time; its consequences are interesting from the point of view of both particle physics and cosmology.

We recall that free massless fermions in two-dimensional space–time are described by a two-component wave function

$$\psi = \begin{pmatrix} \chi \\ \eta \end{pmatrix},$$

satisfying the Dirac equation

$$i\gamma^\mu \partial_\mu \psi = 0, \quad \mu = 0, 1.$$

As before, we shall use the representation (14.15) of the two-dimensional Dirac matrices, $\gamma^0 = \tau^1$, $\gamma^1 = i\tau^2$. In this representation the free equations for χ and η decouple,

$$(i\partial_0 - i\partial_1)\chi = 0, \tag{15.14}$$
$$(i\partial_0 + i\partial_1)\eta = 0. \tag{15.15}$$

The general solution of equation (15.14) is an arbitrary complex function of the variable $x_0 + x_1$, i.e. $\chi = \chi(x_0 + x_1)$. Thus, χ is a wave (generally speaking, not a plane wave), moving to the left. Analogously, $\eta(x_0 - x_1)$, the solution of (15.15), is a wave moving to the right. We shall say that χ describes left fermions (fermions moving to the left), and that η describes right fermions.

For the sequel, it is convenient to suppose that one-dimensional space has a finite length L, at whose ends periodic boundary conditions are imposed. Solutions of the Dirac equation with fixed energy ω have the form

$$
\begin{aligned}
\chi_\omega &= e^{-i\omega x^0 + ikx^1}, \quad \omega = -k \\
\eta_\omega &= e^{-i\omega x^0 + ikx^1}, \quad \omega = +k,
\end{aligned}
\tag{15.16}
$$

where in both cases the momentum k takes a discrete set of values,

$$
k = \frac{2\pi}{L} n, \quad n = 0, \pm 1, \pm 2, \ldots .
\tag{15.17}
$$

The energy spectrum ω is essentially gapless (the first level is separated from zero by a gap of width $2\pi/L$), there are no degeneracies.

Since the equations for left and right fermions decouple, it makes sense to speak of the left fermion number N_L and the right fermion number N_R, which are separately conserved. For example, N_L is the difference between the number of fermions moving to the left and the number of antifermions (holes in the Dirac vacuum), also moving to the left. One state of a fermion system is shown in Figure 15.3. Instead of N_L and N_R, one can introduce the conserved total fermion number $N_F = N_L + N_R$ and chirality $Q^5 = N_L - N_R$.

Suppose now that the fermions interact with the Abelian gauge field A_μ of the group $U(1)$ and carry a charge e. By analogy with four-dimensional electrodynamics, we shall call A_μ the electromagnetic vector potential. In fact, this has nothing to do with magnetism: the only non-zero component of the strength tensor $F_{01} = -F_{10}$ corresponds to the electric field $F_{01} = -E$. The Dirac equation in the external field $A_\mu(x^0, x^1)$ is obtained, as usual, by replacing ordinary derivatives by covariant ones. For the left and right components we have the equations

$$
[i(\partial_0 - ieA_0) - i(\partial_1 - ieA_1)]\chi = 0
\tag{15.18}
$$

$$
[i(\partial_0 - ieA_0) + i(\partial_1 - ieA_1)]\eta = 0.
\tag{15.19}
$$

One might think that N_L and N_R are individually conserved, as before, since the left and right components of the fermion wave functions do not mix with each other. We shall show that that is not always the case.

As a simple example, let us consider a process in which the system of fermions is placed for some period of time in a spatially homogeneous electric field, oriented along x^1 in the positive direction; the electric field vanishes as $x^0 \to \pm\infty$. Suppose that at the beginning of the process the

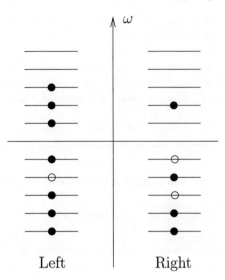

Figure 15.3. State of a fermion system with $N_L = 2$ and $N_R = -1$. The open circles denote empty levels in the Dirac sea, and the closed circles denote occupied levels.

fermions are in the ground state, the Dirac vacuum. When the electric field is switched on, Dirac sea fermions of type η, moving to the right, will acquire energy. Some of them will begin to have positive energy. After the electric field is switched off, these fermions will have positive energy as before, i.e. some positive levels will be filled and real right fermions will appear in the system. Conversely, Dirac sea fermions of type χ, moving to the left, will lose energy in the electric field, i.e. their energy ω will in time become increasingly negative. This means that some negative levels of the left fermions will be empty at the end of the process, i.e. the system will have new holes, left antifermions. The initial and final states of the system are shown in Figure 15.4. Thus, the left fermion number N_L in this process will decrease, and the right fermion number will increase, N_L and N_R will not be individually conserved. For a quantitative description of this process, let us choose the gauge of vector potentials, where $A_0 = 0$ and $A_1 = A_1(x^0)$. The electric field is equal to

$$E = -\partial_0 A_1(x^0). \tag{15.20}$$

In the beginning of the process $A_1(x^0 = -\infty) = 0$, and at its end

$$A_1 = \text{constant} \equiv -\frac{\mu}{e}, \quad x^0 \to +\infty. \tag{15.21}$$

Let us solve equations (15.18) and (15.19) in this background field. Solutions which have energy ω in the beginning of the process, have the form

$$\chi = \text{constant} \cdot \exp\left[-i\omega(x^0 + x^1) - ie\int_{-\infty}^{x^0} A_1(t)dt\right]$$

$$\eta = \text{constant} \cdot \exp\left[-i\omega(x^0 - x^1) + ie\int_{-\infty}^{x^0} A_1(t)dt\right].$$

As $x^0 \to \infty$ these solutions are characterized by energies $(\omega - \mu)$ and $(\omega + \mu)$, respectively,

$$\chi = \text{constant} \cdot e^{-i(\omega - \mu)x^0 - i\omega x^1}, \qquad (15.22)$$

$$\eta = \text{constant} \cdot e^{-i(\omega + \mu)x^0 + i\omega x^1}. \qquad (15.23)$$

The fact that the absolute value of the momentum for these solutions does not coincide with the energy is explained by the presence of the non-zero vector potential (15.21) at the end of the process. This vector potential can be eliminated by the gauge transformation $A_\mu \to A_\mu + \frac{1}{e}\partial_\mu\alpha$, $\psi \to e^{i\alpha}\psi = \psi'$ with $\alpha = \mu x^1$. As a result the gauge transformed wave functions at the end of the process will have the usual form

$$\chi' = \text{constant} \cdot e^{-i(\omega - \mu)(x^0 + x^1)},$$

$$\eta' = \text{constant} \cdot e^{-i(\omega + \mu)(x^0 - x^1)}.$$

At the beginning of the process, all levels with $\omega \leq 0$ are filled. It follows from (15.22) and (15.23) that at the end of the process all left levels with energy less than $(-\mu)$ and all right levels with energy less than $(+\mu)$ will be filled. Left levels in the energy interval $(-\mu, 0)$ will be empty. Thus, equal numbers of right fermions and left antifermions will be created in the system, i.e.

$$\Delta N_L = -\Delta N_R. \qquad (15.24)$$

The total fermion number $N_F = N_L + N_R$ is conserved, but the chirality $Q^5 = N_L - N_R$ changes.

To compute ΔN_R, we take into account the fact that the energies of the free fermions run over the values $\frac{2\pi}{L}n$, $n = 0, \pm 1, \pm 2, \ldots$ (see (15.16) and (15.17)), and that the levels are non-degenerate. At the end of the process, real right fermions fill all levels with energies from 0 to μ, i.e. levels with

$$0 < n \leq \frac{L}{2\pi}\mu.$$

There are $\frac{L}{2\pi}\mu$ of these levels; thus,

$$\Delta N_R = \frac{L}{2\pi}\mu. \qquad (15.25)$$

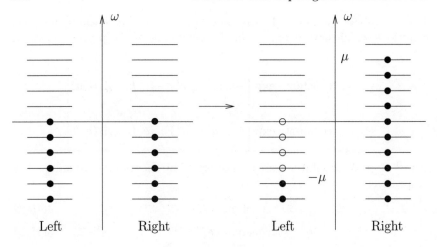

Figure 15.4.

Taking into account (15.20) and (15.21), this equation can be given the following form:

$$\Delta N_R = \frac{e}{2\pi} \int dx^1 dx^0 E \tag{15.26}$$

or

$$\Delta N_R = -\frac{e}{4\pi} \int d^2 x \varepsilon_{\mu\nu} F_{\mu\nu}. \tag{15.27}$$

The changes in the total fermion number and the chirality are given by

$$\Delta N_F = 0 \tag{15.28}$$

$$\Delta Q^5 = \frac{e}{2\pi} \int d^2 x \varepsilon_{\mu\nu} F_{\mu\nu}. \tag{15.29}$$

Thus, in this model one of the fermion quantum numbers, the chirality (the difference between the number of left and right fermions) is not conserved, and the change in the chirality in this process is given by equation (15.29). We note that the right-hand side of (15.29) is proportional to the topological number of the two-dimensional configuration of the gauge field introduced in Section 13.3.

Equations (15.24) and (15.27) or, equivalently, (15.28) and (15.29), are in fact very general. To show this, let us first consider a gauge field $A_1(x^0, x^1)$, varying adiabatically in time and defining an electric field $E(x^0, x^1) = -\partial_0 A_1(x^0, x^1)$, which is switched off as $x^0 \to \pm\infty$. Otherwise the field $A_1(x^\mu)$ is arbitrary; without loss of generality, we may suppose that at the beginning of the process $A_1(x^0, x^1) = 0$, $x^0 \to -\infty$. At the end of the process $A_1(x^0 \to +\infty, x^1)$ is, in general, non-zero.

Solutions of the Dirac equation in an adiabatically slowly varying external field have the form

$$\psi(x^0, x^1) = \exp\left[-i \int_{-\infty}^{x^0} \omega(t) dt\right] \psi_{\omega(x^0)}(x^1), \qquad (15.30)$$

where $\psi_{\omega(x^0)}(x^1)$ is an eigenfunction of the instantaneous Dirac Hamiltonian

$$H_D(x^0)\psi_{\omega(x^0)}(x^1) = \omega(x^0)\psi_{\omega(x^0)}(x^1), \qquad (15.31)$$

where

$$H_D(x^0) = -i\alpha[\partial_1 - ieA_1(x^0, x^1)].$$

As before, $\alpha = \gamma^0\gamma^1 = -\tau^3$. In equation (15.31), the time x^0 must be viewed as a parameter. Equations (15.30) and (15.31) imply that in the course of the process the energies $\omega(x^0)$ of the fermion levels vary (the levels move), while the fermions remain at the same levels as they were initially (there are no jumps from level to level). At the beginning and the end of the process the electric field is switched off, thus the systems of fermion levels coincide; these are the systems of levels of the free Dirac Hamiltonian. At the same time, the movement of the fermion levels may be non-trivial; we shall now see that the left and right levels may move as shown in Figure 15.5. Some levels may cross zero and move from the region of negative[2] ω to the region of positive ω (in Figure 15.5, these are the two right fermion levels). If, for example, the initial state of a fermion system is the Dirac vacuum, as shown in Figure 15.5 for right fermions, then in the final state, there will exist real fermions, and the number of real fermions will be equal to the number of levels crossing zero from below. Some levels may cross zero from above and move from the region of positive ω to the region of negative ω. These levels will be empty in the final state (if the initial state is the Dirac vacuum), i.e. antifermions will appear in the system. Thus, the change in the right fermion number in this process will be equal to

$$\Delta N_R = \left(\begin{array}{c}\text{Number of right levels} \\ \text{crossing zero from below}\end{array}\right) - \left(\begin{array}{c}\text{Number of right levels} \\ \text{crossing zero from above}\end{array}\right)$$

and analogously for ΔN_L. Clearly, this equation does not depend on the initial state, and this is illustrated in Figure 15.5 for left fermions ($\Delta N_L = -2$ in Figure 15.5).

---

[2] We assign the zero level to the Dirac sea; this is a matter of convention, which is not important for what follows. In the limit of infinite spatial size this subtlety is irrelevant.

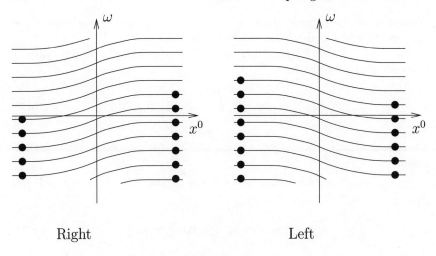

Right   Left

Figure 15.5.

Let us mention one subtlety. The spectra of the fermions will indeed coincide at the beginning and the end of the process if the final field $A_1(x^0 \to +\infty, x^1)$ can be made equal to zero by a non-singular gauge transformation with gauge function $e^{i\alpha(x^1)}$, *periodic in space with period L*. For this we require (see Section 13.3), the quantity

$$n = \frac{e}{2\pi} \int_{-L/2}^{L/2} A_1(x^0 \to \infty, x^1) dx^1,$$

which is the topological number of the gauge vacuum in the final state, to be an integer. We recall that n is related to the topological number of a two-dimensional configuration of the gauge field (we set $A_1 = 0$ for $x^0 \to -\infty$),

$$q = \frac{e}{4\pi} \int d^2 x \varepsilon_{\mu\nu} F_{\mu\nu}.$$

It is precisely the cases in which q is an integer which we shall consider in what follows.

In order to relate ΔN_R and ΔN_L to q, we shall solve equation (15.31). For right and left fermions the solutions have the form

$$\eta_{\omega(x^0)}(x^1) = \exp\left[i\omega(x^0)x^1 + ie\int_0^{x^1} A_1(x^0, x^1) dx^1\right]$$

$$\chi_{\omega(x^0)}(x^1) = \exp\left[-i\omega(x^0)x^1 + ie\int_0^{x^1} A_1(x^0, x^1) dx^1\right].$$

The conditions for periodicity of the wave functions η and χ at the boundaries of the space are satisfied if

$$\text{right}: \qquad \omega_n(x^0)L + e\int_{-L/2}^{L/2} A_1(x^0, x^1)dx^1 = 2\pi n$$

$$\text{left}: \qquad \omega_n(x^0)L - e\int_{-L/2}^{L/2} A_1(x^0, x^1)dx^1 = 2\pi n, \quad n = 0, \pm 1, \pm 2, \ldots.$$

Thus, the energies of fermions in the nth level at time x^0 are equal to:

$$\text{right}: \qquad \omega_n(x^0) = \frac{2\pi n}{L} - \frac{e}{L}\int_{-L/2}^{L/2} A_1(x^0, x^1)dx^1 \qquad (15.32)$$

$$\text{left}: \qquad \omega_n(x^0) = \frac{2\pi n}{L} + \frac{e}{L}\int_{-L/2}^{L/2} A_1(x^0, x^1)dx^1. \qquad (15.33)$$

Hence it is clear that, for $q \neq 0$, the levels indeed move as shown in Figure 15.5. As $x^0 \to -\infty$ we have $\omega_n = \frac{2\pi n}{L}$, and as $x^0 \to +\infty$ this level will occupy the position of the $(n \pm q)$th level of free fermions (the upper sign relates to left fermions, the lower sign to right ones),

$$\omega_n(x^0 \to +\infty) = \frac{2\pi(n \pm q)}{L}.$$

We conclude that the difference between the number of levels crossing zero from below and from above is equal to q for left fermions and to $(-q)$ for right ones. Consequently,

$$\Delta N_R = -q \qquad (15.34)$$
$$\Delta N_L = q, \qquad (15.35)$$

which agrees with the formula for a spatially homogeneous field.

Finally, let us drop the requirement for adiabatic change of the field A_1 with time. In the non-adiabatic situation, jumps of fermions from level to level are possible, but the movement of the levels themselves is given, as before, by formulae (15.32) and (15.33). It is significant that left-hand fermions cannot jump to right levels and vice versa, since the Weyl equations for left and right fermions decouple. All that can happen in the final state, in comparison with the adiabatic case, is the occurrence of a certain number of additional right fermions in the real state (with $\omega > 0$) and *the same number* of holes in the Dirac sea of right fermions (and analogously for left ones). This has no effect on the right (and left) fermion numbers of the final state, since they are defined as the *difference* between the number of fermions with $\omega > 0$ and the number of holes in the

344 Fermions and Topological External Fields

Dirac sea (antifermions). We conclude that formulae (15.34) and (15.35) (or, equivalently, (15.28) and (15.29)) are of a general nature, if the gauge field is in the vacuum state at the beginning and end of the process (and even in the more general situation, when the configurations of the field A_1 at the beginning and end do not depend on time and differ only by a non-singular and periodic gauge transformation).

Several remarks are in order.

1. We have said nothing about the reasons why the gauge field $A_1(x^0, x^1)$ arises, or about whether it interacts with other bosonic fields: the only thing that was important for us was that at the beginning and end its configuration corresponds to topologically distinct vacua. This field may arise in the Abelian Higgs model at zero temperature as a result of the tunneling process (instanton). Thus, the instanton transition in the Abelian Higgs model with massless fermions leads to non-conservation of chirality (for one instanton $q = 1$ and the chirality changes by $\Delta Q_5 = 2$). In the Abelian Higgs model at a finite temperature, the field $A_1(x^0, x^1)$ may occur as a result of a thermal jump (via the sphaleron); this process in the theory with fermions also leads to non-conservation of chirality.

2. In any case, non-conservation of chirality requires "large" fields: small perturbations of A_1 do not lead to transitions between topologically distinct vacua.

3. If the model had only right fields and the left component of ψ were completely absent (two-dimensional Weyl fermions), then we would have to deduce non-conservation of electric charge. Indeed, we have asserted that the electric charge of the Dirac sea is zero, thus, for a system of right Weyl fermions the electric charge is equal to

$$Q = N_R.$$

Non-conservation of N_R would mean non-conservation of electric charge. However, electrodynamics with a non-conserved electric charge is inconsistent (Maxwell's equations are only self-consistent when the current vector is conserved $\partial_\mu j_\mu = 0$). Thus, we conclude that the theory of electrically charged Weyl fermions is inconsistent in two-dimensional space–time.

In quantum field theory one can construct operators of left and right fermion currents

$$j_\mu^L = \bar{\psi}\gamma^\mu \frac{1+\gamma^5}{2}\psi$$
$$j_\mu^R = \bar{\psi}\gamma^\mu \frac{1-\gamma^5}{2}\psi,$$

where $\gamma^5 = -\gamma^0\gamma^1 = \tau^3$ in two-dimensional space–time. The left and right fermion number operators have the form

$$N_{L,R} = \int j_0^{L,R} dx^1.$$

In quantum field theory, it is demonstrated that the currents j_μ^L and j_μ^R are not conserved (analogue of the triangle anomaly of Adler, Bell and Jackiw, which occurs in four-dimensional space–time):

$$\partial_\mu j_\mu^L = \frac{e}{4\pi}\varepsilon_{\mu\nu}F_{\mu\nu} \tag{15.36}$$

$$\partial_\mu j_\mu^R = -\frac{e}{4\pi}\varepsilon_{\mu\nu}F_{\mu\nu}. \tag{15.37}$$

Equations (15.34) and (15.35) are evidently integrals of the identities (15.36) and (15.37) over space–time. It is clear from equation (15.37) that two-dimensional electrodynamics with only right fermions is inconsistent.

4. In the model we considered, chirality is violated and the total fermion number is conserved. However, it is not difficult to construct a model in which the total fermion number $N_F = N_L + N_R$ is not conserved, see the problem at the end of the section.

5. The fact that the difference between the numbers of right levels crossing zero from above and below is equal to the topological number q of the gauge field configuration (or of configurations, differing at the beginning and the end by a gauge transformation) is a very simple version of the Atiyah–Patodi–Singer theorem. The possibility of proving this fact by direct solution of the Dirac equation is a specific property of the simple two-dimensional model considered in this section.

One may have the impression that level crossing is possible only for massless fermions. In fact, this is not the case: level crossing and the non-conservation of fermion quantum numbers can occur also when the fermions acquire a mass as a result of the Yukawa interaction with the Higgs fields (this mechanism of mass generation was discussed at the end of Section 14.3). As an example, let us consider the Abelian Higgs model in two-dimensional space–time and introduce into it fermions interacting with the Higgs field à la Yukawa.

The Lorentz and gauge-invariant action can be written down, by introducing a charged right fermion χ and a neutral left fermion ξ (see the example at the end of Section 14.3). Under gauge transformations, we shall have $\varphi \to e^{i\alpha}\varphi$, $\chi \to e^{i\alpha}\chi$, $\xi \to \xi$ and the fermion action with the Yukawa term can be chosen in the form

$$S_F = \int d^2x[\chi^+ i(D_0 + D_1)\chi + \xi^+ i(\partial_0 - \partial_1)\xi - h(\chi^+ \xi\varphi + \xi^+ \chi\varphi^*)], \quad (15.38)$$

where, as before, $D_\mu = \partial_\mu - ieA_\mu$. Here, we have used the explicit form of the two-dimensional γ-matrices.

The equations, following from the action (15.38), have the form

$$i(D_0 + D_1)\chi - h\varphi\xi \;=\; 0 \qquad\qquad (15.39)$$
$$i(\partial_0 - \partial_1)\xi - h\varphi^*\chi \;=\; 0. \qquad\qquad (15.40)$$

If the external fields take the vacuum values, $A_\mu = 0$, $\varphi = v$, then the system (15.39), (15.40) describes a free massive Dirac fermion in two-dimensional space–time: it reduces to the free Dirac equation

$$i\gamma_\mu \partial_\mu \psi - m\psi = 0$$

for the column

$$\psi = \begin{pmatrix} \xi \\ \chi \end{pmatrix}. \qquad\qquad (15.41)$$

Here, the fermion mass is equal to $m = hv$.

Let us now consider external fields $A_1(x^0, x^1)$, $\varphi(x^0, x^1)$ in the gauge $A_0 = 0$. We shall assume that they interpolate between neighboring topologically distinct vacua, i.e.

$$A_1(x^1) = 0, \quad \varphi(x^1) = v, \quad \text{for } x^0 \to -\infty$$

$$A_1(x^1) = \frac{1}{e}\partial_1\alpha(x^1), \quad \varphi(x^1) = e^{i\alpha(x^1)}v, \quad \text{for } x^0 \to +\infty,$$

where

$$\alpha(x^1 \to +\infty) - \alpha(x^1 \to -\infty) = 2\pi$$

(since all fermions are massive, there is no need to consider the system in space of finite length). We now raise the question as to whether or not level crossing occurs in these external fields. In other words, we need to analyze the spectrum of the instantaneous Dirac Hamiltonian for the system (15.39), (15.40) in the given external fields. This Hamiltonian has the form

$$H_D = \begin{pmatrix} i\partial_1 & h\varphi^* \\ h\varphi & -iD_1 \end{pmatrix}, \qquad\qquad (15.42)$$

since for $A_0 = 0$, equations (15.39), (15.40) are written down for the column (15.41) in the form

$$i\partial_0 \psi = H_D \psi.$$

We note that the Hamiltonian (15.42) is Hermitian, as it should be.

We shall solve only part of the problem, namely, we shall show (and only for a certain class of configurations of fields A_1 and φ) that the instantaneous Hamiltonian (15.42) has eigenvalue zero for some x^0. The corresponding eigenfunction ψ_0 decreases rapidly as $x^1 \to \pm\infty$; thus, we shall not be worried about the boundary conditions for ψ_0. In order to make the problem more symmetric in time, we perform a gauge transformation with gauge function $-\frac{\alpha(x^1)}{2}$ over all the quantities. Then, the external bosonic fields will interpolate between static configurations of the form

$$A_1(x^1) = \frac{1}{e}\partial_1\beta_-(x^1), \quad \varphi(x^1) = e^{i\beta_-(x^1)}v, \quad x^0 \to -\infty$$

and

$$A_1(x^1) = \frac{1}{e}\partial_1\beta_+(x^1), \quad \varphi(x^1) = e^{i\beta_+(x^1)}v, \quad x^0 \to +\infty,$$

where $\beta_\pm = \pm\frac{\alpha(x^1)}{2}$. Without loss of generality, we may suppose that $\alpha(x^1 \to -\infty) = 0$ and $\alpha(x^1 \to +\infty) = 2\pi$. Then, at the beginning of the process as x^1 changes, the phase of the field φ changes from zero (when $x^1 \to -\infty$) to $(-\pi)$ (when $x^1 \to +\infty$). This means that, as x^1 changes, the initial field $\varphi(x^1)$ runs along a semicircle of radius v in the complex plane, located in the lower half-plane. As $x^0 \to +\infty$, the field $\varphi(x^1)$ runs along a semicircle, located in the upper half-plane, and the whole process proceeds as shown in Figure 15.6.

Let us restrict ourselves, for simplicity, to configurations which at some moment of time (let us say $x^0 = 0$) are described by real $\varphi(x^1)$ and have $A_1(x^1) = 0$. At this moment of time (and for arbitrary x^0) the field $\varphi(x^1)$ changes from v to $(-v)$, as x^1 changes from $-\infty$ to $+\infty$. At $x^0 = 0$, the instantaneous Hamiltonian (15.42) is equal to

$$H_D = \begin{pmatrix} i\partial_1 & h\varphi^* \\ h\varphi & -i\partial_1 \end{pmatrix} = -i\alpha\partial_1 + \beta h\varphi(x^1).$$

It coincides with the Hamiltonian (15.1) in the antikink background field and, as shown in Section 15.1, has a zero eigenvalue. Thus, in the system, one of the fermion levels indeed crosses zero. We note that for level crossing, the fact that the field φ takes zero value at some point (x^0, x^1) is important.

It is a good deal more complicated to show that the fermion number changes exactly by unity in the external fields A_1 and φ of the type in

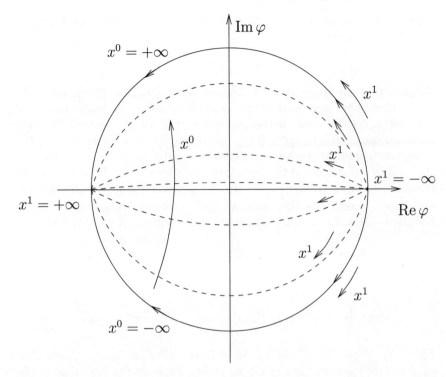

Figure 15.6. Evolution of field φ with increasing x^0. Continuous curves show "trajectories" of $\varphi(x^1)$ for $x^0 = -\infty$ (lower semicircle) and $x^0 = +\infty$ (upper semicircle). Dashed curves show "trajectories" of $\varphi(x^1)$ at intermediate times.

question. We shall not present the corresponding proof here. We note only that at the level of quantum field theory the model contains an anomaly of the type (15.37) in the fermionic current or, in integrated form,

$$\Delta N_F = -q,$$

and that the calculation of the level crossing agrees with this formula.[3]

Problem 4. *Consider massless two-dimensional fermions with axial interaction to the gauge field of the group $U(1)$. The Dirac equation has the form*

$$i\gamma^\mu D_\mu \psi = 0,$$

---

[3]The gauge current in this model also has an anomaly. To make the model self-consistent one has to add other fermions to it.

where ψ is a two-component column and D_μ is defined as $D_\mu = \partial_\mu - ie\gamma^5 A_\mu$. As before, the matrix γ^5 is equal to $\gamma^5 = -\gamma^0\gamma^1 = \tau^3$. Note that the gauge transformations over ψ have the form

$$\psi = e^{i\alpha\gamma^5}\psi.$$

Determine the behavior of the system of fermion energy levels in the external field $A_1(x^0, x^1)$ with $q \neq 0$. Show that the chirality in this model is conserved, but that the total fermion number is not.

Chapter 16

Fermions in Background Fields of Solitons and Strings in Four-Dimensional Space–Time

The purpose of this chapter is to discuss certain dynamical effects associated with the behavior of fermions in background fields of topological objects in four-dimensional space–time, namely 't Hooft–Polyakov magnetic monopoles (Sections 16.1 and 16.2) and strings, i.e. Abrikosov–Nielsen–Olesen vortices (Section 16.3). These objects themselves were described in Sections 9.1, 9.2 and 7.3, and we shall use the notation introduced there. The interest in the matters considered in this chapter is primarily due to the fact that monopoles and strings occur in a natural way in theories of the grand unification of the strong, weak and electromagnetic interactions (monopoles exist in practically all Grand Unified Theories, while strings occur in a relatively narrow class of models). Of course, these theories also contain fermions, at least the known quarks and leptons. Thus, the study of the dynamics of fermions, interacting with monopoles and strings, is important for the experimental search for these objects, and also from the point of view of their possible role in the early Universe.

351

16.1   Fermions in a monopole background field: integer angular momentum and fermion number fractionalization

In this section and in the following sections, we shall consider a simple model, containing 't Hooft–Polyakov monopoles. This model (see Sections 9.1 and 9.2) is based on the gauge group $SU(2)$ and contains a real triplet of Higgs fields φ^a. We recall that the ground state of bosonic fields can be chosen in the form

$$A^a_\mu = 0 \qquad \varphi^a = v\delta_{a3}. \tag{16.1}$$

Here, the subgroup of electromagnetism $U(1)_{\text{e.m.}}$, whose gauge transformations have the form

$$\omega(x) = e^{i\frac{\tau^3}{2}\alpha(x)} \tag{16.2}$$

remains unbroken. In a unitary gauge, fields A^1_μ and A^2_μ are massive, while the massless electromagnetic field is described by the vector potential A^3_μ.

For later reference, we again give the form of the monopole solution (9.10) in a regular gauge:

$$\varphi^a = n^a v(1 - H(r)), \tag{16.3}$$

$$A^a_i = \frac{1}{gr}\varepsilon^{aij}n^j(1 - F(r)), \tag{16.4}$$

where the radial functions satisfy the boundary conditions

$$\begin{aligned} F(0) &= H(0) = 1 \\ F(\infty) &= H(\infty) = 0. \end{aligned} \tag{16.5}$$

We shall again use the fact that in the gauge (16.3) the unbroken subgroup of electromagnetism (which makes sense far away from the core of the monopole) is not defined by formula (16.2), but by the transformations

$$\omega(x) = \exp\left[i\frac{\tau^a n^a}{2}\alpha(x)\right], \tag{16.6}$$

where $n^a = x^a/r$ are components of the unit radius vector in three-dimensional space.

Let us include in the model fermions, which form a doublet with respect to the gauge group $SU(2)$. In this section, we shall consider Dirac fermions, containing both left- and right-handed components, and we shall include the Yukawa interaction with the Higgs field φ^a. The corresponding construction was described in Section 14.3, therefore, we shall immediately write down the Dirac equation (14.54):

$$\left(i\gamma^\mu D_\mu - h\frac{\tau^a}{2}\varphi^a \right)\psi = 0. \tag{16.7}$$

We note that we have set the explicit mass parameter m in (14.54) equal to zero. As usual for the doublet representation of $SU(2)$, the covariant derivative in (16.7) is

$$D_\mu = \partial_\mu - ig\frac{\tau^a}{2}A_\mu. \tag{16.8}$$

Let us in the first place find the masses and electric charges of fermions in the vacuum (16.1). For this, in the covariant derivative (16.8), we retain only the electromagnetic field A_μ^3 and we set $\varphi^a = v\delta^{a3}$. Then equation (16.7) will have the form

$$\left[i\gamma^\mu\left(\partial_\mu - ig\frac{\tau^3}{2}A_\mu^3 \right) - h\frac{\tau^3}{2}v \right]\psi = 0.$$

The equations for the upper and lower components of the $SU(2)$-doublet

$$\psi = \begin{pmatrix} \psi_1 \\ \psi_2 \end{pmatrix}$$

decouple and have the form

$$\begin{aligned} \left[i\gamma^\mu\left(\partial_\mu - i\frac{g}{2}A_\mu^3 \right) - h\frac{v}{2} \right]\psi_1 &= 0 \\ \left[i\gamma^\mu\left(\partial_\mu + i\frac{g}{2}A_\mu^3 \right) + h\frac{v}{2} \right]\psi_2 &- 0. \end{aligned} \tag{16.9}$$

Clearly, the fermions ψ_1 and ψ_2 have the same mass

$$m_F = \frac{hv}{2}, \tag{16.10}$$

the fermion ψ_1 has electric charge $(+\frac{g}{2})$ and the fermion ψ_2 has charge $(-\frac{g}{2})$. We note that the non-standard sign of the mass term in equation (16.9) does not affect the energy spectrum of the fermion ψ_2, as can be seen from the result of the following problem.

Problem 1. *Find a unitary matrix U such that the wave function $\psi_2' = U\psi_2$ satisfies the equation*

$$\left[i\gamma^\mu\left(\partial_\mu + i\frac{g}{2}A_\mu^3 \right) - h\frac{v}{2} \right]\psi_2' = 0,$$

if ψ_2 satisfies the equation (16.9).

Thus, the model has two types of fermions with opposite electric charges and the same mass.

Let us now consider equation (16.7) in the external monopole field (16.3), (16.4). The first notable property of this equation is the existence of an *integer* angular momentum for the fermions (Dereli *et al.* 1975; Jackiw and Rebbi 1976b; Hasenfratz and 't Hooft 1976). Indeed, the external monopole field is invariant under spatial rotations, supplemented by rotations in the "internal" space, see (9.6). Thus, the conserved quantity for the Dirac equation in the background monopole field is not the usual fermion angular momentum, but a combination which includes generators of the gauge group $SU(2)$,

$$J^a = L^a + s^a + \frac{\tau^a}{2}, \quad a = 1, 2, 3, \tag{16.11}$$

where the orbital and spin momenta of the fermion have the usual form

$$
\begin{aligned}
L^a &= -i\varepsilon^{abc} x^b \partial_c, \\
s^a &= \frac{1}{2} \begin{pmatrix} \sigma_a & 0 \\ 0 & \sigma^a \end{pmatrix};
\end{aligned}
$$

here, σ^a are Pauli matrices acting on the Lorentz indices of the left- and right-handed components of the fermion (whereas τ^a are Pauli matrices acting on the internal, isotopic indices of the $SU(2)$-doublet). To check the conservation of J^a explicitly, let us write equation (16.7) in the external fields (16.3), (16.4) in the form

$$i \frac{\partial \psi}{\partial x^0} = H_D \psi.$$

The quantum-mechanical Hamiltonian here is equal to

$$H_D = -i\alpha^a D_a + \beta h \frac{\tau^a}{2} \varphi^a, \tag{16.12}$$

where α^a and β are standard matrices (14.5), acting on the Lorentz indices. The fact that J_a commutes with H_D,

$$[J_a, H_D] = 0, \tag{16.13}$$

is verified by direct computation.

Indeed, let us compute the commutator $[L^a, H_D]$. Carrying the derivatives to the right, we obtain

$$[L^a, H_D] = -i\varepsilon_{abc} x_b \left[(-i) \left(-ig \frac{\tau^d}{2} \alpha^m \partial_c A_m^d \right) + \beta h \frac{\tau^d}{2} \partial_c \varphi^d \right],$$

which for the external fields (16.3), (16.4) gives

$$[L^a, H_D] = \frac{i}{2r}(1-F)(\tau^a \alpha^b n_b - \alpha^a \tau^b n_b) - ih\frac{v}{2}(1-H)\beta\varepsilon_{abc}n_b\tau^c + \varepsilon_{abc}\alpha^b\partial_c.$$

$$(16.14)$$

Now, the commutator of s_a with H_D is equal to

$$[s^a, H_D] = -i[s^a, \alpha^m]D_m + \frac{h}{2}\tau^b\varphi^b[s^a, \beta].$$

Using the explicit form of the matrices s^a, α^m and β, we obtain

$$[s^a, H_D] = -\varepsilon_{abc}\alpha^b\partial_c - \frac{i}{2r}(1-F)(\tau^a\alpha^m n_m - n^a\tau^m\alpha^m).$$

$$(16.15)$$

Finally, the commutator of s^a with H_D is equal to

$$\left[\frac{\tau^a}{2}, H_D\right] = (-i)\left(-ig\alpha^m A_m^b \left[\frac{\tau^a}{2}, \frac{\tau^b}{2}\right]\right)$$

$$(16.16)$$

$$= -\frac{i}{2r}(1-F)(\alpha^a\tau^b n_b - n^a\alpha^m\tau^m) + \frac{h\beta}{2}i\varepsilon_{abc}\varphi^b\tau^c.$$

The sum of the quantities (16.14), (16.15) and (16.16) is equal to zero, which proves (16.13).

The quantum-mechanical operators (16.11) obey the commutational relations of the angular momentum

$$[J^a, J^b] = i\varepsilon_{abc}J^c.$$

$$(16.17)$$

One somewhat unusual property of this angular momentum is that it contains the isospin operator $\frac{\tau^a}{2}$ ("spin from isospin"). This property means, according to the rule for composition of momenta, that the angular momentum J^a is integer valued, despite the fact that we are dealing with fermions.

Problem 2. *As discussed in Section 9.2, far from the center of the monopole, its configuration can be brought to a unitary gauge by a gauge transformation. Then, only the electromagnetic potential A_μ^3 will be non-zero. Find an expression for conserved angular momentum of the fermion far from the center of the monopole in the unitary gauge; express it in terms of the fermion electric charge. Show, by explicit computation, that the commutational relations (16.17) are satisfied for the angular momentum obtained.*

For the sequel, it is convenient to make a change of variables in the Dirac equation. We recall that the wave function ψ contains eight components $\psi_{\kappa i}$, where $\kappa = 1, \ldots, 4$ is the Lorentz index and $i = 1, 2$ is the isotopic index (index, corresponding to the $SU(2)$ gauge symmetry). The Dirac matrices α^a and β act on the index κ, and the isotopic matrices τ^a act on the index i,

$$(\tau^a \psi)_{\kappa i} = \tau^a_{ij} \psi_{\kappa j}.$$

It is convenient to interpret $\psi_{\kappa i}$ as a 4×2 matrix; here, the action τ^a is written in the form $(\psi \tau^{aT})_{\kappa i}$, where the brackets contain the product of a 4×2 and a 2×2 matrix. A convenient change of variables involves the introduction of the matrix $\tilde{\psi}_{\kappa i}$ (also 4×2) according to the rule

$$\psi_{\kappa i} = \tilde{\psi}_{\kappa j} \varepsilon_{ji}, \qquad (16.18)$$

where, as usual, ε_{ij} is an antisymmetric tensor of rank two. Then

$$\psi \tau^{aT} = \tilde{\psi} \varepsilon \tau^{aT} = -\tilde{\psi} \tau^a \varepsilon,$$

where all the products are matrix products, and ε is a 2×2 matrix with matrix elements ε_{ij}. The substitution (16.18) is convenient because we eliminate the transposed Pauli matrices: the Dirac equation in terms of $\tilde{\psi}$ has the form

$$i \frac{\partial \tilde{\psi}}{\partial x^0} = -i\alpha^a \partial_a \tilde{\psi} + \frac{g}{2} A^b_a (\alpha^a \tilde{\psi} \tau^b) - h\varphi^a (\beta \tilde{\psi} \tau^a), \qquad (16.19)$$

where the expressions in brackets have to be understood in the sense of matrix products. Analogously, the action of the operator J^a has the form

$$J^a \tilde{\psi} = -i\varepsilon_{abc} x^b \partial_c \tilde{\psi} + s^a \tilde{\psi} - \tilde{\psi} \frac{\tau^a}{2}.$$

We note that this change of variables and the rewriting of the Dirac equation in matrix form (16.19) are convenient not only in the case of a monopole, but also for other external fields.

The operator J^a does not intermix the left- and right-handed components of the wave function ψ: if the original wave function ψ is represented in the form

$$\psi = \begin{pmatrix} \chi \\ \eta \end{pmatrix},$$

where χ and η are the left- and right-handed components, respectively, then the operator J^a will act as follows:

$$J^a \psi = \begin{pmatrix} j^a \chi \\ j^a \eta \end{pmatrix},$$

where

$$j^a = L^a + \frac{1}{2}\sigma^a + \frac{1}{2}\tau^a. \tag{16.20}$$

In terms of the transformed function $\tilde\psi$ we will have

$$\tilde\psi = \begin{pmatrix} \tilde\chi \\ \tilde\eta \end{pmatrix},$$

where $\tilde\chi$ and $\tilde\eta$ are 2×2 matrices. Here, the operator j^a acts on $\tilde\chi$ and $\tilde\eta$ as

$$j^a\tilde\chi = -i\varepsilon_{abc}x^b\partial_c\tilde\chi + \frac{1}{2}(\tau^a\tilde\chi - \tilde\chi\tau^a). \tag{16.21}$$

Since we are using the matrix formulation, we have unified the notation for the Pauli matrices in (16.21): instead of σ^a we use τ^a. The action of j^a on $\tilde\eta$ coincides literally with (16.21). Taking into account the explicit form (14.5) of the matrices α^a and β, equation (16.19) can be written in terms of the left- and right-handed components in the form of the system

$$i\frac{\partial\tilde\chi}{\partial x^0} = i\tau^a\partial_a\tilde\chi - \frac{g}{2}A_b^a(\tau^b\tilde\chi\tau^a) - \frac{h}{2}\varphi^a(\tilde\eta\tau^a), \tag{16.22}$$

$$i\frac{\partial\tilde\eta}{\partial x^0} = -i\tau^a\partial_a\tilde\eta + \frac{g}{2}A_b^a(\tau^b\tilde\eta\tau^a) - \frac{h}{2}\varphi^a(\tilde\chi\tau^a).$$

Wave functions with zero angular momentum $J^a = 0$ are of particular interest.

As far as their explicit construction is concerned, we note that, according to (16.20), the operator j^a is, formally speaking, a sum of three angular momentum operators. According to the rule for the composition of momenta, there are two eigenfunctions with $j^a = 0$, one of which has $l = 0$, and the other $l = 1$. The first wave function $\chi_{0,0}$ does not depend on angles, and the sum of the spin and isospin momenta for it is equal to zero. In terms of the function $\tilde\chi_{0,0}$, this last property means (see (16.21)) that

$$[\tau^a, \tilde\chi_{0,0}] = 0,$$

i.e.

$$\tilde\chi_{0,0} = \mathbf{1} \cdot u_1(r,t),$$

where $\mathbf{1}$ is the unit 2×2 matrix, and u_1 is a complex function of r and t. The wave function with $l = 1$ is proportional to the unit radius vector n^a; the requirement that it should be invariant under a combination of spatial and isotopic rotations (i.e. $j^a = 0$) immediately leads to the expression

$$\tilde\chi_{0,1} = \tau^a n^a v_1(r,t).$$

Right-handed wave functions with zero angular momentum have precisely the same form:

$$\tilde{\eta}_{0,0} = \mathbf{1} \cdot u_2(r,t); \quad \tilde{\eta}_{0,1} = \tau^a n^a v_2(r,t). \tag{16.23}$$

Thus, in the sector with zero angular momentum, we have, in the general case

$$\tilde{\chi} = \mathbf{1} \cdot u_1(r,t) + \tau^a n^a v_1(r,t) \tag{16.24}$$

and analogously for $\tilde{\eta}$.

Problem 3.   *Show by direct computation that $j^a \tilde{\chi}_{0,1} = 0$.*

Let us show that, for $h \neq 0$, there exists a normalizable eigenfunction of the Dirac Hamiltonian (16.12) with zero eigenvalue; this is the zero mode, completely analogous to that considered in Section 15.1. In other words, we need to find a smooth, time-independent solution of equations (16.22) which decreases rapidly as $|\mathbf{x}| \to \infty$. Clearly, the solution must be sought in the sector with $J^a = 0$. Moreover, near the center of the monopole we have $A_i = 0$, $\varphi = 0$, therefore the Dirac Hamiltonian reduces to the free one. It will not contain a centrifugal barrier, if $\tilde{\chi}$ and $\tilde{\eta}$ are angle independent. These considerations suggest the Ansatz

$$\tilde{\chi} = \mathbf{1} \cdot u_1(r); \quad \tilde{\eta} = \mathbf{1} \cdot u_2(r), \tag{16.25}$$

where $u_1(r)$ and $u_2(r)$ are two complex functions to be determined. The fact that the Ansatz (16.25) "passes through" equation (16.22) is not completely trivial: in the background fields (16.3), (16.4), the right-hand sides of these equations are proportional to the matrix $(\tau^a n^a)$, and the equality to zero of the coefficients leads to precisely two equations for the unknown functions $u_1(r)$ and $u_2(r)$,

$$i\left(u_1' + \frac{1-F}{r} u_1\right) - m_F(1-H)u_2 = 0, \tag{16.26}$$

$$-i\left(u_2' + \frac{1-F}{r} u_2\right) - m_F(1-H)u_1 = 0,$$

where m_F is given by formula (16.10).

Problem 4.   *Show that, for the Ansatz (16.25), equations (16.22) reduce to the system (16.26).*

The only decreasing solution of the system (16.26) is

$$u_2 = -iu_1 = C \exp\left\{-\int_0^r \left[\frac{1-F}{r} + m_F(1-H)\right] dr\right\}, \tag{16.27}$$

where C is a normalization constant. Because of the property (16.5) this solution is smooth as $r \to 0$ and decreases as $e^{-m_F \cdot r}/r$, as $r \to \infty$.

Problem 5. *Show that a second linearly independent solution of the system (16.26) grows as $r \to \infty$.*

Problem 6. *Write down a system of equations for time-independent wave functions with $J^a = 0$ of general form (that is, suppose that $v_1(r)$ and $v_2(r)$ are non-zero, see (16.24), (16.23)). Show that (16.27) is the unique solution which is smooth at $r = 0$ and decreases as $r \to \infty$.*

The result of the last problem means that the Dirac Hamiltonian in the monopole background field has exactly one zero mode, at least in the sector with $\mathbf{J} = 0$. In sectors with $\mathbf{J} \neq 0$ there are no zero modes; the corresponding analysis is quite complex, and we shall not present it here.

In complete analogy with Section 15.1, the existence of a unique fermion zero mode means that in the given model there are two degenerate monopole states (Jackiw and Rebbi 1976b): in one of these the fermion level with zero energy is occupied, in the other it is not. These states have fermion numbers $\left(+\frac{1}{2}\right)$ and $\left(-\frac{1}{2}\right)$, respectively. Thus, the model of this section demonstrates that the fermion number of topological solitons in four-dimensional space–time can be fractional.

Problem 7. *Show that the Dirac equation in arbitrary real external fields is invariant under the following analogue of C-conjugation,*

$$\psi_{\alpha i} \to \psi^c_{\alpha i} = (\gamma^5 C)_{\alpha\beta} \varepsilon_{ij} \psi^*_{\beta j},$$

where $\alpha = 1,\ldots,4$ and $i = 1, 2$ are the Lorentz and internal indices, C is the usual C-conjugation matrix (14.14). This operation clearly changes the sign of the fermion energy. Show that the fermion zero mode found is invariant under this operation. (Note that here we have a complete analogy with Section 15.1).

16.2 Scattering of fermions off a monopole: non-conservation of fermion numbers

In this section, we consider the asymptotic scattering states of massless s-wave fermions (i.e. fermions with $\mathbf{J} = 0$) in the presence of a magnetic monopole. Taking into account additional considerations relating to the conservation of electric charge, we show that the scattering of fermions off a monopole must lead to non-conservation of the fermion numbers (Rubakov 1981, 1982, Callan 1982). However, our simple analysis does not allow us to identify the mechanisms for this non-conservation; in fact, these mechanisms are highly complex and depend on the model, and their description, in any case, requires an analysis in the framework of quantum field theory. In many Grand Unified Theories, the interaction of fermions with monopoles leads to monopole catalysis of the proton decay, a process of type

$$p + \text{monopole} \rightarrow e^+ + \text{monopole},$$

which must take place with a high probability (Rubakov 1981, Callan 1982). This is a reason for the considerable interest in the questions discussed in this section.

The model which we shall use differs from the model of Section 16.1 only in the fact that the Yukawa interaction with the Higgs field is switched off, i.e. $h = 0$. In the vacuum (16.1) the fermions have no mass. Since the equations for the left- and right-handed components of the wave function decouple, it makes sense to consider the Weyl equation for the left-handed fermions in the field of a magnetic monopole. In terms of the matrix $\tilde{\chi}(x)$, it has the form (see (16.22))

$$i\frac{\partial \tilde{\chi}}{\partial x^0} = i\tau^a \partial_a \tilde{\chi} - \frac{g}{2} A_b^a \tau^b \tilde{\chi} \tau^a. \tag{16.28}$$

We shall be interested in the asymptotic ($|\mathbf{x}| \rightarrow \infty$) states of fermions.[1] For large $|\mathbf{x}|$ the field A_b^a has the form

$$A_b^a(\mathbf{x}) = \frac{1}{gr}\varepsilon^{abc}n_c. \tag{16.29}$$

Fermions with zero angular momentum are most interesting: their wave functions have the form (16.24).

Furthermore, for large $|\mathbf{x}|$ the monopole field is purely electromagnetic. This means that fermions with electric charges $(+\frac{1}{2}g)$ and $(-\frac{1}{2}g)$ behave independently. In a regular gauge the fermion charge operator is equal to

$$Q = \frac{1}{2}\tau^a n^a,$$

as can be seen from (16.6). In terms of the function $\tilde{\chi}$ we have

$$Q\tilde{\chi} = -\frac{1}{2}(\tilde{\chi}\tau^a)n^a.$$

The states of the s-wave fermions with specific electric charges have the form

$$Q = +\frac{1}{2}: \quad \tilde{\chi}_+ = (1 - \tau^a n^a)u_+, \tag{16.30}$$

$$Q = -\frac{1}{2}: \quad \tilde{\chi}_- = (1 + \tau^a n^a)u_-, \tag{16.31}$$

---

[1] For $h = 0$, there is no normalizable fermion zero mode of the Dirac Hamiltonian. This is clear from (16.27): for $m_F = 0$, the right-hand side of (16.27) behaves as $1/r$ as $r \rightarrow \infty$.

where u_+ and u_- are, for the moment, arbitrary complex functions. In the general case the s-wave function is equal to

$$\tilde{\chi} = \tilde{\chi}_+ + \tilde{\chi}_-, \tag{16.32}$$

and we expect that the equations for u_+ and u_-, following from (16.28), will decouple in the external field (16.29). Indeed, by direct substitution of expressions (16.32), (16.30) and (16.31) in equation (16.28), we obtain in the field (16.29), separate equations for u_+ and u_-:

$$\dot{u}_+ = -\left(u_+' + \frac{1}{r}u_+\right), \tag{16.33}$$

$$\dot{u}_- = u_-' + \frac{1}{r}u_-, \tag{16.34}$$

where the dots and the primes denote differentiation with respect to time and with respect to r, respectively.

Problem 8. *Show that in the external field (16.4) with $F \neq 0$, the equations for u_+ and u_- have the form*

$$\dot{u}_+ = -\left(u_+' + \frac{1}{r}u_+\right) - \frac{1}{r}Fu_-,$$

$$\dot{u}_- = u_-' + \frac{1}{r}u_- + \frac{1}{r}F'u_+.$$

The last terms in these equations mix components of the wave function with different electric charges (this occurs because of the presence of charged vector fields in the monopole core).

Solutions of equations (16.33) and (16.34) with a fixed energy have the form

$$u_+ = \frac{1}{r}e^{-i\omega(x_0-r)}, \tag{16.35}$$

$$u_- = \frac{1}{r}e^{-i\omega(x_0+r)}. \tag{16.36}$$

We see that the wave functions of the left-handed fermions with positive electric charge contain only outgoing waves, while the wave functions of negatively charged left-handed fermions have only incoming waves!

One immediate consequence of the result (16.36) is the fact that s-wave negatively charged fermions reach the core of the monopole with unit probability. Even if the size of the monopole core is small in comparison with the fermion wavelength (in other words, even for $m_v \gg \omega$, where m_v is the vector boson mass), s-wave fermions interact strongly with the monopole core and probe its structure. This phenomenon is quite unusual:

as a rule, the probability of the interaction of particles with an object of small size is suppressed by this size. The reason why the opposite situation occurs in our case ultimately lies in the strong interaction of charged fermions with the long-ranged monopole magnetic field, which leads to the effect of a type of fall on the center.

Let us now discuss in more detail the possible final state of the process, when the negatively charged fermion in s-wave state is incident on the monopole. Let us suppose that in the final state there is also one particle, fermion or antifermion. First, we note that a negatively charged s-wave fermion cannot occur in the final state, since the wave function (16.36) does not contain outgoing waves. One option would be to convert the negatively charged fermion to a positively charged one with wave function (16.35). At the level of quantum mechanics in the external monopole field, precisely this possibility is realized: the Dirac equation near the monopole core contains terms which mix u_- and u_+ (see Problem 8). It is clear, however, that here we have a case where approximation of the external bosonic field is not appropriate. Indeed, because of the conservation of electric charge, the transition $u_- \to u_+$ must be accompanied by the appearance of an electric charge on the monopole, the monopole must transfer to the dyon[2] with charge $(-g)$. The mass of this dyon is greater than the monopole mass, and the scale of the mass difference is set by the mass of the vector boson m_v. Consequently, for low (in comparison with m_v) energies of the colliding fermion, the process

$$u_- + \text{monopole} \to u_+ + \text{dyon}$$

is impossible because of conservation of energy. We have to search for other versions of the final state.

In principle, an s-wave fermion may transfer to a fermion with the same charge, but a different angular momentum. This possibility must also be discarded at low energies of the colliding fermion: in such a process the monopole would acquire angular momentum, which would also require energy of scale m_v (in addition, fermions with non-zero angular momentum experience a centrifugal barrier, while the process must take place near the monopole core).

It remains to search for negatively charged fermions or antifermions with zero angular momentum, whose wave functions have the form of outgoing spherical waves. If there are no right-handed fermions in the model then the only candidate is a negatively charged *antifermion*, the antiparticle to the fermion u_+. We conclude that the properties of the asymptotic states (16.35), (16.36), together with the laws of the conservation of energy, angular momentum and electric charge require violation of the law of

---

[2]Precisely at this point, the assumption that the back reaction of fermions on the bosonic fields can be neglected is violated.

conservation of the fermion number (in transmutation of the fermion into the antifermion) with unit probability in the s-wave sector!

In fact, the situation is even more complex. A model with gauge group $SU(2)$ and one left-handed fermion doublet does not exist at the level of quantum field theory because of a global anomaly (Witten 1982). If we add a second left-handed doublet $\binom{u'_+}{u'_-}$, then there is no global anomaly, and the theory is self-consistent. In this model, there is a possibility of anomalous non-conservation of the fermion numbers, which we shall discuss in the next chapter. Corresponding selection rules permit essentially a unique final state for the process, in which a charged fermion of the first type u_- is incident in the s-wave state on the monopole; the process has the form

$$u_- + \text{monopole} \to \overline{(u'_+)} + \text{monopole},$$

where $\overline{(u'_+)}$ denotes the antiparticle to the positively charged fermion of the second type.

If right-handed fermions are included in the model, then their wave functions in the s-wave state will contain only incoming waves for a positive electric charge and outgoing waves for a negative charge. Thus, the process will take place without a change in the total fermion number

$$u_-^L + \text{monopole} \to u_-^R + \text{monopole}. \tag{16.37}$$

In this process, however, the number of left-handed fermions N_L and the number of right-handed fermions N_R, which are conserved in weak fields, are not conserved individually.[3]

Problem 9. *Find asymptotic s-wave functions of right-handed fermions in the background monopole field as $|\mathbf{x}| \to \infty$.*

Thus, one characteristic property of the interaction of massless fermions with a magnetic monopole is the non-conservation of the fermion numbers. Even for low fermion energies and small monopole core sizes this happens with a high probability: the cross section is determined by the probability of having colliding fermions in an s-wave state, which is, roughly speaking, of the order of unity (in fact, the cross section grows as the energy decreases). As previously mentioned, in realistic Grand Unified Theories this phenomenon leads to decay of the proton, induced by the monopole; the cross section of this process is of the order of the hadronic cross sections (and even greater), despite the fact that the size of the monopole core is fantastically small (of the order of 10^{-30} cm).

---

[3]We note that the selection rule $\Delta N_R = -\Delta N_L = 1$ for the process (16.37) is analogous to the properties (15.34), (15.35). This analogy is not accidental: non-conservation of the fermion numbers occurs in the models of this section and of Section 15.2, by virtue of one and the same mechanism.

16.3   Zero modes in a background field of a vortex: superconducting strings

In Sections 15.1 and 16.1, we saw that in external fields of topological solitons, there may exist fermion zero modes, the eigenfunctions of the Dirac Hamiltonian with zero energy. In this section, we shall show that for a specific choice of gauge and Yukawa interactions, fermion zero modes exist in the field of a vortex[4] (Jackiw and Rossi 1981). The situation where the vortex is viewed as an extended one-dimensional object, a string, in four-dimensional space–time is of particular interest. It is precisely this situation which we consider in this section. In this case, the existence of zero modes of the transverse part of the Dirac operator leads to the occurrence of fermion states localized on the string, which may move freely along the string. Strings in such models may be superconducting (Witten 1985); this property is interesting from the point of view of possible astrophysical and cosmological applications.

As an example, let us consider the vortex discussed in Section 7.3. The model has a gauge symmetry $U(1)_R$ and a Higgs field φ with R-charge e_R (we have introduced the index R to distinguish the gauge group $U(1)_R$ from the unbroken group of electromagnetism, which we shall introduce later; the gauge field of the group $U(1)_R$ will be denoted by B_μ). The structure of the vortex field has the form

$$
\begin{aligned}
\varphi(r,\theta) &= v e^{i\theta} F(r), \\
B_\alpha(r,\theta) &= -\frac{1}{e_R r}\varepsilon_{\alpha\beta} n_\beta B(r) \\
B_0 = B_3 &= 0.
\end{aligned}
\tag{16.38}
$$

Here $\alpha, \beta = 1, 2$, r and θ are the radius and polar angle in the plane $x^3 = $ constant, and $n_\alpha = x_\alpha/r$. The vortex field does not depend on x^3, the vortex is an infinite straight string, directed in space along the third axis. The functions $F(r)$ and $B(r)$ have the following boundary values:

$$
\begin{aligned}
F(0) &= B(0) = 0, \\
F(\infty) &= B(\infty) = 1.
\end{aligned}
$$

We note that we have changed the notation for the radial function of the vector field from that of Section 7.3.

Let us now introduce fermions, interacting with bosonic fields. In fact, we have already constructed a model with a non-trivial Yukawa interaction

---

[4]Fermion zero modes in the vortex field have been discussed in the context of two-dimensional instantons in Nielsen and Schroer (1977a), Kiskis (1977), Ansourian (1977) and by other authors.

at the end of Section 14.3. Namely, suppose the left-handed fermion component χ has R-charge $(-\frac{1}{2}e_R)$, and the right-handed component η has charge $(+\frac{1}{2}e_R)$. Then the Dirac equation has the form of the system (14.59), which we write down again for further reference:

$$i\tilde{\sigma}^\mu D_\mu^{(-)}\chi - h\varphi^*\eta \;=\; 0 \tag{16.39}$$

$$i\sigma^\mu D_\mu^{(+)}\eta - h\varphi\chi \;=\; 0, \tag{16.40}$$

where $D_\mu^{(\pm)} = \partial_\mu \mp i\frac{1}{2}e_R B_\mu$, and the constant h is real. In the background of the vacuum state $\varphi = v$, $B_\mu = 0$, this system describes fermions with mass $m_F = hv$. We note that systems of this type arise in certain Grand Unified Theories.

In arbitrary external fields, the system (16.39), (16.40) has the property of C-symmetry: if

$$\psi = \begin{pmatrix} \chi \\ \eta \end{pmatrix} \tag{16.41}$$

satisfies this system, then $\psi^C = C\psi^*$ also satisfies this system of equations, where the matrix C, as before, is defined by formula (14.14). In terms of left- and right-handed components, the C-conjugation operation has the form

$$\chi \to -\varepsilon\eta^*, \qquad \eta \to \varepsilon\chi^*. \tag{16.42}$$

Problem 10. *Show by explicit computation that the system (16.39), (16.40) is invariant under the C-conjugation operation (16.42).*

Let us now discuss solutions of the system of equations (16.39), (16.40) in the external field of the string. Since the string fields (16.38) do not depend on x^0 and x^3, a solution may be sought in the form

$$\psi(x^0, x^3; x^\alpha) = e^{-i\omega x^0 + ik_3 x^3}\psi_T(x^\alpha). \tag{16.43}$$

Our next task is to find properties of the energy spectrum ω and of the corresponding eigenfunctions. Since far from the string the fields B_μ and φ tend to their values in the ground state (up to a gauge transformation), the part of the spectrum with $|\omega| > m_F$ coincides with the spectrum of the free Dirac equation, and the corresponding wave functions far from the string are (gauge transformed) conventional plane waves. Besides these, there may exist states with $|\omega| < m_F$, whose transverse wave functions $\psi_T(x^i)$ are localized near the string. We shall be interested in precisely this part of the spectrum in what follows.

Taking into account the fact that for the string $B_0 = B_3 = 0$, we obtain the following equation for ψ_T from (16.39), (16.40),

$$(k_3 C_T + D_T)\psi_T = \omega\psi_T, \tag{16.44}$$

where

$$C_T = \begin{pmatrix} -\sigma^3 & 0 \\ 0 & \sigma^3 \end{pmatrix},$$

$$D_T = \begin{pmatrix} i\tilde{\sigma}^\alpha D_\alpha^{(-)} & h\varphi^* \\ h\varphi & i\sigma^\alpha D_\alpha^{(+)} \end{pmatrix} \qquad (16.45)$$

are 4×4 matrices. The transverse part of the Dirac operator D_T has a number of important properties. The first of these has to do with the fact that the string configuration is invariant under spatial rotations around the third axis, $\theta \to \theta + \alpha$, supplemented by phase transformations $\varphi \to e^{-i\alpha}\varphi$. Thus, the operator analogous to the third component of the angular momentum is conserved (compare with section 16.1),

$$J_3 = L_3 + s_3 - R, \qquad (16.46)$$

where, as usual, $L_i = -i\varepsilon_{ijk}x_j\partial_k$ $(i,j,k = 1,2,3)$, i.e.

$$L_3 = -i\varepsilon_{\alpha\beta}x_\alpha\partial_\beta$$

is the third component of the orbital angular momentum,

$$s_3 = \frac{1}{2}\begin{pmatrix} \sigma^3 & 0 \\ 0 & \sigma^3 \end{pmatrix}$$

is the third component of the spin and

$$R = \frac{1}{2}\begin{pmatrix} -1 & 0 \\ 0 & 1 \end{pmatrix}$$

is the R-charge operator.

Problem 11. *Show by direct computation that the operator J_3 commutes with the Hamiltonian $(k_3 C_T + D_T)$ entering (16.44), if the external bosonic fields have the form (16.38).*

The second property is that the operators C_T and D_T anticommute,

$$C_T D_T + D_T C_T = 0. \qquad (16.47)$$

As we now see, this property enables us to relate the energy spectrum ω to the spectrum of the operator D_T.

Problem 12. *Verify that equation (16.47) holds.*

Now suppose that $\psi_{T,\lambda}(x^\alpha)$ are eigenfunctions of the operator D_T with positive eigenvalues λ,

$$D_T \psi_{T,\lambda}(x^\alpha) = \lambda \psi_{T,\lambda}(x^\alpha).$$

Then, by virtue of (16.47), the functions

$$\psi_{T,\lambda}^{C_T} = C_T \psi_{T,\lambda} \tag{16.48}$$

will be eigenfunctions of the operator D_T with eigenvalues $(-\lambda)$,

$$D_T \psi_{T,\lambda}^{C_T} = -\lambda \psi_{T,\lambda}^{C_T}.$$

This correspondence between the eigenfunctions is one to one[5] (for $\lambda \neq 0$), since $C_T^2 = 1$. To find the energy spectrum ω we shall seek a solution of equation (16.44) in the form

$$\psi_T = u\psi_{T,\lambda} + v\psi_{T,\lambda}^{C_T}.$$

Substituting this expression in (16.44) and using the linear independence of $\psi_{T,\lambda}$ and $\psi_{T,\lambda}^{C_T}$, we obtain for u and v the equations

$$
\begin{aligned}
(\omega - \lambda)u - k_3 v &= 0, \\
-k_3 u + (\omega + \lambda)v &= 0.
\end{aligned}
$$

These equations are soluble for

$$\omega^2 = \lambda^2 + k_3^2, \tag{16.49}$$

which provides the relation between the eigenvalues of the operator D_T and the fermion energies.

According to what has already been said, the spectrum of the values of λ is continuous for $\lambda > m_F$, and the wave functions $\psi_{T,\lambda}(x^\alpha)$ are not localized near $x^\alpha = 0$ for such λ. At the same time, there may exist a discrete set of eigenvalues λ of the operator D_T with eigenfunctions $\psi_T(x^\alpha)$, confined near $x^\alpha = 0$. From dimensional analysis, it is clear that the distance between these levels is of the order of m_F. Fermions lying in these levels are localized near the string and propagate freely along the string (along the third axis): their wave functions have the form (16.43).

Until now we have assumed that $\lambda \neq 0$. However, the possibility that the operator D_T may have a *zero* eigenvalue is of particular interest. We shall show that this possibility is indeed realized and we shall find the corresponding eigenfunction in explicit form. By virtue of (16.47), the eigenfunction $\psi_{T,0}(x^\alpha)$ is at the same time an eigenfunction of the operator C_T. From physical considerations, it is clear that it must have zero "angular momentum," J_3, and also zero orbital momentum L_3 (the last property comes from the fact that near the center of the string, i.e. for $x^\alpha = 0$, the vortex fields are equal to zero, so that for $L_3 \neq 0$, there exists a centrifugal

---

[5] The operator C_T may be called the C-conjugation operator in the space of transverse functions, since its action changes the sign of an eigenvalue of the operator D_T.

barrier). Thus, for the zero mode we have $s_3 = R$, which is equivalent to $C_T = 1$; moreover, the function $\psi_{T,0}(x^\alpha)$ is a function of r only, i.e. it does not depend on the polar angle θ.

From the equality

$$C_T\psi_{T,0} = \psi_{T,0},\tag{16.50}$$

it follows that the left- and right-handed components of the function $\psi_{T,0}$ have the structure

$$\chi_{T,0} = \begin{pmatrix} 0 \\ f_1(r) \end{pmatrix}, \qquad \eta_{T,0} = \begin{pmatrix} f_2(r) \\ 0 \end{pmatrix},\tag{16.51}$$

where $f_1(r)$ and $f_2(r)$ are yet unknown complex functions. Using the explicit form (16.45) of the operator D_T and the string field (16.38), we obtain for the functions $f_1(r)$ and $f_2(r)$ the equations

$$f_1' - \frac{B}{2r}f_1 + ihvFf_2 = 0,$$

$$f_2' - \frac{B}{2r}f_2 - ihvFf_1 = 0.\tag{16.52}$$

It is important that the dependence on the angle θ in these equations drops out. Equations (16.52) can be easily solved: the solution, decreasing for $r \to \infty$, has the form

$$f_1(r) = -if_2(r) = C\exp\left[-\int_0^r\left(\frac{B(r)}{2r} + hvF(r)\right)dr\right],\tag{16.53}$$

where C is a normalization constant. For small r this solution is regular, and for large r it behaves as

$$f_1(r) \sim f_2(r) \sim \frac{e^{-m_F r}}{\sqrt{r}}.$$

The second linearly independent solution of the system (16.52) grows as $r \to \infty$.

Thus, the operator D_T indeed has a zero mode. Its explicit form is given by formulae (16.51) and (16.53). We note that the zero mode is invariant (up to a phase factor) under the usual C-transformations (compare with (16.42))

$$\chi_T \to -\varepsilon\eta_T^*, \qquad \eta_T \to \varepsilon\chi_T^*,$$

as one might expect.

Problem 13. *Show that for the fermion wave functions (16.51), the equation $D_T\psi_T = 0$ reduces to the system (16.52).*

Problem 14. *Consider the equation $D_T\psi_T = 0$ in the sector with $J_3 = 0$ and $C_T = -1$. Find the angular dependence and the structure of the fermion wave function (analogue of formula (16.51)). Write down corresponding radial equations and show that they do not have normalizable solutions.*

Let us turn to a discussion of the wave function (16.43), with the transverse part $\psi_T(x^\alpha)$ being the zero mode. Taking into account (16.50), equation (16.44) implies that

$$k_3 = \omega, \tag{16.54}$$

i.e. the fermions (with $\omega > 0$) move along the string *in one direction only*. Moreover, ω can take values arbitrarily close to zero, i.e. the spectrum has no gap, unlike the states with $\lambda \neq 0$, see (16.49). In other words, fermions lying at a level with $\lambda = 0$ behave as massless particles with a specific chirality in $(1+1)$-dimensional space–time. As we shall now see, when an electromagnetic field is included in the model, this property means that the string behaves as a superconducting thin wire.

Interaction of the fermions with the electromagnetic field can be introduced, by assuming that the field φ is electrically neutral and attaching the same electric charge e to the left- and right-handed components χ and η. Then equations (16.39) and (16.40) will have the previous form and, taking into account the electromagnetism, the covariant derivatives in them will be replaced by

$$D^{(\pm)}_\mu = \partial_\mu \mp i\frac{e_R}{2}B_\mu - ieA_\mu, \tag{16.55}$$

where the A_μ are electromagnetic vector potentials.

We shall show that there exist states of the string with an electric current, and this current does not dissipate (superconductivity). For this, let us consider a state of the fermion system in which all negative energy levels with $\lambda \neq 0$ are occupied, all positive energy levels with $\lambda \neq 0$ are empty, and levels with $\lambda = 0$ are filled up to some energy $\omega = \mu > 0$ (Fermi energy). This situation is illustrated in Figure 16.1. We assign zero fermion number, zero electric charge, zero electric current, etc. to the Dirac sea[6] (the state with all levels with $\omega < 0$ occupied and those with $\omega > 0$ empty). Thus, the state of the fermion system shown in Figure 16.1 has a finite linear fermion density in the string. Since fermions with $\lambda = 0$ move along the string in one direction, electric current flows along the string. Moreover, the minimal non-zero value $|\lambda|_{\min}$ of the quantity $|\lambda|$ is finite,

---

[6]As mentioned in Section 15.1, in the presence of the external bosonic field, the Dirac sea may be polarized, i.e. it may have a non-zero charge and also, generally speaking, a non-zero electric current. However, they do not depend on μ, unlike the corresponding quantities which arise thanks to the real fermions; thus, the arguments which follow remain valid.

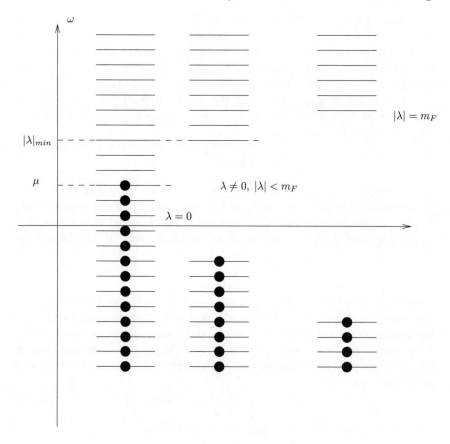

Figure 16.1.

and the continuous spectrum of the operator D_T begins at $|\lambda| = m_F$. As mentioned earlier, $|\lambda|_{\min} \sim m_F$. Because of this, fermion energy levels with $\lambda \neq 0$ are separated from zero by a gap of finite height, as shown in Figure 16.1. If $\mu < |\lambda|_{\min}$, real fermions with $\lambda = 0$ and $0 < \omega < \mu$ cannot lose energy or momentum by jumping to a lower level, since all lower levels are occupied. Consequently, the state of the fermion system shown in Figure 16.1 is absolutely stable. The electric current cannot dissipate, the string is superconducting. We note that if the levels with $\lambda = 0$ are occupied up to $\omega = \mu > |\lambda|_{\min}$, then dissipation of the electric current is possible because of fermion jumps from levels with $\lambda = 0$ to levels with $\lambda \neq 0$ (including levels in the continuous spectrum for $\mu > m_F$). Thus, there exists a maximum value of the electric current, up to which the string remains superconducting.

We shall find the linear density of fermions and the electric current in the state shown in Figure 16.1 using the results of Section 15.2. Since the fermions with zero D_T behave as one-dimensional massless fermions with a specific chirality, their linear density is equal to (see (15.25))

$$\frac{N}{L} = \frac{\mu}{2\pi}.$$

The fermions move along the string with the speed of light; therefore, their electric current is equal to

$$j_3^z = \frac{e\mu}{2\pi}.$$

According to what has already been said, the string remains super-conducting for $\mu < \mu_{\max}$, where $\mu_{\max} \sim m_F$. Hence, it follows that the maximal superconducting current is of order

$$j_3^{\mathrm{em,max}} \sim \frac{e m_F}{2\pi}.$$

Problem 15. *Estimate the maximum value (in Amperes) of the superconducting current flowing along the string, assuming that e and m_F are the electron charge and mass, respectively.*

If an electric field (not too strong) is imposed along the string, then fermions with $\lambda = 0$ will be created. The corresponding analysis repeats the arguments of Section 15.2 word for word; in particular, the number of fermions created is given by formula (15.26). Thus, here we have an explicit example of level crossing in a four-dimensional model. In the model of this section, as it stands, the electric charge is not conserved either (the effective one-dimensional fermions have a specific chirality). This points to an internal consistency of the model, which is indeed confirmed by a triangle anomaly computation in quantum field theory.

The model ceases to be internally inconsistent if another type of fermion

$$\hat{\psi} = \begin{pmatrix} \hat{\chi} \\ \hat{\eta} \end{pmatrix}$$

is added to it; here the R-charges of the left- and right-handed components are $(\frac{1}{2}e_R)$ and $(-\frac{1}{2}e_R)$, respectively (i.e. the components of ψ and $\hat{\psi}$ have opposite R-charges). The electric charge of the fermion $\hat{\psi}$ can be chosen equal to $(-e)$. Then fermions $\hat{\psi}$ with $\lambda = 0$ will have dispersion law $k_3 = -\omega$, i.e. they would move along the string in the opposite direction to ψ. The electric field, as before, will create fermions $\hat{\psi}$ (they will have a negative electric charge and will accelerate toward negative x^3, as permitted by the dispersion law); the number of these fermions will be equal to the number of fermions ψ created. Hence, the total fermion number will not be conserved, but the electric charge is conserved, as is required for consistency of the model. The total electric current along the string will be non-zero (this is generated by both fermions ψ with charge $(+e)$ moving along the electric field, and by fermions $\hat{\psi}$ with charge $(-e)$ moving in the opposite direction). This current, if it is not too large, will be superconducting.

Thus, in the model considered in this section, the string has the superconductivity property, and the application of a longitudinal electric field leads to non-conservation of the fermion numbers as a result of fermion level crossing. From the theoretical point of view the direct analogy between the fermions localized in the string and the one-dimensional fermions of Section 15.2 is interesting.

We note that superconductivity of a string is not necessarily associated with fermions. Witten (1985) revealed the possibility that superconducting strings can exist in purely bosonic theories. We stress that the super-conductivity property is in no way general for strings which arise in unified gauge theories of elementary particles.

Chapter 17

Non-Conservation of Fermion Quantum Numbers in Four-dimensional Non-Abelian Theories

In Section 15.2, for the example of a two-dimensional Abelian model, we considered a mechanism for the non-conservation of fermion quantum numbers, associated with the occurrence of fermion level crossing. In this chapter, we shall see that this mechanism also works in non-Abelian four-dimensional theories. In the Standard Model, it leads to electroweak non-conservation of the baryon and lepton numbers ('t Hooft 1976a,b). Under the usual conditions (at low temperatures and densities or in collisions of particles whose energies are not too high) the probabilities of electroweak processes with non-conservation of the baryon number are extremely small, since they are due to instantons and are strongly suppressed by the tunneling exponential. However, the probability of these processes ceases to be small at sufficiently high temperatures, which is important for cosmology (Kuzmin *et al.* 1985).

The quantity which is broken in strong interactions is the chirality of quarks; its breaking has direct experimental consequences (the absence in nature of the ninth light pseudo-Goldstone boson, analogous to the pions, kaons and the η-meson). Thus, the results studied in this chapter are also of great interest for particle physics.

The direct study of the movement of fermion levels in external fields is technically very complicated when it comes to four-dimensional theories. Thus, in Section 17.1, we formulate an approach which enables us to reduce the problem to the study of Euclidean fermion zero modes. In many cases, Euclidean zero modes are quite easy to find, as we shall see in Section 17.2. We note that it was the Euclidean approach (and the use of the formalism of the functional integral, see Appendix) which was adopted by 't Hooft (1976a,b) in his pioneering works on anomalous non-conservation of fermion quantum numbers.

17.1 Level crossing and Euclidean fermion zero modes

Let us briefly recall (see Section 15.2) what the phenomenon of fermion level crossing and the associated non-conservation of the fermion quantum numbers involve. Suppose we have a uniparametric family of bosonic field configurations (path in the space of static configurations), where the corresponding parameter τ varies from $-\infty$ to $+\infty$. Let us suppose, to be specific, that the bosonic fields take vacuum values both as $\tau \to -\infty$ and as $\tau \to \infty$; we shall see in what follows that the situation in which the vacua for $\tau = -\infty$ and $\tau = +\infty$ are topologically distinct (see Chapter 13) is of interest. For gauge fields, we shall assume that the gauge $A_0 = 0$ is chosen. For each fixed value of τ, there exists a fermion Hamiltonian $H_D(\tau)$; its dependence on τ is determined by the external bosonic fields. The explicit form of H_D depends on the model and we shall not yet define it concretely. The eigenvalues of $H_D(\tau)$ determine the system of fermion levels; as τ changes it also changes (the levels "move"). As $\tau \to -\infty$ and $\tau \to +\infty$ the systems of fermion levels are the same and coincide with the system of levels of the free fermion Hamiltonian. However, situations are possible in which as τ varies, the fermion levels cross zero and the total number of levels crossing zero from below N_+ is not equal to the total number of fermion levels crossing zero from above N_-. The quantity

$$\Delta N_F = N_+ - N_-$$

is of interest. As discussed in Section 15.2, as the bosonic fields evolve along a given path in the space of bosonic field configurations, the fermion quantum number of the system changes by ΔN_F, and this change does not depend on how quickly the bosonic fields vary or on the state in which the fermion system was at the beginning of the process (how many fermions and antifermions it had). Thus, we have to compute $(N_+ - N_-)$ for different paths in the space of configurations of bosonic fields.

It is convenient to reformulate this problem. We shall initially suppose that the bosonic fields vary slowly with τ. Let us consider the auxiliary equation

$$-\frac{\partial \psi}{\partial \tau} = H_D(\tau)\psi. \tag{17.1}$$

This equation can be formally interpreted as a Euclidean Dirac equation, since it is obtained from the usual Dirac equation $i\frac{\partial \psi}{\partial x^0} = H_D(\tau)\psi$ by the formal substitution $x^0 = -i\tau$ (compare with Chapters 11 and 13). In the case of slowly varying bosonic fields, the solutions of equation (17.1) have the form

$$\psi(\tau) = \exp\left[-\int \omega(\tau)d\tau\right]\psi_{\omega(\tau)}, \tag{17.2}$$

where $\psi_{\omega(\tau)}$ are eigenfunctions of the instantaneous Hamiltonian

$$H_D(\tau)\psi_{\omega(\tau)} = \omega(\tau)\psi_{\omega(\tau)}.$$

If some level crosses zero from below as the bosonic fields change, then for that level we have

$$\begin{aligned} \omega(\tau = -\infty) &= -\omega_i < 0 \\ \omega(\tau = +\infty) &= \omega_f > 0. \end{aligned}$$

This means that the solution of (17.2) decreases both as $\tau \to -\infty$ and as $\tau \to +\infty$,

$$\begin{aligned} \psi &= \text{constant} \cdot e^{\omega_i \tau}, \quad \tau \to -\infty, \\ \psi &= \text{constant} \cdot e^{-\omega_f \tau}, \quad \tau \to +\infty. \end{aligned}$$

In other words, in Euclidean space with coordinates $(\tau, \mathbf{x})$, the operator

$$D = \frac{\partial}{\partial \tau} + H_D \tag{17.3}$$

has zero mode $\psi_0(\tau, \mathbf{x})$, i.e. there exists a solution of the equation

$$D\psi_0 = 0 \tag{17.4}$$

which decreases as $\tau \to \pm\infty$. We shall not discuss here the boundary conditions for the fermion wave functions as $|\mathbf{x}| \to \infty$ since we shall see that in the interesting cases the Euclidean zero modes ψ_0 decrease in all directions in the Euclidean space $(\tau, \mathbf{x})$. In what follows, we shall consider the case of four-dimensional space–time, so that $(\tau, \mathbf{x})$ will be a four-dimensional Euclidean space. We shall call the decreasing solutions of equation (17.4) Euclidean fermion zero modes.

Since any solution of equation (17.1) in the adiabatic case is a linear combination of functions (17.2), the aforementioned correspondence between the levels of the Hamiltonian $H_D(\tau)$ crossing zero from below and the zero modes of the operator D is one to one. Thus

$$N_+ = N_0(D),$$

where $N_0(D)$ is the number of (Euclidean) zero modes of the operator D.

Analogously, we obtain a correspondence between the levels of the operator H_D crossing zero from above and the zero modes of the operator

$$D^\dagger = -\frac{\partial}{\partial \tau} + H_D(\tau). \tag{17.5}$$

We have

$$N_- = N_0(D^\dagger).$$

The operators D and $D^\dagger$ are Hermitian conjugates of each other, if we introduce the scalar product for functions $\psi(\tau, \mathbf{x})$,

$$(\psi, \psi') = \int d^3x \, d\tau \, \psi^\dagger(\tau, \mathbf{x}) \psi'(\tau, \mathbf{x}), \tag{17.6}$$

where the integration is over the whole of four-dimensional Euclidean space.[1]

Thus, the problem of computing the change in the fermion quantum number reduces to that of computing

$$\Delta N_F = N_0(D) - N_0(D^\dagger). \tag{17.7}$$

Until now, we have assumed that the external bosonic fields vary adiabatically slowly as τ varies. Now, we shall see that this assumption is unnecessary; moreover, we shall see that the quantity (17.7) does not depend on the specific choice of family of boson configurations; in that sense, it is a topological invariant.

Therefore, we shall consider the operators (17.3) and (17.5) in external bosonic fields, depending on $\mathbf{x}$ and on τ, i.e. in external fields in four-dimensional Euclidean space. Our next task is to show that the right-hand side of (17.7) is unchanged under smooth changes of the external Euclidean bosonic fields which do not affect the values of the fields at infinity in the Euclidean space. It is convenient to move to the Hermitian (in the sense of the scalar product (17.6)) operators $D^\dagger D$ and $DD^\dagger$. This is possible due to the fact that the zero modes of the operator D are zero modes of the

---

[1] We shall not dwell on subtleties relating to the rate of decrease of the functions $\psi(\tau, \mathbf{x})$ here.

operator $D^\dagger D$ *and conversely.* The first assertion is evident; to prove the second assertion we note that if $D^\dagger D\psi_0 = 0$, then

$$(\psi_0, D^\dagger D\psi_0) = (D\psi_0, D\psi_0) = 0,$$

i.e. $D\psi_0 = 0$, as required. Thus,

$$N_0(D) = N_0(D^\dagger D).$$

Analogously,

$$N_0(D^\dagger) = N_0(DD^\dagger)$$

and

$$\Delta N_F = N_0(D^\dagger D) - N_0(DD^\dagger). \tag{17.8}$$

The right-hand side of (17.8) is easier to study than (17.7). The set of zero modes of the (elliptic) operator $D^\dagger D$ is a linear space, called the kernel of this operator and denoted by $\mathrm{Ker}\,(D^\dagger D)$. The number of linearly independent zero modes is the dimension of the kernel, $N_0(D^\dagger D) = \dim \mathrm{Ker}(D^\dagger D)$. The difference

$$I(D^\dagger D) = \dim \mathrm{Ker}(D^\dagger D) - \dim \mathrm{Ker}(DD^\dagger)$$

is called the index of the operator $D^\dagger D$, so that (17.8) can be written in the form

$$\Delta N_F = I(D^\dagger D).$$

The majority of this section represents a study of the elements of Atiyah–Singer theory of the index of the elliptical operator and its connection with Atiyah–Patodi–Singer theorem.

In order to show that the right-hand side of (17.8) is unchanged under smooth variation of the external fields in four-dimensional Euclidean space, we shall show that the sets of *non-zero* eigenvalues of the operators $D^\dagger D$ and $DD^\dagger$ coincide. Indeed, let ψ_λ be an eigenfunction of the operator $D^\dagger D$ with eigenvalue $\lambda \neq 0$,

$$D^\dagger D\psi_\lambda = \lambda\psi_\lambda.$$

Acting on this equality from the left by the operator D, we obtain that the function $\tilde{\psi}_\lambda = D\psi_\lambda$ is an eigenfunction of the operator $DD^\dagger$ with the same eigenvalue:

$$DD^\dagger\tilde{\psi}_\lambda = \lambda\tilde{\psi}_\lambda.$$

This correspondence between the eigenfunctions of the operators $D^\dagger D$ and $DD^\dagger$ is invertible (for $\lambda \neq 0$), $\psi_\lambda = \lambda^{-1}D^\dagger\tilde{\psi}_\lambda$. Thus, the sets of non-zero eigenvalues (including possible degeneracy) coincide for the operators $D^\dagger D$ and $DD^\dagger$.

Let us now consider smooth variation of the external bosonic fields in four-dimensional Euclidean space. Under this variation, one of the non-zero eigenvalues of the operator $D^\dagger D$ may evolve smoothly to zero, $\lambda \to 0$, i.e. $N_0(D^\dagger D)$ may change by one. But, as $\lambda \to 0$, one of the eigenvalues of the operator $DD^\dagger$ also tends to zero, so that the difference $N_0(D^\dagger D) - N_0(DD^\dagger)$ is unchanged. This demonstrates the invariance of (17.8) under smooth variations of the external Euclidean fields.

In the next section, we shall apply the result studied here to compute the change in the fermion quantum number induced by external gauge fields, interpolating between topologically distinct vacua. According to the above discussion, the value of ΔN_F is the same for all external fields of the same homotopy class, so it will suffice to compute the number of zero modes of the operators (17.3) and (17.5) in *one* particular field with a given topological number, the choice of this field being a matter of convenience.

17.2 Fermion zero mode in an instanton field

To be specific, we shall consider a model with gauge group $SU(2)$ and massless fermions, transforming according to the fundamental representation of the group $SU(2)$. For the moment, we shall assume that there are no scalar fields in the model. Since the equations for left- and right-handed fermions decouple, we can consider left- and right-handed fermions separately. Our task is to show that external gauge fields, interpolating between topologically inequivalent vacua (see Section 13.3) lead to non-conservation of the number of left-handed fermions N_L and of the number of right-handed fermions N_R. Moreover,

$$\Delta N_L = n(\Omega_f) - n(\Omega_i) = \Delta n, \tag{17.9}$$

$$\Delta N_R = -\Delta N_L \tag{17.10}$$

for each left- and right-handed fermion doublet, where $n(\Omega_i)$ and $n(\Omega_f)$ are the topological numbers of the initial and final classical vacua of the gauge field. Without loss of generality, we can set $\Omega_i = 1$, i.e. in the initial vacuum $\mathbf{A} = 0$. For the moment we shall use the gauge $A_0 = 0$.

As shown in the previous section, the problem of computing ΔN_L reduces to counting the zero modes of the operators

$$D_L = \frac{\partial}{\partial \tau} + H_L,$$

and

$$D_L^\dagger = -\frac{\partial}{\partial \tau} + H_L,$$

where H_L is the Dirac (more precisely, Weyl) Hamiltonian in the external field $A_i(\tau, \mathbf{x})$. This external field can be chosen arbitrarily, provided the following conditions are satisfied:

$$
\begin{aligned}
A_i(\tau, \mathbf{x}) &\rightarrow 0 \quad \text{as } \tau \rightarrow -\infty \\
A_i(\tau, \mathbf{x}) &\rightarrow \Omega_f(\mathbf{x})\partial_i[\Omega_f(\mathbf{x})]^{-1} \quad \text{as } \tau \rightarrow +\infty.
\end{aligned} \tag{17.11}
$$

The difference between the numbers of zero modes of the operators D_L and $D_L^\dagger$, whence also the value of ΔN_L, does not depend on the specific choice of configuration $A_i(\tau, \mathbf{x})$ in the class (17.11).

The Hamiltonian H_L, as usual, is equal to (as before, we use the representation (14.4) for the γ-matrices and the representation (14.5) for the matrices α^i)

$$
H_L = i\sigma^i D_i = i\sigma^i[\partial_i + A_i(\tau, \mathbf{x})].
$$

For right-handed fermions, we need to consider the operator

$$
D_R = \frac{\partial}{\partial \tau} + H_R,
$$

where

$$
H_R = -i\sigma^i[\partial_i + A_i(\tau, \mathbf{x})].
$$

It is useful to note that

$$
D_R = -D_L^\dagger. \tag{17.12}
$$

From the last equation, there immediately follow the relations for the numbers of zero modes:

$$
\begin{aligned}
N_0(D_R) &= N_0(D_L^\dagger), \\
N_0(D_R^\dagger) &= N_0(D_L).
\end{aligned} \tag{17.13}
$$

Since changes in the numbers of left- and right-handed fermions are related to the number of zero modes by the equations

$$
\begin{aligned}
\Delta N_L &= N_0(D_L) - N_0(D_L^\dagger), \\
\Delta N_R &= N_0(D_R) - N_0(D_R^\dagger),
\end{aligned}
$$

equation (17.13) leads immediately to the selection rule (17.10).

Thus, we need to solve the equation

$$
D_L \chi_0 = \left[\frac{\partial}{\partial \tau} + i\sigma^i(\partial_i + A_i)\right]\chi_0 = 0, \tag{17.14}
$$

and also the equation

$$-D_L^\dagger \eta_0 = \left[\frac{\partial}{\partial \tau} - i\sigma^i(\partial_i + A_i)\right]\eta_0 = 0. \tag{17.15}$$

Here, $\chi_0(\tau, \mathbf{x})$ and $\eta_0(\tau, \mathbf{x})$ are two-component columns. As previously mentioned, equation (17.14) can be interpreted as a Euclidean Weyl equation for left-handed fermions in a (Euclidean) external gauge field with $A_0 = 0$. By virtue of (17.12), equation (17.15) is a Euclidean Weyl equation for right-handed fermions. To find explicit solutions it is convenient to omit the requirement that the Euclidean external gauge field has $A_0 = 0$. For this, we use the fact that equation (17.14) for the function $\chi_0(\tau, \mathbf{x})$ in the external field $A_0 = 0$, $A_i(\tau, \mathbf{x})$ is equivalent to the equation

$$\left[\left(\frac{\partial}{\partial \tau} + A_0'\right) + i\sigma^i(\partial_i + A_i')\right]\chi_0' = 0 \tag{17.16}$$

for the function $\chi_0'(\tau, \mathbf{x}) = \omega(\tau, \mathbf{x})\chi_0(\tau, \mathbf{x})$ in the external field

$$\begin{aligned}
A_0' &= \omega \partial_\tau \omega^{-1}, \\
A_i' &= \omega A_i \omega^{-1} + \omega \partial_i \omega^{-1},
\end{aligned}$$

where $\omega(\tau, \mathbf{x})$ is an arbitrary function with values in $SU(2)$. Essentially, we have merely performed a gauge transformation (in four-dimensional Euclidean space) from the gauge $A_0 = 0$ to an arbitrary gauge. Let us introduce the matrices

$$\begin{aligned}
\sigma_E^\mu &= (1, -i\vec{\sigma}), \\
\sigma_E^{\mu\dagger} &= (i, i\vec{\sigma}),
\end{aligned}$$

which are Euclidean analogues of the matrices entering (14.4) and coincide with the matrices used in Section 13.2 (see (13.12)) but now they act on the Lorentz indices. Omitting the primes in (17.16), we finally write the Euclidean Weyl equation in the external field as

$$D_L \chi_0 \equiv \sigma_E^{\mu\dagger} D_\mu \chi_0 = 0, \tag{17.17}$$

where, as usual, $D_\mu = \partial_\mu + A_\mu$, and $x^0 = \tau$ emerges as the Euclidean time. Analogously, equation (17.15) in an arbitrary gauge has the form

$$-D_L^\dagger \eta_0 = D_R \eta_0 \equiv \sigma_E^\mu D_\mu \eta_0 = 0. \tag{17.18}$$

We are interested in the difference between the numbers of decreasing solutions of these two equations.

Problem 1. *Define Euclidean γ-matrices by analogy with (14.4)*

$$\gamma_E^\mu = \begin{pmatrix} 0 & \sigma_E^\mu \\ \sigma_E^{\mu\dagger} & 0 \end{pmatrix}.$$

Show that the following anti-commutation relation holds:

$$\{\gamma_E^\mu, \gamma_E^\nu\} = \delta^{\mu\nu}.$$

Problem 2. *Consider equation (17.3) for massive fermions (so that ψ is a four-component column). Show that it is equivalent to the Euclidean Dirac equation*

$$(\gamma_E^\mu D_\mu + m)\psi = 0,$$

with the matrices γ_E^μ defined in the previous problem.

The only requirement imposed upon the Euclidean gauge field $A_\mu(x)$ is the condition that after transition to the gauge $A_0 = 0$, equation (17.11) is satisfied. This requirement is equivalent to the equation (see Section 13.3, in particular, formula (13.45))

$$Q = \Delta n, \tag{17.19}$$

where, as in Section 13.3,

$$Q = -\frac{1}{16\pi^2} \int d^4x \,\mathrm{Tr}\, (F_{\mu\nu} \tilde{F}_{\mu\nu}). \tag{17.20}$$

Since condition (17.19) is gauge invariant, there is no longer any need to turn to the gauge $A_0 = 0$.

Thus, in order to verify equation (17.9) for the non-conservation of the fermion quantum number, we need to show that

$$N_0(D_L) - N_0(D_L^\dagger) = Q$$

for the Euclidean Weyl equations (17.17), (17.18) in any (it does not matter which) external gauge field with topological number Q. Let us consider the case $Q = 1$ when, for the external field, one can take the instanton solution of the Euclidean Yang–Mills equations.

We shall show, in the first place, that the operator D_R (or, what amounts to the same thing, $D_L^\dagger$) does not have zero modes in the instanton field. For this, we act on the left-hand side of equation (17.18) by the operator $D_R^\dagger = -D_L = -\sigma_E^{\mu\dagger} D_\mu$. We take into account the fact that $\sigma_E^{\mu\dagger}\sigma_E^\nu =$

$\delta^{\mu\nu} + i\bar{\eta}_{\mu\nu a}\sigma_a$ (see (13.31)), together with the antisymmetry of the 't Hooft symbols in the indices μ and ν. We obtain

$$D_R^\dagger D_R = -D_\mu D_\mu - \frac{1}{2} i\bar{\eta}_{\mu\nu a}[D_\mu, D_\nu].$$

Moreover, $[D_\mu, D_\nu] = F_{\mu\nu}$ (see the problem in Section 4.2). For the instanton, $F_{\mu\nu}$ is proportional to $\eta_{\mu\nu a}\tau_a$, and for the 't Hooft symbols we have $\bar{\eta}_{\mu\nu a}\eta_{\mu\nu b} = 0$. Thus,

$$D_R^\dagger D_R = -D_\mu D_\mu.$$

This is a positive definite operator, since the operator D_μ is anti-Hermitian in the space of functions with scalar product (17.6). Consequently, the operator $D_R^\dagger D_R$ has no zero modes; thus, the operator D_R has none either. Therefore, for the instanton external field,

$$N_0(D_R) = N_0(D_L^\dagger) = 0. \tag{17.21}$$

Let us now consider equation (17.17) in the instanton field. It is convenient to use the method described in Section 16.1 and move from the variables $\chi_{\alpha i}$ (here, $\alpha = 1,2$ is the Lorentz index and $i = 1,2$ is the isotopic index) to the variables $\tilde{\chi}_{\alpha i}$ according to the formula (see (16.18))

$$\chi_{\alpha i} = \tilde{\chi}_{\alpha j}\varepsilon_{ji}.$$

Then, equation (17.17) can be written in matrix form

$$\sigma_\mu^\dagger \partial_\mu \tilde{\chi}_0 - \sigma_\mu^\dagger \tilde{\chi}_0 A_\mu = 0, \tag{17.22}$$

where we have omitted the index E on the σ-matrices. The instanton field A_μ, given by expression (13.27), can be conveniently written in the form

$$A_\mu = -(\sigma_\mu \sigma_\nu^\dagger - \delta_{\mu\nu})\frac{n_\nu}{r} f(r), \tag{17.23}$$

where n_ν and r are, respectively, the unit radius vector and the radial coordinate in four-dimensional space. In writing down the expression (17.23), we have employed equation (13.20) and used the notation $f(r)$ for the function (13.26),

$$f(r) = \frac{r^2}{r^2 + \rho^2}.$$

Equation (17.22) in the instanton field takes the form

$$\sigma_\mu^\dagger \partial_\mu \tilde{\chi}_0 - \sigma_\mu^\dagger n_\mu \frac{f(r)}{r}\tilde{\chi}_0 + \frac{f(r)}{r}\sigma_\mu^\dagger \tilde{\chi}_0 \sigma_\mu \sigma_\nu^\dagger n_\nu = 0, \tag{17.24}$$

where $\tilde{\chi}_0$ is to be interpreted as a 2×2 matrix and its products with $\sigma_\mu^\dagger$ and σ_μ are matrix products.

The form of equation (17.24) suggests the evident Ansatz

$$\tilde{\chi}_0 = \mathbf{1} \cdot a(r),$$

where $a(r)$ is a function which is as yet unknown, and $\mathbf{1}$ is the unit 2×2 matrix. Taking into account the fact that $\sigma_\mu^\dagger \sigma_\mu = 4$, we obtain that the left-hand side of (17.24) is proportional to $\sigma_\mu^\dagger n_\mu$, and equation (17.24) leads to the single equation for $a(r)$:

$$a' + 3\frac{f(r)}{r}a = 0.$$

Hence,

$$a(r) = \text{constant} \cdot \exp\left[-3\int \frac{dr}{r}f(r)\right] = \frac{\text{constant}}{(r^2 + \rho^2)^{3/2}}.$$

This is precisely the zero mode in the instanton field ('t Hooft 1976a,b). In terms of the original function $\chi(x)$ it has the form

$$(\chi_0)_{\alpha i} = \text{constant} \cdot \varepsilon_{\alpha i}\frac{1}{(r^2 + \rho^2)^{3/2}}. \tag{17.25}$$

The zero mode (17.25) is smooth everywhere; it decreases sufficiently rapidly as $r \to \infty$, so that its norm (17.6) is finite.

The fact that the operator D_L in the instanton field has a zero mode, i.e. $N_0(D_L) \neq 0$, already implies, together with (17.21), that the left fermion quantum number is not conserved in backgrounds of gauge configurations starting and ending in adjacent topologically distinct vacua. In fact, in the instanton field, there is exactly one fermion zero mode. The proof of this last statement is quite complicated, and we shall not present it here. The absence of zero modes, other than (17.25), depending solely on r is the subject of the following problem.

Problem 3. *Consider fermion functions $\chi(r)$ depending solely on r. Show that*

1. *for these functions equation (17.17) leads to a closed system of equations;*

2. *the zero mode (17.25) is the only decreasing solution of equation (17.17).*

Hint: use the fact that any 2×2 matrix $\tilde{\chi}(r)$ can be represented in the form
$\tilde{\chi} = \mathbf{1} \cdot a(r) + \sigma^a b_a(r)$, where the σ^a are Pauli matrices; use the equation
for the zero mode in the form (17.24).

Thus, for transitions between adjacent gauge vacua ($\Delta n = 1$), equation (17.9) is indeed satisfied. That this equation is satisfied for arbitrary Δn can be seen from the following argument. Let us consider, for example, the case $\Delta n = 2$. The external gauge field $A_i(\tau, \mathbf{x})$ (in the gauge $A_0 = 0$) can be chosen in such a way that it corresponds to a sequence of two transitions with $\Delta n = 1$, i.e. for $\tau = -\infty$, $\tau = 0$ and $\tau = +\infty$, it is a vacuum field with $n = 0, n = 1$ and $n = 2$, respectively. As τ changes from $(-\infty)$ to 0, the fermion level crossing corresponds to a change in the number of left-handed fermions by $\Delta N_L = 1$; the fermion quantum number changes by the same amount as τ changes from 0 to $(+\infty)$. The total number of level crossings thus corresponds to $\Delta N_L = 2$. The movement of levels in this case is illustrated schematically in Figure 17.1.

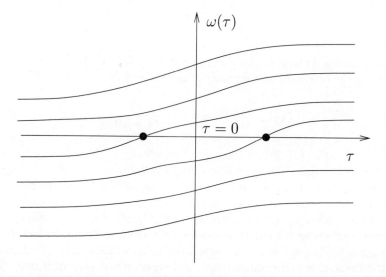

Figure 17.1. Movement of left-handed fermion levels in the gauge field described by sequential transitions between neighboring vacua. The fermion levels crossing zero are shown.

Since the change in the fermion quantum number does not depend on the specific choice of field, for a given Δn, we conclude that $\Delta N_L = 2$ for all fields with $\Delta n = 2$. Clearly, this consideration can be easily generalized to any $\Delta n > 0$. An evident modification of this argument (a sequence of transitions with $\Delta n = 1$ and $\Delta n = -1$ is a topologically trivial field, leading to $\Delta N_L = 0$) shows that equation (17.9) is also valid for $\Delta n < 0$.

Problem 4. *Show that in the anti-instanton field the operator D_L does not have zero modes. Find a zero mode of the operator $D_L^\dagger$, or, what amounts to the same thing, of the operator D_R. This result gives an explicit confirmation of formula (17.9) for $\Delta n = -1$.*

The result studied here is a particular case of the Atiyah–Singer index theorem (the connection between non-conservation of fermion quantum numbers and the Atiyah–Singer theorem was explained by Schwarz (1977), Nielsen and Schroer (1977b), Brown *et al.* (1977) and Jackiw and Rebbi (1977)). Applied to left-handed fermions, it gives

$$\dim \mathrm{Ker}\,(D_L^\dagger D_L) - \dim \mathrm{Ker}\,(D_L D_L^\dagger) = Q, \qquad (17.26)$$

where Q is the topological number of a Euclidean configuration of the gauge field (17.20). Equation (17.9) is a reformulation of equation (17.26). This equation is also connected with the Adler–Bell–Jackiw anomaly, which exists in quantum field theory. Namely, in quantum field theory in Minkowski space, one can introduce the left-handed fermion current operator

$$j_L^\mu = \chi^\dagger \bar\sigma^\mu \chi$$

(in the notation of Section 14.1). Naively this current is conserved, i.e. naively, the left-handed fermion quantum number

$$N_L = \int j_L^0 d^3x$$

is conserved. However, quantum effects lead to an anomaly in the divergence of the current j_L^μ,

$$\partial_\mu j_L^\mu = -\frac{1}{16\pi^2} \mathrm{Tr}\,(F_{\mu\nu} \tilde F_{\mu\nu}). \qquad (17.27)$$

Equation (17.9) can be interpreted as the anomalous identity (17.27) integrated over space–time.

17.3 Selection rules

As a matter of fact, we have already formulated, in the previous section, a selection rule for the processes with non-conservation of the fermion quantum numbers discussed in this chapter (in the theory with gauge group $SU(2)$ and massless fermions in the fundamental representation). Namely, if a process takes place in which the topological number of the gauge vacuum changes by Δn, then the fermion quantum number *of every left-handed doublet* changes by $\Delta N_L = \Delta n$ and the fermion quantum number of every right-handed doublet changes by $\Delta N_R = -\Delta n$. In specific models, it is convenient to reformulate this selection rule in more physical terms, which we shall do in this section for the theory of strong interactions (quantum chromodynamics) and for the Standard Model of electroweak interactions.

Let us begin with quantum chromodynamics, the theory with gauge group $SU(3)_C$ and fermion triplets (quarks) with both left- and right-handed components. Three types of quark u, d, s have very small masses and a reasonable approximation is the limit of massless u, d, s. In this limit, the numbers of left- and right-handed quarks of each type would be conserved individually: $N_{u_L}, N_{u_R}; \ldots; N_{s_L}, N_{s_R}$, if there did not exist processes with transition between distinct gauge vacua.

To establish selection rules, which are valid when processes of the instanton type are taken into account, we note that the considerations of the previous section, based on the identity $D_L^\dagger = -D_R$, leading to equation (17.10), go over directly to the case of the gauge group $SU(3)_C$. Thus, the following equations are satisfied:

$$\begin{aligned}
\Delta N_{u_R} &= -\Delta N_{u_L}, \\
\Delta N_{d_R} &= -\Delta N_{d_L}, \\
\Delta N_{s_R} &= -\Delta N_{s_L}.
\end{aligned} \qquad (17.28)$$

Furthermore, the components u_L, d_L and s_L (in the limit of massless quarks) behave in exactly the same way in the external gauge fields of the group $SU(3)_C$. Thus, level crossing takes place in the same way for each type of quark and we have

$$\Delta N_{u_L} = \Delta N_{d_L} = \Delta N_{s_L}. \qquad (17.29)$$

Equations (17.28) and (17.29) represent five selection rules, valid in all processes in chromodynamics with three types of massless quarks. They mean that the overall chirality

$$Q^5 = (N_{u_L} + N_{d_L} + N_{s_L}) - (N_{u_R} + N_{d_R} + N_{s_R}) \qquad (17.30)$$

may not be conserved, while at the same time, five other linear combinations of the numbers $N_{u_L}, \ldots, N_{s_R}$ are conserved.

In order to see that chirality is actually not conserved due to transitions between topologically distinct gauge vacua, let us consider the gauge fields of the $SU(2)$ subgroup embedded in $SU(3)_C$ as follows:

$$\begin{pmatrix} SU(2) & \mathbf{0} \\ \mathbf{0} & 1 \end{pmatrix}. \qquad (17.31)$$

Every $SU(3)_C$ quark triplet decomposes into a doublet and a singlet with respect to this $SU(2)$. The singlets do not interact with the gauge fields of this $SU(2)$ subgroup and level crossing in gauge fields with this structure does not occur for them. For doublets, level crossing does occur in topologically non-trivial external gauge fields, corresponding to the subgroup (17.31). Here, as is clear from the results of the previous section, the selection rules (17.28), (17.29) are indeed valid, but $N_{u_L}, \ldots, N_{s_R}$ are not conserved individually, i.e.

$$\Delta Q^5 \neq 0.$$

Thus, the chirality (17.30) is not conserved in chromodynamics.

Physically, non-conservation of chirality (17.30) in quantum chromodynamics is manifest in the absence in nature of a ninth light pseudoscalar boson, analogous to the octet $\pi^\pm, \pi^0, K^\pm, K^0, \bar{K}^0, \eta$. The fermionic part of the quantum chromodynamic action in the limit of massless u, d, s has the form

$$S_F = \int d^4x(\bar{u}i\gamma^\mu D_\mu u + \bar{d}i\gamma^\mu D_\mu d + \bar{s}i\gamma^\mu D_\mu s), \tag{17.32}$$

where, as usual, $D_\mu = \partial_\mu - ig_s t^a A^a_\mu$; $a = 1, \ldots, 8$; g_s is the coupling constant of the group $SU(3)_C$, t^a are generators of $SU(3)_C$, A^a_μ is an octet of gauge (gluon) fields. The action (17.32) is naively invariant under global transformations from the group

$$SU(3)_L \times SU(3)_R \times U(1)_L \times U(1)_R, \tag{17.33}$$

where the subgroups $SU(3)_L \times U(1)_L$ and $SU(3)_R \times U(1)_R$ act on left- and right-handed quark components, respectively. Here, for example, the quarks (u_L, d_L, s_L) form a triplet under $SU(3)_L$ and transform with a common phase for transformations from $U(1)_L$. The naive conservation of the six numbers $N_{u_L}, \ldots, N_{s_R}$ corresponds to the diagonal subgroup of the group (17.33). In particular, the chirality (17.30) corresponds to transformations from the subgroup $U(1)_{L-R}$,

$$\begin{pmatrix} u_L \\ d_L \\ s_L \end{pmatrix} \to e^{i\alpha} \begin{pmatrix} u_L \\ d_L \\ s_L \end{pmatrix} ; \quad \begin{pmatrix} u_R \\ d_R \\ s_R \end{pmatrix} \to e^{-i\alpha} \begin{pmatrix} u_R \\ d_R \\ s_R \end{pmatrix}.$$

Non-conservation of the chirality (17.30) means that, in fact, the global symmetry subgroup is

$$SU(3)_L \times SU(3)_R \times U(1)_{L+R}. \tag{17.34}$$

Strong interactions lead to spontaneous breaking of the symmetry (17.34) down to $SU(3)_{L+R} \times U(1)_{L+R}$ (there is still no reliable theoretical explanation for this experimental fact). According to Goldstone's theorem, this leads to the occurrence of eight massless bosons, eight being the number of broken generators. In fact, u, d, s quarks have small masses, therefore the masses of the Nambu–Goldstone bosons are small, but non-zero. These are the $\pi^{\pm}, \pi^0, K^{\pm}, K^0, \bar{K}^0$ and η-mesons. If the symmetry group were the group (17.33) then there would have to be a ninth pseudo-Goldstone boson. Its absence is direct experimental proof of the non-conservation of chirality (non-invariance under $U(1)_{L-R}$) in quantum chromodynamics.

We stress that in quantum chromodynamics, the coupling constant is not small (at large distances), therefore, the studied effects leading to breaking of chirality are not small.

Continuing the discussion of chromodynamics, here is a remark concerning quarks with non-zero (and sufficiently large) mass. In this case, the numbers of left- and right-handed quarks are not conserved, even when there are no external gauge fields. Thus, it only makes sense to talk about conservation of the total number of quarks of each type:

$$N_q = N_{qL} + N_{qR}. \tag{17.35}$$

These numbers are conserved for arbitrary masses (including zero, see (17.28)) and in arbitrary external gauge fields of the group $SU(3)_C$. Thus, in the case of massive quarks, topologically non-trivial gauge fields do not lead to any change in the naive selection rules.

Let us now turn to the theory of electroweak interactions. We point out that as far as selection rules are concerned, the subgroup $U(1)$ of the electroweak gauge group $SU(2) \times U(1)$ is unimportant, since the gauge vacua of four-dimensional Abelian theories do not have a complex structure. Thus, in what follows, we shall consider only gauge fields corresponding to the group $SU(2)$. Furthermore, we shall temporarily turn off the Yukawa interactions of fermions with the Higgs field; we shall see in what follows that taking into account the Yukawa interactions essentially does not change the selection rules. Neglecting the gauge group $U(1)$ and the Yukawa interactions, the right-handed components of leptons and quarks are free (strong interactions are also of no interest to us here), but left-handed leptons and quarks interact with the gauge field of the group $SU(2)$, exactly as described in the previous section. Since there are no right-handed doublets in the model, the relation (17.10) is not relevant; whereas the equality (17.9) is satisfied for *every* left-handed doublet. The number of right-handed fermions does not change, and we obtain that, in the case of transitions between topologically distinct vacua, the following selection rule holds:

$$\Delta L_e = \Delta L_\mu = \Delta L_\tau = 3\Delta B, \tag{17.36}$$

where the lepton numbers L_e, L_μ and L_τ are defined by equations (14.79), (14.80) and (14.81), and B is the baryon number. In the last equation we took account of the fact that the number of left-handed quark doublets in the model is equal to the product of the number of generations ($N_{\text{gen}} = 3$) and the number of colors ($N_c = 3$), and the baryon number of a single quark is equal to $1/3$; hence, we obtain the coefficient $N_{\text{gen}} N_c \cdot \frac{1}{3} = 3$. Thus, processes of the instanton type lead in electroweak theory to non-conservation of the baryon and lepton numbers, although at the same time, the selection rule (17.36) is satisfied.

The conclusion that the baryon and lepton numbers are not conserved, and that the selection rule (17.36) holds remains valid when one takes into account the Yukawa interactions of fermions with the Higgs field, which ultimately lead to generation of quark and lepton masses (Krasnikov *et al.* 1979). This can be shown in the following, albeit not completely rigorous, way. Let us consider, for simplicity, a set of fermions which includes a left-handed doublet L and two right-handed singlets R_1 and R_2 under the group $SU(2)$. In this simplified case, the fermion action has the form (see (14.74))

$$
\begin{aligned}
S_F \;=\; \int d^4x \Bigg\{ & L^\dagger i \tilde{\sigma}^\mu D_\mu L + R_1^\dagger i \sigma^\mu \partial_\mu R_1 + R_2^\dagger i \sigma^\mu \partial_\mu R_2 \\
& -h_1 \left[R_1^\dagger (\varphi^\dagger L) + (L^\dagger \varphi) R_1 \right] \\
& -h_2 \left[R_2^\dagger (\tilde{\varphi}^\dagger L) + (L^\dagger \tilde{\varphi}) R_2 \right] \Bigg\},
\end{aligned}
\tag{17.37}
$$

where we have neglected the gauge interaction corresponding to the group $U(1)$. The system of Dirac *Euclidean* equations, analogous to equation (17.17) has the form

$$
\begin{aligned}
\sigma_\mu^\dagger D_\mu L - h_1 \varphi R_1 - h_2 \tilde{\varphi} R_2 &= 0, \\
\sigma_\mu \partial_\mu R_1 - h_1 (\varphi^\dagger L) &= 0, \\
\sigma_\mu \partial_\mu R_2 - h_2 (\tilde{\varphi}^\dagger L) &= 0.
\end{aligned}
\tag{17.38}
$$

By introducing the column

$$
\psi = \begin{pmatrix} L \\ R_1 \\ R_2 \end{pmatrix},
$$

this system can be written in symbolic form

$$
\mathcal{D}\psi = 0.
$$

The change in the fermion quantum number in the external gauge and Higgs fields of instanton type is determined by the difference between the numbers of zero modes of the operators $\mathcal{D}$ and $\mathcal{D}^\dagger$,

$$\Delta N_F = N_0(\mathcal{D}) - N_0(\mathcal{D}^\dagger). \tag{17.39}$$

Here, as usual, the fermion quantum number is understood to be the difference between the number of fermions and the number of antifermions (both left- and right-handed).

Problem 5. *Using the arguments of Section 17.1, show that equation (17.39) remains valid with the operator $\mathcal{D}$ defined by the system (17.38), for the theory with the fermion action (17.37). Write down the explicit form of the operator $\mathcal{D}$, without suppressing the group indices corresponding to the gauge group $SU(2)$.*

As we showed in Section 17.1, the difference $(N_0(\mathcal{D}) - N_0(\mathcal{D}^\dagger))$ does not change under continuous variations of the operator $\mathcal{D}$. In particular, it should not change under changes of the Yukawa constants h_1 and h_2. But when $h_1 = h_2 = 0$, the equations for the left- and right-handed fermions decouple, and we return to the case considered in the previous section. In this limit, the right-handed fermions are free and do not have zero modes, but for left-handed fermions, we have

$$N_0(\mathcal{D}) - N_0(\mathcal{D}^\dagger) = \Delta n,$$

where Δn is the change in the topological number of the classical vacuum of the gauge fields. Since this formula does not depend on h_1 and h_2, we obtain

$$\Delta N_F = \Delta n$$

for arbitrary Yukawa constants.

In order to apply this argument to leptons, it suffices to set $h_2 = 0$. The generalization of this argument to the case of quarks (with non-trivial mixing, briefly considered in Section 14.4) is quite evident. Thus, we conclude that the inclusion of Yukawa interactions leaves the selection rule (17.36) valid in the electroweak theory.

The structure of the Euclidean fermion zero mode in external gauge and Higgs fields, having the symmetries of an instanton solution, is fairly easy to find in the case of identical Yukawa constants (Krasnikov *et al.* 1979). Indeed, let us consider Euclidean gauge and Higgs fields of the form (13.56),

$$\begin{aligned} A_\mu(x) &= f(r)\omega\partial_\mu\omega^{-1}, \\ \varphi(x) &= H(r)\omega\varphi^{\text{vac}}, \end{aligned}$$

where

$$\omega(x) = n_\alpha\sigma_\alpha, \quad \varphi^{\text{vac}} = \frac{1}{\sqrt{2}}\begin{pmatrix} 0 \\ \varphi_0 \end{pmatrix},$$

and let us use the notation of Chapter 13 (in particular, $r = \sqrt{\mathbf{x}^2 + \tau^2}$). For $h_1 = h_2 = h$ (in other words, for fermions of equal mass), equation (17.38) can be solved using the Ansatz

$$
\begin{aligned}
L_{\alpha i}(x) &= \varepsilon_{\alpha i} l(r), \\
R_{1\alpha}(x) &= \varepsilon_{\alpha j} \varphi_j^{\text{vac}} \rho(r), \\
R_{2\alpha}(x) &= \varepsilon_{\alpha j} \tilde{\varphi}_j^{\text{vac}} \rho(r),
\end{aligned} \tag{17.40}
$$

where $\alpha = 1, 2$ and $i, j = 1, 2$ are Lorentz and weak isospin indices, respectively. As a result of computations analogous to those performed in section 14.2, we obtain that the system (17.38) reduces to two equations for $l(r)$ and $\rho(r)$,

$$
\begin{aligned}
\rho' - m_F H l &= 0, \\
l' + \frac{3f}{r} l - m_F H \rho &= 0,
\end{aligned} \tag{17.41}
$$

where $m_F = h\varphi_0/\sqrt{2}$ is the mass of fermions in the vacuum φ^{vac}. In order to show the existence of smooth solutions of the system (17.41), decreasing for $r \to \infty$, we use the following argument, which is also useful in other cases. Let us consider first the behavior of solutions of the system (17.41) as $r \to \infty$. In this region we have

$$
\begin{aligned}
f(r) &\to 1, \\
H(r) &\to 1, \quad r \to \infty.
\end{aligned} \tag{17.42}
$$

It is not difficult to show that *one* solution of the equations (17.41) decreases exponentially as $r \to \infty$, namely, $\rho, l \propto \exp(-m_F r)$, while the other grows exponentially. Let us now determine the behavior of solutions of equations (17.41) as $r \to 0$. In this region,

$$
\begin{aligned}
m_F H(r) &= \alpha_1 r, \\
f(r) &= \alpha_2 r^2, \quad r \to 0,
\end{aligned} \tag{17.43}
$$

where α_1 and α_2 are some constants. (The behavior of (17.43) is due to the requirement that the fields A_μ and φ be smooth at the point $x^\mu = 0$). Taking into account (17.43), one can show that *both* solutions of the system (17.41) are smooth as $r \to 0$. Indeed, we shall look for a solution in the form of a series

$$
\begin{aligned}
l &= C_1 + C_2 r^2 + \cdots, \\
\rho &= D_1 + D_2 r^2 + \cdots
\end{aligned}
$$

with constant coefficients $C_1, \ldots, D_1, \ldots$. Equations (17.41) reduce to recurrence relations for the coefficients $C_1, \ldots, D_1, \ldots$, the first of which has the form

$$2C_2 + 3\alpha_2 C_1 - \alpha_1 D_1 = 0,$$
$$2D_2 - \alpha_1 C_1 = 0.$$

It is significant that the constants D_1 and C_1 are arbitrary and the remaining coefficients are expressed in terms of them. Hence it follows that both solutions of the equations (17.41) are smooth as $r \to 0$, and there do not exist solutions which are singular at the origin. In particular, the solution of (17.41), which is decreasing as $r \to \infty$ is (like all others) smooth as $r \to 0$. This is the Euclidean zero mode.

Problem 6. *Using the Ansatz (17.40), show that the system (17.38) reduces to the two equations (17.41).*

Problem 7. *Assuming that the functions $H(r)$ and $f(r)$ tend exponentially to values (17.42) as $r \to \infty$, find asymptotics of $l(r)$ and $\rho(r)$ for $r \to \infty$, including the power of r in front of the exponential.*

Problem 8. *Find an explicit solution for $l(r)$ and $\rho(r)$ in the field of the 't Hooft configuration*

$$f(r) = \frac{r^2}{r^2 + \rho^2},$$
$$H(r) = \frac{r}{\sqrt{r^2 + \rho^2}}.$$

17.4   Electroweak non-conservation of baryon and lepton numbers at high temperatures

In the previous section, we showed that theories of electroweak interactions predict non-conservation of the baryon and lepton numbers, and that the selection rule has the form (17.36). This non-conservation arises due to transitions between topologically distinct classical vacua of gauge theory and is due to the phenomenon of the fermion level crossing.

Under usual conditions, at low temperatures and particle densities and for energies which are not too high, transitions between topologically distinct vacua are tunneling processes and are described by instantons.[2]

---

[2]More precisely, by configurations of the instanton type, constrained instantons; see Appendix.

And it is precisely instanton processes that generate gauge and Higgs fields in which the fermion level crossing takes place. Since the gauge coupling constant of the non-Abelian subgroup $SU(2)$ of the electroweak group $SU(2) \times U(1)$ is small,

$$\alpha_W = \frac{g^2}{4\pi} = \frac{\alpha}{\sin^2 \theta_W} \simeq \frac{1}{30},$$

the probability of electroweak instanton processes is strongly suppressed:

$$\Gamma \propto \exp\left(-\frac{4\pi}{\alpha_W}\right) \sim 10^{-160}. \tag{17.44}$$

Thus, the electroweak breaking of the baryon and lepton number cannot be detected experimentally. We again note that its theoretical description requires one to go beyond the standard method of quantum field theory, namely expansion in the small coupling constant.

The situation changes radically if we consider systems experiencing electroweak interactions in extremal conditions. The case of high temperatures is most interesting, since it was realized at early stages in the evolution of the Universe. As we have mentioned a number of times, at high temperatures, thermal jumps are possible across the potential barrier separating topologically distinct vacua. The most naive estimate of the rate of these jumps at moderate temperatures is obtained by estimating the probability that the system emerges at a saddle point separating topologically distinct vacua (sphaleron), by the Boltzmann exponent

$$\Gamma \propto \exp\left(-\frac{E_{\text{sph}}}{T}\right). \tag{17.45}$$

In the electroweak theory, we have (see (13.58))

$$E_{\text{sph}} = \frac{2m_W}{\alpha_W} B\left(\frac{m_H}{m_W}\right) \sim 10 \text{ TeV},$$

and even the estimate (17.45) shows that at high temperatures $T \geq 1$ TeV, the rate of thermal jumps ceases to be small. Thus, we deduce an intense non-conservation of the baryon and lepton numbers at sufficiently high temperatures.

In fact, the estimate (17.45) is a gross underestimate in the temperature range of interest. The probability that a particular state of the system is realized at high temperatures is determined not by the energy of that state E, but by its free energy F, which is itself a function of temperature. This means, in particular, that in order to estimate the rate of thermal jumps across the potential barrier, we need to seek not the extremum of the static energy functional $E(\mathbf{A}, \varphi)$, but the extremum of the free energy $F(\mathbf{A}, \varphi)$. This extremum, the thermal sphaleron, may not coincide with the usual sphaleron configuration, and its free energy F_{sph} may be significantly less than E_{sph}. This is what is obtained in the electroweak theory, thus the rate of thermal jumps, leading to non-conservation of the baryon and lepton numbers

$$\Gamma \sim T^4 \exp\left(-\frac{F_{\text{sph}}(T)}{T}\right) \tag{17.46}$$

is very much greater than that given by the estimate (17.45). In the estimate (17.46), we introduced the factor T^4 on dimensional grounds; Γ is the probability of processes with non-conservation of the baryon and lepton numbers per unit time per unit spatial volume.

A reasonable estimate of F_{sph} is obtained if one takes account of the fact that the expectation value of the Higgs field depends on the temperature. This temperature dependence (Kirzhnits 1972, Kirzhnits and Linde 1972, Dolan and Jackiw 1974, Weinberg 1974) is studied by methods which go far beyond the scope of this book. The expectation value of the Higgs field $\varphi_0(T)$ decreases as the temperature grows and above some temperature T_{crit} becomes zero. This critical temperature is known as the electroweak phase transition temperature; roughly speaking, for $T > T_{\text{crit}}$ the electroweak group $SU(2) \times U(1)$ is not broken.[3] Since the masses of the vector and Higgs bosons are proportional to the expectation value φ_0, they also vary with temperature; in particular, $m_W(T) = \frac{1}{2}g\varphi_0(T)$. The free energy of the thermal sphaleron can be estimated by setting $m_W = m_W(T)$, $m_H = m_H(T)$ in formula (13.58), i.e.

$$F_{\text{sph}}(T) = \frac{2m_W(T)}{\alpha_W} B\left(\frac{m_H}{m_W}\right).$$

This quantity decreases as the temperature grows, and for $T > T_{\text{crit}}$, it is equal to zero. This last property means that for $T > T_{\text{crit}}$, processes with non-conservation of the baryon and lepton numbers cease to be exponentially suppressed. At the same time they can no longer be described by semiclassical methods, which presents a difficulty from the theoretical point of view.

In the electroweak theory, the temperature of the phase transition depends on an (as yet!) unknown parameter, the mass of the Higgs boson in the vacuum. Its value is around $T_{\text{crit}} \sim 200$ GeV. Thus, electroweak processes with non-conservation of the baryon and lepton numbers are rapid

---

[3]In fact, the situation in the electroweak theory at high temperatures is even more subtle. The order parameter in this theory is actually missing (Fradkin and Shenker 1979, Banks and Rabinovici 1979) and there may be no phase transition at all (Kajantie *et al.* 1996).

for $T \geq 200$ GeV. This conclusion is important for cosmology, namely for an explanation of the observed excess of baryons over antibaryons in the present-day Universe (problem of the generation of baryon asymmetry in the Universe (Sakharov 1967, Kuzmin 1970)).

In conclusion, we note that rapid non-conservation of the baryon and lepton numbers due to electroweak interactions is also possible in other, more exotic cases, namely at high fermion densities (Matveev *et al.* 1987) or in the decays of heavy particles (Ambjorn and Rubakov 1985). The cross sections of such processes, occurring in particle *collisions*, are suppressed by the factor (17.44) at low energies, but increase rapidly at energies in the several TeV range (Ringwald 1990, Espinosa 1990), most likely remaining exponentially suppressed at all accessible energies. Instanton processes at high energies are discussed in brief in the Appendix.

Supplementary Problems
for Part III

Problem 1. *Symmetry restoration in cold dense fermionic matter (Kirzhnits and Linde 1976).*

Consider the four-dimensional theory of a single real scalar field φ, interacting with a single type of Dirac fermion. The action of the scalar field itself is chosen in the form

$$S_\varphi = \int d^4x \left[\frac{1}{2} (\partial_\mu \varphi)^2 - \frac{\lambda}{4} (\varphi^2 - v^2)^2 \right], \tag{S3.1}$$

and the fermion action in the form

$$S_\psi = \int d^4x (i\bar{\psi}\gamma^\mu \partial_\mu \psi - f\varphi\bar{\psi}\psi).$$

The coupling constants λ and f are assumed to be small.

1. Show that the theory has a discrete symmetry $\varphi \to -\varphi$. In the absence of real fermions this discrete symmetry is spontaneously broken: there are two ground states with $\varphi = \pm v$. Consider next the system with a finite density of fermions n_F (and temperature zero).

2. Find the energy density of the fermionic matter $\varepsilon_F(\varphi)$ in an arbitrary homogeneous external field φ, if the fermion number density is equal to n_F.

3. Sketch the dependence of the total energy density of the system[1]

$$V_{\text{eff}}(\varphi) = \frac{\lambda}{4} (\varphi^2 - v^2)^2 + \varepsilon_F(\varphi)$$

---

[1] In fact, there are quantum corrections to this expression. However, they are small for small λ and f.

397

on the field φ for various n_F. Estimate the value of n_F at which symmetry is restored, i.e. $V_{\text{eff}}(\varphi)$ has a unique minimum at the point $\varphi = 0$.

4. Find the corresponding critical density for $f^4 \ll \lambda$.

Problem 2. *Non-topological soliton in a theory with fermions (Friedberg and Lee 1977).*
Consider the four-dimensional theory of a single real scalar field φ, interacting with N types of fermions à la Yukawa. The action of the scalar field is chosen in the form (S3.1). Suppose, for simplicity, that the Yukawa coupling constants of all fermions are the same, i.e. the fermionic action has the form

$$S_\psi = \int d^4x \sum_{i=1}^{N} (i\bar{\psi}_i \gamma^\mu \partial_\mu \psi_i - f\varphi\bar{\psi}_i\psi_i).$$

The constants λ and f are assumed to be small, but $\lambda \gg f^4$. Using considerations analogous to those in Chapter 10, show that for sufficiently large N, the theory possesses non-topological solitons (for example, those in which the number of fermions of each type is equal to 1). Estimate the corresponding minimal value of N. The polarization of the vacuum (including the contribution of the Dirac sea to the total energy) is neglected.[2]

Problem 3. *Fermion zero modes in a monopole background field in a theory with a fermion triplet (Jackiw and Rebbi 1976b).*
Consider the model of Section 16.1, but instead of the fermion isodoublet, introduce the fermion isotriplet ψ^a, $a = 1, 2, 3$, with the action

$$S_\psi = \int d^4x (i\bar{\psi}^a \gamma^\mu (D_\mu \psi)^a - f\varepsilon^{abc}\bar{\psi}^a\psi^b\varphi^c),$$

where, as usual, $(D_\mu \psi)^a = \partial_\mu \psi^a + g\varepsilon^{abc}A_\mu^b\psi^c$.

1. Find the analogue of the conserved fermion angular momentum in the monopole background field.

2. Considering only fermions with minimal angular momentum, show that there exist fermion zero modes in the monopole field (eigenstates of the Dirac Hamiltonian with zero energy). Find the number of these.

---

[2]For large N, corrections to the total energy associated with the polarization of the fermion vacuum are not small. This difficulty can be avoided at the cost of introducing additional boson fields.

Problem 4. *Euclidean fermion zero modes in $U(1)$-theory (Nielsen and Schroer 1977a, Ansourian 1977, Kiskis 1977).*
Consider a two-dimensional theory with gauge group $U(1)$ and massless Dirac fermions with unit charge.

1. Using the considerations of Section 17.1, write down the Euclidean Dirac equation in an arbitrary gauge.

2. Consider the gauge $\partial_\mu A_\mu = 0$ in the Euclidean theory. Show that in this gauge the smooth Euclidean field A_μ can be represented in the form $A_\mu = \varepsilon_{\mu\nu}\partial_\nu\sigma$, where $\sigma(x)$ is a smooth function of the coordinates. Considering only spherically symmetric asymptotics, $\sigma = \sigma(r)$ as $r \to \infty$ (where $r = \sqrt{x_0^2 + x_1^2}$), determine the behavior of $\sigma(r)$ for large r for fields with a fixed and integer topological number

$$ q = \frac{e}{4\pi} \int d^2x\, \varepsilon_{\mu\nu} F^{\mu\nu} $$

   (see the end of Section 13.3).

3. Find the general non-singular solution of the Dirac equation in the external field $A_\mu = \varepsilon_{\mu\nu}\partial_\nu\sigma$.

4. For external fields with spherically symmetric asymptotics and integer q, find all Euclidean fermion zero modes, i.e. solutions of the Euclidean Dirac equation, which decrease as $r \to \infty$ (including[3] fermion zero modes, which decrease as r^{-1} for large r). Which fermion quantum numbers are conserved in this model and which are not?

5. Show that in the theory without Higgs fields (two-dimensional massless quantum electrodynamics, model due to Schwinger (1962)), the bosonic action of Euclidean configurations of the field A_μ can be arbitrarily small for $q \neq 0$.

   Thus, in the Schwinger model, transitions between topologically distinct gauge vacua take place without exponential suppression and cannot be described by the usual semiclassical methods. In fact, the Schwinger model reduces at the quantum level to the theory of some free field, but in terms of the original fields ψ and A_μ, its vacuum has a complex structure (Lowenstein and Swieca 1972).

---

[3]The norm of these zero modes diverges logarithmically, however, they play the same role as normalizable zero modes (Patrascioiu 1979).

Problem 5.   *Fermion zero modes in a sphaleron field.*
Let us consider a four-dimensional gauge theory with gauge group $SU(2)$
and a Higgs doublet, as in Section 13.4. We include in it the massless Dirac
fermion isodoublet with the action

$$S_\psi = \int d^4x\, i\bar\psi\gamma^\mu D_\mu\psi.$$

Consider the Dirac equation in the external sphaleron field (13.57).

1. Using the symmetry of the sphaleron external field, find the analogue
   of the conserved fermion angular momentum.

2. Considering only fermions with minimal angular momentum, show
   that in the sphaleron field there exist zero modes of the Dirac
   Hamiltonian (eigenfunctions with zero energy). Find the number of
   zero modes of each chirality.

3. Give an interpretation of the fermion zero modes in terms of level
   crossing.

Problem 6.   *The same as in the previous problem, but in a model with
fermion action (17.37) with $h_1 = h_2$ (Ringwald 1988).*

Problem 7.   *Anomalous non-conservation of fermion quantum numbers
in the presence of a monopole.*
Consider the model of Section 16.2.

1. Choose the following class of Euclidean configurations of the boson
   fields

$$A_0 = 0 \qquad A_i = \Omega A_i^{\mathrm{mon}}\Omega^{-1} + \Omega\partial_i\Omega^{-1} \qquad \varphi = \varphi^{\mathrm{mon}}, \qquad (S3.2)$$

   where $A_i^{\mathrm{mon}}, \varphi^{\mathrm{mon}}$ is the classical static monopole solution (16.3),
   (16.4), $\Omega(\mathbf{x}, t)$ is a spherically symmetric function with values in
   $SU(2)$,

$$\Omega = \exp[i\tau^a n^a f(r, t)].$$

   Find an expression for the topological number Q (see (17.20)) and the
   Euclidean action $(S_{A,\varphi} - S_M)$ for configurations of the form (S3.2).
   Here, $S_M = E_M T$, E_M is the energy of the classical monopole
   solution, T is normalization time. The expressions (S3.2) describe
   some class of perturbations about the monopole background; in the
   monopole sector, the Euclidean action is referenced from the action

of the unperturbed monopole S_M. Show that, for $Q \neq 0$, the action $(S - S_M)$ can be arbitrarily small.

Thus, in the monopole sector of the four-dimensional theory we have a situation analogous to the Schwinger model (see Problem 4); the exponential suppression, characteristic for processes of the instanton type in the vacuum sector of four-dimensional theories, is absent.

2. Considering only *s*-wave fermions, find the Euclidean fermion zero modes in fields of the form (S3.2) in the limit when the size of the monopole core is small in comparison with all parameters of the problem, including the length scale characterizing the function Ω.

The results of this problem indicate that anomalous non-conservation of fermion quantum numbers takes place without exponential suppression in the monopole sector. A consistent description of this phenomenon can only be given in the framework of quantum field theory.

Problem 8. *Non-conservation of the fermion number in cold fermionic matter in a two-dimensional model (Rubakov 1986).*
Let us consider the Abelian Higgs model in two-dimensional Minkowski space–time. Let us include in it massless fermions with the action

$$S_\psi = \int d^2 x i \bar{\psi} \gamma^\mu (\partial_\mu - ie\gamma^5 A_\mu)\psi$$

(see Problem 15.4). Consider first the state of the system in which the fields A_μ and φ take vacuum values $A_\mu = 0$, $\varphi = v$, and there is a finite fermion density n_F (here, the densities of the numbers of left- and right-handed fermions are equal, so that the total charge of the system is equal to zero). Let us now vary the vector field $A_1(x^1)$ adiabatically (using the gauge $A_0 = 0$). Neglecting the polarization of the Dirac vacuum (which does not depend on n_F), find the change in the energy of the fermionic matter, i.e. the contribution of the fermions to the static energy as a functional of $A_1(x^1)$. Using the expression for the sphaleron energy in the Abelian Higgs model (see Supplementary Problems for Part II), estimate the critical value of n_F at which non-conservation of the fermion number occurs without tunneling, i.e. without exponential suppression.

Appendix
Classical Solutions and
the Functional Integral

One of the most suitable approaches, capable of providing a consistent interpretation of *classical* solutions in *quantum* field theory, is the formalism of functional integration. Within this formalism the role of classical solutions of the field equations is that they represent non-trivial saddle points of a functional integral. Integration near saddle points leads to a semiclassical decomposition of the functional integral, which is usually legitimate in weakly coupled theories. One example of the use of this approach is the semiclassical quantization of solitons, which, however, can also be performed by operator methods. In this Appendix, we shall consider another class of problems, for whose solution the functional integration method is indispensable, namely, computation of instanton effects in quantum field theory. The use of the functional integral in these cases allows one not only to find the leading semiclassical tunneling exponential (which was the main subject of our interest in Chapters 11–13), but also to compute, at least in principle (and sometimes explicitly), the pre-exponential factor and the subsequent corrections of the semiclassical decomposition. Furthermore, using the functional integration method, it is possible to reveal a number of non-trivial properties of the instanton contributions to the Green's functions of quantum fields.

In this Appendix, we shall not attempt any form of systematic study. Our aim is to give an initial idea of semiclassical methods in quantum field theory, based on the use of classical solutions.

A.1 Decay of the false vacuum in the functional integral formalism

In this section we shall consider one of the simplest examples of the use of the functional integral in problems relating to tunneling in quantum field theory, namely, the decay of the false vacuum (Callan and Coleman 1977), discussed in Chapter 12. As before, we shall consider the model of a single scalar field with a scalar potential, shown in Figure 12.1, in d-dimensional space–time. The main quantity of interest will be the energy of the false vacuum with

$$\langle \phi \rangle = \phi_-.$$

Since the false vacuum is unstable, we may expect its energy $E(\phi_-)$ to contain an imaginary part, associated with its decay width,

$$\mathrm{Im}\, E(\phi_-) = -\frac{1}{2}\Gamma(\phi_-).$$

We shall not consider gravitational interactions, thus the value of the real part of the energy, $\mathrm{Re}\, E(\phi_-)$, is unimportant to us; our purpose is a semiclassical computation of the width $\Gamma(\phi_-)$. We already know, from Chapter 12, that in a weakly coupled theory the width $\Gamma(\phi_-)$ is exponentially small.

We can write down an expression for the energy of the false vacuum in the form of a Euclidean functional integral

$$\mathrm{e}^{-E(\phi_-)T} = \int \mathcal{D}\phi\, \mathrm{e}^{-S[\phi]}, \tag{A.1}$$

where

$$S[\phi] = \int d^d x \left[\frac{1}{2}(\partial_\mu \phi)^2 + V(\phi) \right]$$

is the Euclidean action of the model (the summation over the index μ here and in what follows is performed with the Euclidean metric $g_{\mu\nu} = \mathrm{diag}\,(+1,+1,+1,\ldots)$), T is normalization time. Since we are interested in the state with $\langle \phi \rangle = \phi_-$, the fields over which the integration in (A.1) is performed have the asymptotics

$$\phi(|x| \to \infty) = \phi_-. \tag{A.2}$$

The semiclassical computation of the integral (A.1) involves finding its saddle points and taking into account their contributions to the integral.

The saddle points of the integral (A.1) are extrema of the Euclidean action $S[\phi]$. They satisfy the classical Euclidean field equation

$$-\partial_\mu \partial_\mu \phi + \frac{\partial V}{\partial \phi} = 0$$

and the boundary condition (A.2).

The results of Chapter 12 show that the width $\Gamma(\phi_-)$ is determined by the contribution of a non-trivial saddle point, the bounce (Euclidean bubble), $\phi_B(x^\mu)$. Before considering the contribution of this saddle point, let us consider the contribution of the trivial solution

$$\phi(x) = \phi_-, \tag{A.3}$$

which is uniform throughout Euclidean space–time. The action for the trivial solution is equal to zero; thus, according to the standard rules for saddle-point integration, the contribution of this point is determined by a Gaussian integral over the perturbations about it. Writing

$$\phi(x) = \phi_- + \eta(x),$$

we obtain a quadratic action for the fluctuations

$$S_2^{(0)}(\eta) = \int \left[\frac{1}{2}(\partial_\mu \eta)^2 + \frac{1}{2} V''(\phi_-)\eta^2 \right] d^d x. \tag{A.4}$$

The contribution of the saddle point (A.3) to the integral (A.1) is equal to

$$I_0 = \int \mathcal{D}\eta \, e^{-S_2^{(0)}(\eta)} \tag{A.5}$$

up to corrections, which are small in the weakly coupled theory.

The formal computation of the Gaussian integral (A.5) involves the following. We shall suppose that Euclidean space–time is a box with a large but finite volume (on whose boundaries, for example, periodicity conditions are imposed). Using the fact that the quadratic action (A.4) can be represented in the form

$$S_2^{(0)} = \int d^d x \frac{1}{2} \eta [-\partial^2 + V''(\phi_-)]\eta,$$

we introduce orthonormalized eigenfunctions of the corresponding operator,

$$[-\partial_\mu^2 + V''(\phi_-)]\eta_\lambda = \lambda \eta_\lambda. \tag{A.6}$$

We note that $V''(\phi_-) > 0$, thus the eigenvalues λ are positive. We decompose the integration variable $\eta(x)$ in terms of these eigenfunctions,

$$\eta(x) = \sum_\lambda a_\lambda \eta_\lambda(x). \tag{A.7}$$

In terms of the new integration variables a_λ, we have for the action

$$S_2^{(0)} = \sum_\lambda \frac{1}{2} \lambda a_\lambda^2,$$

so that instead of (A.5), we obtain

$$I_0 = \int \prod_\lambda \left[\frac{da_\lambda}{\sqrt{2\pi}} \right] \exp \left[-\frac{1}{2} \sum_\lambda \lambda a_\lambda^2 \right] \qquad (A.8)$$

(the factor $\prod_\lambda [\frac{1}{\sqrt{2\pi}}]$ is introduced for convenience and corresponds to the specific choice of normalization for the integral (A.1), i.e. the choice of an additive constant in the energy). The expression (A.8) is the product of Gaussian integrals, and it is equal to

$$I_0 = \prod_\lambda \lambda^{-1/2}. \qquad (A.9)$$

The quantity $\prod_\lambda \lambda$ can be interpreted as the determinant of the operator $(-\partial^2 + V''(\phi_-))$, since it is the product of its eigenvalues. Thus,

$$I_0 = [\text{Det}\,(-\partial^2 + V''(\phi_-))]^{-1/2}, \qquad (A.10)$$

and the contribution of the saddle point $\phi = \phi_-$ to the energy of the false vacuum equals, up to higher-order perturbative corrections,

$$E_0(\phi_-) = \frac{1}{2T} \ln \text{Det}\,[-\partial^2 + V''(\phi_-)].$$

This expression is real and, as mentioned earlier, its value is not important to us.

In the case of a homogeneous external field, a functional determinant of type (A.10) can be computed without particular difficulty (Coleman and Weinberg 1973). In a space–time box of size $L = T$, the eigenfunctions of the operator $(-\partial^2 + V''(\phi_-))$ for homogeneous ϕ are plane waves $\exp(ik_n x)$ with $k_n^\mu = \frac{2\pi}{L}(n_0, n_1, n_2, n_3)$, where the n_μ are integers (to be specific, we shall consider the case $d = 4$). Here, the eigenvalues are equal to $(k_n^2 + V'')$, thus the first quantum correction to the energy density of the homogeneous field ϕ (the effective potential) is equal to

$$\Delta V(\phi) = \frac{\Delta E(\phi)}{L^3} = \frac{1}{2TL^3} \sum_n \ln[k_n^2 + V''(\phi)].$$

In the limit of large $L = T$ we have

$$\Delta V(\phi) = \frac{1}{2} \int_{k^2 < \Lambda^2} \frac{d^4 k}{(2\pi)^4} \ln(k^2 + V''(\phi)).$$

The integral on the right-hand side diverges, thus, we have introduced the ultraviolet cut-off Λ. As a result, for the field energy (taking into account the classical energy $V(\phi)$) in a Gaussian (one-loop) approximation we have, up to a divergent constant, independent of ϕ,

$$
\begin{aligned}
V(\phi) + \Delta V(\phi) &= V(\phi) + \frac{\Lambda^2}{32\pi^2} V''(\phi) \\
&\quad - \frac{1}{64\pi^2} [V''(\phi)]^2 \ln \Lambda^2 + \frac{1}{64\pi^2} [V''(\phi)]^2 \left[\ln V''(\phi) - \frac{1}{2} \right].
\end{aligned}
$$

$$(A.11)$$

If $V(\phi)$ is a polynomial of degree at most four (renormalizable theory), then the divergent terms in (A.11) are also polynomials of degree at most four. These divergences are removed by renormalizing the parameters of the original potential $V(\phi)$, i.e. by standard renormalization of the mass and the coupling constant. Thus, the renormalized correction to the energy density of the homogeneous field is equal to

$$
\Delta V(\phi) = \frac{1}{64\pi^2} [V''(\phi)]^2 \ln \frac{V''(\phi)}{\mu^2} + P_4(\phi),
$$

where μ is the renormalization parameter, and $P_4(\phi)$ is a polynomial of degree four, whose form depends on the normalization conditions for the mass and the coupling constant of the theory.

Another saddle point of the integral (A.1) is the bounce solution. Denoting its contribution to the integral by I_B, and the associated contribution to the energy by E_B, we write

$$
e^{-(E_0 + E_B)T} = I_0 + I_B + \ldots = e^{-E_0 T} + I_B + \ldots . \tag{A.12}
$$

Here, I_B and E_B are suppressed by the tunneling exponent; the dots in (A.12) denote corrections, suppressed by higher orders of the tunneling exponent (they arise due to the contributions of multiple bounces). To the first order in the tunneling exponent we have

$$
-e^{-E_0 T} E_B T = I_B
$$

or

$$
E_B = -\frac{1}{T} \frac{I_B}{I_0}. \tag{A.13}
$$

To compute the contribution of a bounce to the integral (A.1) we again proceed in the manner standard for saddle-point computations. Near this solution we write

$$
\phi(x) = \phi_B(x) + \eta(x),
$$

and retain terms of up to second-order in η in the action,

$$S^{(2)} = S_B + \int d^d x \frac{1}{2} \eta [-\partial_\mu^2 + V''(\phi_B)] \eta,$$

where S_B is the Euclidean action of the bounce. Consequently, up to higher-order perturbative corrections

$$I_B = e^{-S_B} \int \mathcal{D}\eta \exp\left[-\frac{1}{2} \int dx \eta [-\partial_\mu^2 + V''(\phi_B)] \eta\right]. \qquad (A.14)$$

The contribution of a bounce is indeed suppressed by the quantity e^{-S_B}, which agrees with the results of Section 12.2. The pre-exponential factor is given by a Gaussian integral over perturbations η.

To compute the Gaussian integral in (A.14) we need to consider the eigenvalue problem

$$[-\partial_\mu^2 + V''(\phi_B(x))] \eta_\lambda = \lambda \eta_\lambda, \qquad (A.15)$$

analogous to the problem (A.6). If the eigenvalues λ were positive, we would have a formula analogous to (A.9) or (A.10). However, in the case of a bounce there exist zero and even negative eigenvalues λ; thus, further analysis is required.

Let us begin with a discussion of zero eigenvalues. The occurrence of zero modes of equation (A.15) is associated with the fact that the center of the bounce may be located at any point in Euclidean space–time, i.e. we in fact have a whole family of solutions of the Euclidean field equations $\phi_B(x - x_0)$, parametrized by d parameters x_0^μ (we study the contribution of a spherically symmetric bounce, considered in Chapter 12). By complete analogy with Section 7.1, this leads to the existence of d zero modes around the bounce (with center at $x_0^\mu = 0$),

$$\eta_0^\mu = S_B^{-1/2} \partial_\mu \phi_B(x), \quad \mu = 0, \ldots, d-1.$$

The normalization factor here is chosen such that

$$\int d^d x \eta_0^\mu \eta_0^\nu = \delta^{\mu\nu},$$

as one can see using the virial theorem, obtained according to the considerations of Section 7.2.

If we were to treat zero modes in the same way as non-zero modes, then the corresponding integral with respect to $\prod_\mu da_0^\mu$ in the expression of type (A.8) would diverge. For an interpretation of this divergence it is convenient to move from integration with respect to da_0^μ to integration with respect to the "collective coordinates," the position of the bounce x_0^μ. To this end, we shall use a procedure of the Faddeev–Popov type and insert unity in the original functional integral (A.1):

$$1 = \Delta[\phi(x + x_0)] \int \prod_\nu dx_0^\nu \prod_\mu \delta \left\{ \int d^d x \partial_\mu \phi_B(x)[\phi(x + x_0) - \phi_-] \right\},$$

where

$$\Delta[\phi(x + x_0)] = \text{Det}_{\mu,\nu} \left[\int d^d x \partial_\mu \phi_B(x) \partial_\nu \phi(x + x_0) \right]. \tag{A.16}$$

After interchanging the order of integration with respect to $\mathcal{D}\phi$ and $\prod_\nu dx_0^\nu$ and replacement of the variable $\phi(x + x_0) \rightarrow \phi(x)$, we obtain

$$\int \mathcal{D}\phi e^{-S[\phi]} \tag{A.17}$$

$$= \int \prod_\nu dx_0^\nu \int \mathcal{D}\phi \Delta[\phi(x)] \prod_\mu \delta \left\{ \int d^d x \partial_\mu \phi_B(x)[\phi(x) - \phi_-] \right\} e^{-S[\phi]}.$$

Since the integrand is now independent of x_0, the integral with respect to $d^d x_0$ gives the volume of space–time VT, where V is the volume of $(d-1)$-dimensional space. Moreover, thanks to the δ-function in the integral with respect to $\mathcal{D}\phi$, the only contribution to it comes from a neighborhood of the bounce with center at the origin. In this neighborhood, we can write

$$\phi(x) = \phi_B(x) + \sum_\mu a_0^\mu \eta_0^\mu(x) + \sum_{\lambda \neq 0} a_\lambda \eta_\lambda(x). \tag{A.18}$$

Then the δ-function in (A.17) will be equal to

$$\prod_\mu \delta \left\{ \sum_\nu \left(\int d^d x \partial_\mu \phi_B \eta_0^\nu \right) a_0^\nu \right\} = S_B^{-d/2} \prod_\nu \delta(a_0^\nu).$$

Thus, integration with respect to da_0, taking into account the measure in (A.8), leads to the occurrence of a factor $(\frac{1}{S_B 2\pi})^{d/2}$. Finally, for fields of the form (A.18), expression (A.16) is equal to

$$\Delta[\phi(x)] = (S_B)^d.$$

As a result, for the contribution of the bounce to the energy we have

$$I_B = VT \left(\frac{S_B}{2\pi} \right)^{d/2} e^{-S_B} \int \mathcal{D}' \eta \exp \left[-\frac{1}{2} \int dx \eta [-\partial_\mu^2 + V''(\phi_B)]\eta \right],$$

where the prime denotes no integration over zero modes.

In addition to zero eigenvalues, the operator $[-\partial_\mu^2 + V''(\phi_B)]$ in which we are interested has a (single) negative eigenvalue. Indeed, this operator is spherically symmetric (we recall that ϕ_B depends only on $r = \sqrt{x_\mu x_\mu}$), and the zero modes have a non-trivial angular dependence, i.e. non-zero angular momentum. Clearly, the lowest s-wave state has the least eigenvalue, i.e. for that state $\lambda < 0$. Naively, the existence of a negative eigenvalue λ points to the divergence of the integral with respect to the corresponding coefficient a_-,

$$\int \frac{da_-}{\sqrt{2\pi}} e^{-\frac{1}{2}\lambda_- a_-^2} = \infty \quad \text{for } \lambda_- < 0.$$

In fact, one has to view the integral (A.1) as the analytic continuation of some well-defined quantity. Corresponding considerations were presented by Callan and Coleman (1977); the result is that a_- should be viewed as a purely imaginary quantity, and also that the factor $\frac{1}{2}$ occurs in the final answer. Thus, we come to the expression

$$I_B = \frac{i}{2} VT \left(\frac{S_B}{2\pi}\right)^{d/2} e^{-S_B} \left|\text{Det}'\left(-\partial_\mu^2 + V''(\phi_B)\right)\right|^{-1/2}.$$

Here, Det' denotes the determinant (product of the eigenvalues) of the operator in which the zero eigenvalues are omitted.

Taking into account (A.13), the contribution of the bounce to the energy of the false vacuum has the form

$$E_B = -iV\frac{1}{2}\left(\frac{S_B}{2\pi}\right)^{d/2} e^{-S_B} \left|\frac{\text{Det}'\left(-\partial_\mu^2 + V''(\phi_B)\right)}{\text{Det}\left(-\partial_\mu^2 + V''(\phi_-)\right)}\right|^{-1/2}.$$

First and foremost, this contribution is purely imaginary, i.e. the bounce actually describes the decay of the false vacuum with width

$$\Gamma = 2\text{Im}\, E_B = V\left(\frac{S_B}{2\pi}\right)^{d/2} e^{-S_B} \left|\frac{\text{Det}'\left(-\partial_\mu^2 + V''(\phi_B)\right)}{\text{Det}\left(-\partial_\mu^2 + V''(\phi_-)\right)}\right|^{-1/2}. \tag{A.19}$$

Moreover, this width is proportional to the volume of $(d-1)$-dimensional space. In the limit $V \to \infty$ the value of Γ/V, the probability of decay per unit time per unit volume is finite. In fact, it is precisely this quantity that must be finite, since a bubble of a false vacuum may be formed spontaneously at any point of $(d-1)$-dimensional space.

Thus, the functional integration technique allows us to find not only the main semiclassical exponent, but also the pre-exponential factor for the probability of decay of the false vacuum per unit spatial volume, per unit time. The functional determinants in (A.19) can be computed analytically only in the simplest models; if necessary, this can be done numerically. In fact, these determinants are ultraviolet divergent (compare with (A.11)), however, these divergences are removed by the usual renormalization of the parameters of the Lagrangian (mass, coupling constant, wave function).

To conclude this section, we note that the method of treating zero modes, studied in this section, is of a very general nature and, with corresponding modifications, is applicable in the semiclassical quantization of solitons, the computation of instanton contributions to the functional integral in theories with gauge fields, the determination of the probability of sphaleron transitions at finite temperatures, and in other situations.

A.2 Instanton contributions to the fermion Green's functions

In Chapter 17, we saw that instanton transitions lead to anomalous non-conservation of fermion quantum numbers in a number of four-dimensional gauge theories. In this section, we discuss how this result can be obtained in the formalism of functional integration and describe an approach to computing the corresponding fermion Green's functions. In addition, we consider the occurrence of the effective θ-parameter in theories with massive fermions. The approach to these questions, based on the functional integral, was historically the first approach ('t Hooft 1976a,b) and is most appropriate for quantitative analysis.

To be specific, let us consider a four-dimensional model with gauge group $SU(2)$ and a single Dirac $SU(2)$ fermion doublet. The Euclidean action of the model has the form

$$S = S_A + S_\psi,$$

where S_A is the action of the gauge fields

$$S_A = \int d^4x \frac{1}{4} F^a_{\mu\nu} F^a_{\mu\nu} + g.f. \qquad (A.20)$$

(we do not include the θ-term in it for the moment), and

$$S_\psi = \int d^4x (\bar{\psi}\gamma^\mu D_\mu \psi + m\bar{\psi}\psi) \qquad (A.21)$$

is the fermion part of the action (we shall omit throughout the index E for Euclidean objects). In (A.20) the component "*g.f.*" includes a term fixing the gauge and the action of Faddeev–Popov ghosts. In (A.21), Euclidean γ-matrices

$$\gamma^\mu = \begin{pmatrix} 0 & \sigma^\mu \\ \sigma^{\mu\dagger} & 0 \end{pmatrix}$$

are used (compare with Section 17.2), where

$$\sigma^\mu = (1, -i\boldsymbol{\sigma}),$$
$$\sigma^{\mu\dagger} = (1, i\boldsymbol{\sigma}).$$

We shall be interested in Euclidean Green's functions of the type

$$G_{km}(x_1,\ldots,x_k;y_1,\ldots,y_m) = \langle\psi(x_1)\ldots\psi(x_k)\bar\psi(y_1)\ldots\bar\psi(y_m)\rangle.$$

They are given by the functional integral

$$G_{km} = \int \mathcal{D}A\mathcal{D}\bar\psi\mathcal{D}\psi e^{-S}\psi(x_1)\ldots\bar\psi(y_m), \tag{A.22}$$

where integration over ghost fields is implicit; it is understood that ψ and $\bar\psi$ are Grassmann integration variables. In particular, for $k = m = 0$, the integral (A.22) determines the vacuum energy of the system

$$e^{-E_{vac}T} = \int \mathcal{D}A\mathcal{D}\bar\psi\mathcal{D}\psi e^{-S}, \tag{A.23}$$

and for $k = m = 1$, it is an exact two-point Green's function

$$G_{11}(x,y) = \langle\psi(x)\bar\psi(y)\rangle. \tag{A.24}$$

In what follows, we shall, to be specific, consider these two cases.

The action S_A has a non-trivial saddle point (local minimum), which is an instanton solution of the Euclidean field equations. The contribution of its neighborhood to the integral (A.22) will be of primary interest to us.

Let us first consider the case of massless fermions,

$$m = 0.$$

In this case, *in perturbation theory*, chirality is conserved, i.e. vacuum averages of operators invariant under (global) transformations

$$\psi \to e^{i\alpha\gamma^5}\psi, \quad \bar\psi \to \bar\psi e^{i\alpha\gamma^5} \tag{A.25}$$

are non-zero only. For example, the vacuum average of the operator $\bar\psi\psi$ (convolution with respect to the Lorentz and isotopic indices is understood) is equal to zero in perturbation theory.

As follows from the results of Chapter 17, chirality ceases to be conserved when instanton transitions are taken into account. To explain how this phenomenon arises in the formalism of functional integration, we consider the instanton contribution to the integral (A.22). Near an instanton solution, we can write

$$A_\mu = A_\mu^{inst} + a_\mu,$$

so that in a Gaussian approximation we will have

$$G_{km} = e^{-S_{\text{inst}}} \left[\int \mathcal{D}a_\mu e^{-S_A^{(2)}(a_\mu)} \right] G_{km}^{\text{inst}}(x_1, \ldots, y_m), \qquad (A.26)$$

where $S_A^{(2)}$ is the part of the boson action which is quadratic in the perturbations a_μ (including terms fixing the gauge and the action of ghosts; as before, integration over ghost fields is understood but not explicitly indicated), and G_{km}^{inst} is the integral over fermions in the external instanton field

$$G_{km}^{\text{inst}} = \int \mathcal{D}\bar{\psi}\mathcal{D}\psi \exp\left[-\int dx \bar{\psi}\gamma^\mu D_\mu^{\text{inst}}\psi \right] \psi(x_1)\ldots\bar{\psi}(y_m), \qquad (A.27)$$

where

$$D_\mu^{\text{inst}} = \partial_\mu + A_\mu^{\text{inst}}.$$

Integration over the boson fluctuations a_μ in (A.26) is performed using methods essentially analogous to those considered in the previous section, although there are additional complications associated with the large number of bosonic zero modes, gauge invariance, etc. This Gaussian integral was computed explicitly by 't Hooft (1976b). Here, we shall be primarily interested in the fermionic integral (A.27) in the instanton field.

It is useful to discuss the computation of an integral of the type (A.27) in an arbitrary external field A_μ and to consider, for specificity, the cases $k = m = 0$ and $k = m = 1$, i.e. the matrix elements of the unit operator and of the operator $\psi(x)\bar{\psi}(y)$:

$$\langle 1 \rangle^A = \int \mathcal{D}\bar{\psi}\mathcal{D}\psi \exp\left[-\int dx \bar{\psi}\gamma^\mu D_\mu \psi \right], \qquad (A.28)$$

$$\langle \psi(x)\bar{\psi}(y) \rangle^A = \int \mathcal{D}\bar{\psi}\mathcal{D}\psi \exp\left[-\int dx \bar{\psi}\gamma^\mu D_\mu \psi \right] \psi(x)\bar{\psi}(y). \ (A.29)$$

To compute these Gaussian integrals over the Grassmann variables, we diagonalize the quadratic form in the exponential (analogously to the bosonic case considered in the previous section). Since the operator $\gamma^\mu D_\mu$ is anti-Hermitian, its eigenvalues are purely imaginary, and the eigenfunctions ψ_λ satisfy the equation

$$\gamma^\mu D_\mu \psi_\lambda = i\lambda\psi_\lambda.$$

Let us first suppose that there are no Euclidean fermion zero modes, as is the case for topologically trivial (and sufficiently weak) external fields. In other words, we suppose that all the eigenvalues λ are non-zero. From the commutation relations for the γ-matrices, it follows that in addition to the eigenvalue λ, there exists an eigenvalue $(-\lambda)$, where the corresponding eigenfunction is equal to

$$\psi_{-\lambda} = \gamma^5 \psi_\lambda.$$

Let us now decompose the integration variables $\psi(x)$ and $\bar{\psi}(x)$ in terms of the set ψ_λ:

$$\psi(x) = \sum_\lambda b_\lambda \psi_\lambda(x), \qquad (A.30)$$

$$\bar{\psi}(x) = \sum_\lambda \bar{b}_\lambda \psi_\lambda^\dagger(x), \qquad (A.31)$$

where $b_\lambda, \bar{b}_\lambda$ are Grassmann integration variables. Then the integrals (A.28) and (A.29) take the form

$$\langle 1 \rangle^A = \int \prod_\lambda d\bar{b}_\lambda db_\lambda \exp\left[-i \sum_\lambda \lambda \bar{b}_\lambda b_\lambda\right], \qquad (A.32)$$

$$\langle \psi(x)\bar{\psi}(y) \rangle^A = \int \prod_\lambda d\bar{b}_\lambda db_\lambda \exp[-i \sum_\lambda \lambda \bar{b}_\lambda b_\lambda]$$

$$\times \sum_{\lambda',\lambda''} b_{\lambda'} \psi_{\lambda'} \bar{b}_{\lambda''} \psi_{\lambda''}^\dagger(y). \qquad (A.33)$$

These integrals are computed according to the rules for Grassmann integration:

$$\int db = 0 \qquad (A.34)$$

$$\int bdb = 1, \qquad (A.35)$$

and are equal to

$$\langle 1 \rangle^A = \prod_\lambda (i\lambda) = \text{Det}(\gamma D),$$

$$\langle \psi(x)\bar{\psi}(y) \rangle^A = \sum_{\lambda'} \left[\left(\prod_{\lambda \neq \lambda'} i\lambda\right) \psi_{\lambda'}(x)\psi_{\lambda'}^\dagger(y)\right]$$

$$= \text{Det}(\gamma D) \sum_\lambda \frac{\psi_\lambda(x)\psi_\lambda^\dagger(y)}{i\lambda}.$$

The quantity

$$G_0^A(x,y) = \sum_\lambda \frac{\psi_\lambda(x)\psi_\lambda^\dagger(y)}{i\lambda}$$

is a Green's function of the operator $\gamma^\mu D_\mu$. Taking into account (A.30), it can also be written in the form

$$G_0^A(x,y) = \sum_{\lambda>0} \frac{1}{i\lambda} \left[\psi_\lambda(x)\psi_\lambda^\dagger(y) - \gamma^5 \psi_\lambda(x)\psi_\lambda^\dagger(y)\gamma^5 \right].$$

Its chiral invariance (which is evident also from the anti-commutation of $\gamma^\mu D_\mu$ and γ^5) follows immediately from this representation,

$$e^{i\alpha\gamma^5} G_0^A e^{i\alpha\gamma^5} = G_0^A.$$

Thus, the matrix element (A.29) in the topologically trivial external field A_μ also has the property of chiral invariance

$$\langle e^{i\alpha\gamma^5} \psi(x)\bar\psi(y) e^{i\alpha\gamma^5} \rangle^A = \langle \psi(x)\bar\psi(y)\rangle^A.$$

This result generalizes directly to higher fermion Green's functions in any topologically trivial external fields, including the case of strong external fields, in which there may exist fermion zero modes. We again deduce the conservation of chirality in conventional perturbation theory.

The situation is fundamentally different in the case of an external instanton field (in general, of external fields with non-zero topological number (17.20)). In this case, the operator $\gamma^\mu D_\mu$ has one *left-handed* zero mode, i.e. a unique eigenfunction with $\lambda = 0$, having the structure

$$\psi_0 = \begin{pmatrix} \chi_0 \\ 0 \end{pmatrix}. \tag{A.36}$$

For the instanton external field, an explicit form for $\chi_0(x)$ is given by formula (17.25). This zero mode and its conjugate are involved in the decompositions (A.30) and (A.31). The first consequence of the existence of the zero mode is the equality to zero of the integral (A.32). Indeed, for that integral we have

$$\langle 1 \rangle^{\text{inst}} = \int db_0 d\bar b_0 \int \prod_{\lambda\neq 0} db_\lambda d\bar b_\lambda \exp\left[-i\sum_{\lambda\neq 0} \lambda b_\lambda \bar b_\lambda \right],$$

and the integrals with respect to db_0 and $d\bar b_0$ are equal to zero by virtue of the property (A.34) of the Grassmann integral. Thus, in the theory with massless fermions, instantons (and, in general, gauge field configurations with non-zero topological numbers) do not contribute to the vacuum energy (A.23).

Unlike the integral (A.28), the integral (A.29) is non-zero in the instanton background. Here, in the sum in (A.33), only the term with

$\lambda' = \lambda'' = 0$ is significant and, taking into account (A.35), we have

$$
\begin{aligned}
\langle \psi(x)\bar{\psi}(y)\rangle^{\text{inst}} &= \int d\bar{b}_0 db_0 [b_0 \psi_0(x)][\bar{b}_0 \psi_0^\dagger(y)] \\
&\quad \times \int \prod_{\lambda \neq 0} db_\lambda d\bar{b}_\lambda \exp\left[-i\sum \lambda \bar{b}_\lambda b_\lambda\right] \\
&= \psi_0(x)\psi_0^\dagger(y)\text{Det}'(\gamma^\mu D_\mu),
\end{aligned}
\tag{A.37}
$$

where the prime denotes omission of the zero mode. Because of (A.36), this matrix element breaks chirality; under the transformation (A.25) we have

$$
\langle \psi(x)\bar{\psi}(y)\rangle^{\text{inst}} \rightarrow e^{2i\alpha}\langle \psi(x)\bar{\psi}(y)\rangle^{\text{inst}}.
\tag{A.38}
$$

It follows from the results of Chapter 17 that the given discussion is valid for any external gauge fields with a unit topological number. Moreover, this discussion generalizes to arbitrary Green's functions and arbitrary values of the topological number Q: *in the general case, a sector with topological number Q contributes to the vacuum averages of operators with chirality $2Q$ and only to these operators.* This is precisely the mechanism for non-conservation of fermion quantum numbers from the point of view of the functional integral, and the instanton computation described in this section provides an efficient technique for finding anomalous Green's functions in weakly coupled theories.

The analysis presented here can be generalized to other gauge groups and/or fermions in other representations. In particular, the number and structure of the fermion zero modes in the instanton field in QCD and the electroweak theory are such that the selection rules of Section 17.3 are satisfied.

Let us now discuss the dependence of physical quantities on the parameter θ introduced in Section 13.3. Let us first see that this dependence does not exist in theories with massless fermions. As mentioned in Section 13.3, the parameter θ can be introduced using an additional term in the action of the gauge fields; in a Euclidean theory it has the form

$$
S_\theta = \frac{i\theta}{16\pi^2} \int \text{Tr}\,(F_{\mu\nu}\tilde{F}_{\mu\nu})d^4x.
$$

Its role is to make the contributions of sectors with topological charge Q to the functional integral proportional to $\exp(iQ\theta)$. For example, the instanton contribution to the two-point Green's function (A.24) is equal to

$$
\langle \psi(x)\bar{\psi}(y)\rangle_\theta^{\text{inst}} = e^{i\theta}\langle \psi(x)\bar{\psi}(y)\rangle_{\theta=0}^{\text{inst}}.
\tag{A.39}
$$

By virtue of property (A.36) the parameter θ can be eliminated from (A.39), by redefining the fields according to formula (A.25) with $\alpha = -\theta/2$. This chiral rotation eliminates θ from *all* Green's functions, which follows, in the general case, from the previous statement about selection rules. Thus, the parameter θ is unphysical in theories with massless fermions (and also in theories of the type of the standard electroweak model).

Let us now consider the model (A.20), (A.21) with *massive* fermions and $\theta \neq 0$. In this case, the chirality is not conserved even in perturbation theory, and the instanton contributions to the Green's function are not characterized by any special selection rules. The fermion mass term in the action can be generalized somewhat by replacing the last term in (A.21) by the expression

$$S_m = \int d^4x m\bar{\psi}e^{i\beta\gamma^5}\psi. \tag{A.40}$$

This term in the action has the necessary properties of Hermiticity (in Minkowski space–time) and Lorentz and gauge invariance. In perturbation theory the parameter β is unimportant since it can be eliminated by chiral transformation of the fields (A.25) with $\alpha = -\beta/2$. Things are different when instanton contributions are taken into account.

Let us consider, by way of example, a one-instanton contribution to the vacuum energy (A.23), and let us for simplicity suppose that the fermion mass is small. For small m, the mass term (A.40) can be taken into account perturbatively. To zeroth order in m the instanton contribution to the vacuum energy is absent, while to first order in m we have, for $\theta \neq 0$,

$$\langle 1 \rangle^{\text{inst}} = me^{i\theta}\langle\bar{\psi}e^{i\beta\gamma^5}\psi\rangle^{\text{inst}}_{\theta=0;m=0}.$$

Taking into account (A.37) we have

$$\langle\bar{\psi}e^{i\beta\gamma^5}\psi\rangle^{\text{inst}}_{\theta=0;m=0} \sim \psi_0^\dagger e^{i\beta\gamma^5}\psi_0,$$

and it follows from (A.36) that

$$\langle 1 \rangle^{\text{inst}} = me^{i(\theta+\beta)}\langle\bar{\psi}\psi\rangle^{\text{inst}},$$

where the factor $\langle\bar{\psi}\psi\rangle^{\text{inst}}$ does not depend on θ or on β. Thus, the contribution of the instanton to the vacuum energy depends on θ and β, but only in the combination

$$\theta_{\text{eff}} = \theta + \beta.$$

This result holds for all Green's functions; the assumption that the fermion mass is small is also unimportant. In the general case in theories with massive fermions there is a single parameter θ_{eff} (combination of the parameter θ and the phase of the fermion mass matrix), which is not important in perturbation theory, but which manifests itself in physical quantities when instanton contributions are taken into account.

The possible presence of the effective θ-parameter is particularly important in quantum chromodynamics, since it leads to a considerable violation of CP-invariance in strong interactions provided θ_{eff} is not too small. One possible explanation of the experimental fact of the absence of such CP-violation (Peccei and Quinn 1977) is that the phases of quark masses are dynamical variables. In other words, the parameter β is in fact the vacuum expectation value of some scalar field (axion). The scalar Lagrangian is chosen so that, in the absence of QCD instanton effects, there is global $U(1)$ symmetry and the vacuum value of β is arbitrary; the axion in this approximation is a massless Goldstone boson. Instanton effects of QCD break this $U(1)$ symmetry explicitly, and the axion field takes a vacuum value corresponding to $\theta_{\text{eff}} = 0$ (also, the axion acquires a small mass). This hypothesis, which has not yet been experimentally confirmed, is of great interest both from the point of view of particle physics and from the point of view of cosmology.

A.3   Instantons in theories with the Higgs mechanism. Integration along valleys

We have already mentioned in Section 13.4 that the scale argument of Section 7.2 rules out the existence of instanton solutions in four-dimensional theories with the Higgs mechanism. Nevertheless, Euclidean configurations with non-zero topological number Q and a finite action exist and one can ask about the contribution of sectors with $Q \neq 0$ to the functional integral. Here, we present simple considerations ('t Hooft 1976b), which lead quickly to a legitimate answer in the most interesting case of configurations of small size with $|Q| = 1$; a systematic approach to this question was given by Affleck (1981).

Let us consider, for specificity, a model with gauge group $SU(2)$ and Higgs field doublet. This was discussed in Sections 6.2 and 13.4; we shall use the notation introduced there. The Euclidean action of the model has the form

$$S = \int d^4x \left[-\frac{1}{2g^2} \operatorname{Tr} F_{\mu\nu}^2 + (D_\mu\phi)^\dagger D_\mu\phi + \lambda \left(\phi^\dagger\phi - \frac{1}{2}\phi_0^2 \right)^2 \right]. \quad \text{(A.41)}$$

In the sector with $Q = 1$, smooth configurations of fields have the following asymptotics:

$$
\begin{aligned}
A_\mu &= \omega\partial_\mu\omega^{-1}, \\
\phi &= \omega\phi^{\text{vac}}, \quad r = \sqrt{x_\mu x_\mu} \to \infty,
\end{aligned}
\quad \text{(A.42)}
$$

where

$$\omega = \sigma_\alpha n_\alpha, \quad \phi^{\text{vac}} = \frac{1}{\sqrt{2}} \begin{pmatrix} 0 \\ \phi_0 \end{pmatrix},$$

and $n_\alpha = x_\alpha / r$ is a unit radius vector in four-dimensional Euclidean space–time.

In pure Yang–Mills theory (without Higgs fields) the action is scale invariant and its minima in the sector with $Q = 1$ (instantons) are characterized by the parameter ρ (size of instantons), which can take arbitrary values. The action does not depend on ρ, thus ρ emerges as a collective coordinate, analogous to the instanton position x_0^μ, discussed in Section A.1. To compute the contributions of all instantons to the functional integral of pure Yang–Mills theory, one has to integrate with respect to ρ using the techniques described in Section A.1. Loosely speaking, in the space of all Euclidean field configurations of pure Yang–Mills theory, the profile of the action is a valley with a flat bottom, with ρ being the parameter along the valley; this is illustrated schematically in Figure A.1. Integration along the valley must be carried out exactly, while in perpendicular directions it is performed in a Gaussian approximation.[1]

When Higgs fields are included, the Euclidean action begins to depend on the scale of the configuration. As we now see, for small ρ, the contribution of the Higgs fields to the Euclidean action is small in comparison with the contribution of the gauge field. In other words, the valley shown in Figure A.1 ceases to be flat-bottomed, but remains gently sloping for small ρ. This situation is shown in Figure A.2. It is clear that, on the one hand, there is no minimum of the action (it corresponds to the singular limit $\rho \to 0$), while, on the other hand, integration with respect to ρ must be carried out as before without using the Gaussian approximation.

For small ρ the action along the valley may be found from the following considerations. Consider the family of configurations of the form

$$A_\mu(x) = \frac{1}{\rho} B_\mu \left(\frac{x}{\rho} \right), \quad \phi(x) = \phi_0 f \left(\frac{x}{\rho} \right),$$

where B_μ and f do not depend on g and ϕ_0, and depend on ρ through the combination x/ρ. For such configurations, three contributions to the action (A.41) are estimated as follows:

$$S_A = \int d^4x \left(-\frac{1}{2g^2} \operatorname{Tr} F_{\mu\nu}^2 \right) \propto \frac{1}{g^2},$$

---

[1] Of course, there are other flat directions associated with the instanton position, and also its orientation in space–time and isotopic space. These directions are not shown in Figure A.1.

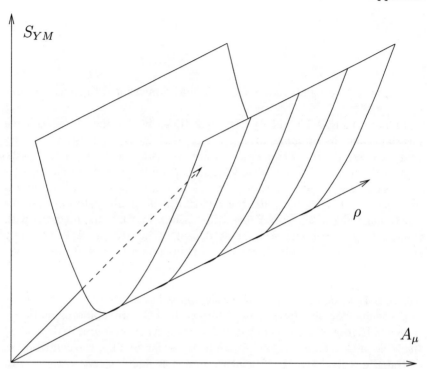

Figure A.1.

$$S_k = \int d^4x (D_\mu\phi)^\dagger D_\mu\phi \propto \rho^2\phi_0^2,$$

$$S_p = \int d^4x \lambda \left(\phi^\dagger\phi - \frac{1}{2}\phi_0^2\right)^2 \propto \lambda\rho^4\phi_0^4.$$

For small ρ, the contribution of S_k is suppressed in comparison with S_A by a factor

$$g^2\phi_0^2\rho^2 \sim m_v^2\rho^2,$$

while the contribution of S_p is even smaller, being suppressed in comparison with S_k by a factor

$$\lambda\phi_0^2\rho^2 \sim m_\chi^2\rho^2.$$

We shall consider the case $m_\chi \lesssim m_v$, and shall be interested in configurations with scales $\rho \ll m_v^{-1}$. For the latter,

$$S_A \gg S_k \gg S_p.$$

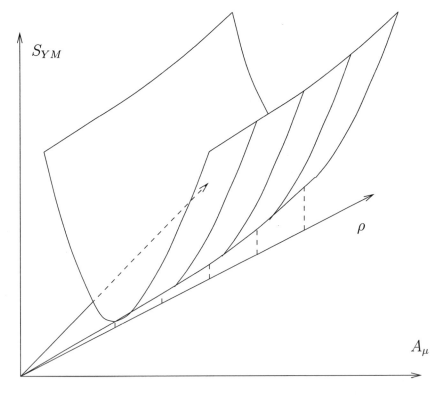

Figure A.2.

Thus, to compute the action on the valley floor we first have to minimize S_A for fixed ρ. This minimum is nothing other than the instanton

$$A_\mu = A_\mu^{\text{inst}} = \frac{r^2}{r^2 + \rho^2}\omega\partial_\mu\omega^{-1}. \qquad (\text{A.43})$$

The first term in the action then does not depend on ρ and is equal to $S_A = 8\pi^2/g^2$, but the field configuration ϕ is not yet fixed. To compute the total action to the first non-trivial order in ρ, one has to find the field ϕ, minimizing the contribution of S_k for a given gauge field (A.43). This leads to the equation

$$D_\mu^{\text{inst}} D_\mu^{\text{inst}} \phi = 0,$$

where $D_\mu^{\text{inst}} = \partial_\mu + A_\mu^{\text{inst}}$. A solution of this equation with the boundary condition (A.42) is

$$\phi = \frac{r}{\sqrt{r^2 + \rho^2}} \omega(\vec{n}) \begin{pmatrix} 0 \\ \phi_0 \end{pmatrix}. \tag{A.44}$$

Here the value of S_k for the configuration (A.43), (A.44) is equal to $\pi^2 \rho^2 \phi_0^2$. Thus, for the total action along the valley, we have

$$S = \frac{8\pi^2}{g^2} + \pi^2 \phi_0^2 \rho^2 + \frac{1}{g^2} O(m_v^2 m_\chi^2 \rho^4). \tag{A.45}$$

As a result, the contribution of configurations with scale $\rho \ll m_v^{-1}$ to the functional integral (in the sector with $Q = 1$) can be written as follows:

$$\int \mu(\rho) d\rho d^4 x_0 \exp\left(-\frac{8\pi^2}{g^2} - \pi^2 \phi_0^2 \rho^2\right). \tag{A.46}$$

Here, we have explicitly written down the integral with respect to the position of the instanton, but not the integral with respect to the collective coordinates associated with the orientations. The dependence of the pre-exponential (measure $\mu(\rho)$) on ρ arises from the functional determinant ('t Hooft 1976b); this dependence is not too important[2] for small g^2.

Since instantons with $\rho \gtrsim m_v^{-1}$ are suppressed by a factor of $\exp(-\frac{4\pi^2}{g^2}\rho^2 m_v^2)$ in comparison with small instantons, the studied approach enables one to find the instanton contribution to the functional integral in the region of values of ρ of greatest interest for the majority of applications. At the same time, it does not allow one to systematically find corrections in ρ and is not suitable for calculating contributions of instantons with $\rho \gtrsim m_v^{-1}$. Moreover, it is not completely satisfactory also from the following points of view. The fields A_μ and ϕ are massive in the given model, thus one might expect them to tend exponentially to the asymptotics (A.42). Configurations (A.43), (A.44) do not have this property.

These disadvantages are removed in the formalism of constrained instantons (Affleck 1981). The idea of this formalism is similar to the idea of the treatment of zero modes (see Section A.1) and involves introducing into the functional integral a constraint, parametrized by the scale ρ (δ-function with corresponding determinant of Faddeev–Popov type), in such a way that the total action has an exact minimum when this constraint is taken into account. The integration with respect to ρ is performed at the very end of the computations. In this formalism the minimum of the constrained action is achieved for field configurations which agree with (A.43), (A.44) for $r \ll m_v^{-1}$ (which is the most interesting region), but tend exponentially to the asymptotics (A.42) as $r \to \infty$. Of course, the result (A.46) is also obtained (for $\rho \ll m_v^{-1}$) in the constrained instanton formalism.

---

[2]In fact, the appropriate quantity is the value of the running coupling at the scale ρ, i.e. $g^2(\rho^{-1})$. Precisely, this quantity occurs in the exponential in (A.46).

A.4 Growing instanton cross sections

In Section 13.4, we saw that the height of the energy barrier between topologically distinct vacua in four-dimensional non-Abelian gauge theories with the Higgs mechanism is finite and proportional to $E_{\text{sph}} \propto m_v/\alpha_g$, where m_v is the mass of the vector boson, and $\alpha_g = g^2/4\pi$. In the standard electroweak theory, $E_{\text{sph}} \sim 10$ TeV. The finiteness of the barrier height leads to a high rate of transitions between gauge vacua at finite temperature (see Section 17.4).

It is natural to ask whether transition processes between topologically distinct vacua can be induced by collisions of particles with energies (in the center-of-mass frame) of the order of E_{sph}, i.e. whether the exponential suppression of the probabilities of these processes disappears. In the case of the electroweak theory, this would imply rapid non-conservation of the baryon and lepton numbers at energies in principle accessible to future accelerators.

At the present time, accurate analysis of this problem has only been performed for the case of relatively low energies ($E \ll E_{\text{sph}}$). It has been shown (Ringwald 1990, Espinosa 1990) that instanton cross sections actually grow rapidly with energy. In this section we briefly discuss the reason for this interesting phenomenon. At the same time, the question of the behavior of cross sections at center-of-mass energies $E \gtrsim E_{\text{sph}}$ remains open, although there are a number of arguments pointing to the exponential suppression of these cross sections at all energies (Banks *et al.* 1990, Zakharov 1991, Maggiore and Shifman 1992, Rebbi and Singleton 1996, Kuznetsov and Tinyakov 1997).

Let us consider, for specificity, the model with gauge group $SU(2)$ and Higgs field doublet, discussed in the previous section, but without fermions.[3] By way of example, we shall study a process in which two colliding vector bosons with center-of-mass energy E are transformed into n vector bosons, and at the same time the system moves from one gauge vacuum to a neighboring one. The topological number Q of the field configurations responsible for this process must be equal to unity. We shall consider the case when the energies of both the initial and the final vector bosons are large in comparison with their mass, i.e.

$$\frac{E}{n} \gg m_v. \tag{A.47}$$

In this case the mass of the vector bosons in the constrained instanton configuration can be neglected, i.e. formula (A.43) for the instanton field

---

[3]We recall that this model may be viewed as a limiting case of the electroweak theory at $\sin\theta_W = 0$. Fermions of the Standard Model play a minor role in this problem and we shall consider a purely bosonic theory.

can be used. To compute the amplitude of the process, we begin with the $(2 + n)$-point Euclidean Green's function

$$G_{n+2}(x_1, x_2, y_1, \ldots, y_n) = \int \mathcal{D}A\mathcal{D}\phi e^{-S[A]} A(x_1)A(x_2)A(y_1)\ldots A(y_n),$$

$$(A.48)$$

where the spatial and group indices are omitted. We perform the semi-classical computation of this integral in the sector with $Q = 1$, assuming that $A(x_1)\ldots A(y_n)$ is the pre-exponential factor. In other words, we calculate only the contribution of configurations of the valley floor, discussed in the previous section. In this approximation, integration with respect to $\mathcal{D}A\mathcal{D}\phi$ is replaced by integration with respect to the collective coordinates of the valley floor (taking into account the measure arising from the determinant of the fluctuations); the action here is (A.45) and the fields $A(x_1)\ldots A(y_n)$ in the pre-exponential are replaced by instanton fields.[4]

As a result we have

$$G_{n+2}^{\text{inst}}(x_1, \ldots, y_n) \qquad\qquad\qquad\qquad\qquad\qquad\qquad (A.49)$$
$$= \int d^4x_0 \frac{d\rho}{\rho^5} \mu(\rho) e^{-\frac{8\pi^2}{g^2} - \pi^2\rho^2\phi_0^2} A^{\text{inst}}(x_1 - x_0; \rho)\ldots A^{\text{inst}}(y_n - x_0; \rho).$$

There is a subtlety here. Asymptotically, as $r \to \infty$, the instanton configuration tends to the purely gauge field (A.42). At the same time, the fields of vector and Higgs bosons are identified with the fields A_μ^a and ϕ in the *unitary* gauge, where

$$\phi = \frac{1}{\sqrt{2}}\begin{pmatrix} 0 \\ \phi_0 + \eta \end{pmatrix}.$$

Thus, it is convenient to perform a gauge transformation with gauge function $\omega^\dagger(x) = \sigma_\alpha^\dagger n_\alpha$ over the configuration (A.43), i.e. to bring the configuration to the unitary gauge. In the unitary gauge,

$$A_\mu^{a,\text{inst}} = \frac{\rho^2}{r^2 + \rho^2}\omega^{-1}\partial_\mu\omega$$

or, in components,

$$A_\mu^{a,\text{inst}} = \frac{1}{g}\frac{2\rho^2}{r^2(r^2 + \rho^2)}\bar{\eta}_{\mu\nu a}x_\nu. \qquad\qquad\qquad (A.50)$$

This configuration is singular at $r = 0$, but this is purely gauge singularity and should not disturb us. It is precisely the configuration (A.50) that appears in (A.49) as A^{inst}.

---

[4]In fact, one has to use the constrained instanton configurations mentioned in Section A.3. This does not alter the result (A.55).

In (A.49), although not written down explicitly, there is the integration with respect to instanton orientations. In further estimations, we shall ignore the fact that an instanton may have different orientations in space–time and isotopic space. Because of this, we shall not be able to find the numerical constants in the expression for the cross section. In fact, integration with respect to instanton orientations is technically very complex and is performed using methods which go far beyond the scope of this short review.

We note immediately that the dependence on coordinates in (A.49) factorizes modulo integration with respect to the instanton position x_0^μ, guaranteeing the translational invariance of the Green's function. This means that, in the approximation used, the corresponding amplitude is pointlike. Hence it is clear that the instanton contribution to the cross section will grow with the energy according to a power law.

To obtain the contribution to the amplitude of the process $2 \to n$ in which we are interested, we move to the momentum representation, i.e. we find $G(k_1, k_2, p_1, \ldots, p_n)$ initially for Euclidean momenta. From (A.49) it is clear that the Green's function in the momentum representation is expressed in terms of Fourier transform of the instanton field $\tilde{A}^{\text{inst}}(q; \rho)$. Integration with respect to x_0^μ in (A.49) leads to a δ-function, ensuring the four-dimensional momentum conservation, and we obtain

$$G_{n+2}^{\text{inst}}(k_1, k_2, p_1, \ldots, p_n) = \delta(k_1 + k_2 + p_1 + \ldots + p_n) \qquad \text{(A.51)}$$
$$\times \int d\rho\, \mu(\rho) e^{-\frac{8\pi^2}{g^2} - \pi^2 \rho^2 \phi_0^2}\, \tilde{A}^{\text{inst}}(k_1) \tilde{A}^{\text{inst}}(k_2) \tilde{A}^{\text{inst}}(p_1) \ldots \tilde{A}(p_n).$$

The Fourier transform of the instanton field (A.50) is not difficult to compute explicitly. The function $\tilde{A}^{\text{inst}}(q)$ is analytic in q^0 and has a pole at $q^2 = 0$. The Green's function (A.51) has the same analytic structure in all its arguments. In fact, the Green's function (A.51) must have a pole on the mass shell of the vector bosons, i.e. at $k_1^2 = k_2^2 = p_1^2 = \ldots = p_n^2 = -m_v^2$, however, we shall consider the case (A.47), in which it is impossible to distinguish between the points[5] $q^2 = 0$ and $q^2 = -m_v^2$. Then we can use the Lehman–Symanzik–Zimmermann formalism, according to which the amplitude of the process $2 \to n$ is determined by the residue of the Green's function G_{n+2} on the mass shell with respect to all momenta. From (A.51) we obtain the instanton contribution to this amplitude in the form

$$A_{2 \to n}^{\text{inst}}(\mathbf{k}_1, \mathbf{k}_2, \mathbf{p}_1, \ldots, \mathbf{p}_n)$$
$$= \int d\rho\mu(\rho) e^{-\frac{8\pi^2}{g^2} - \pi^2 \rho^2 \phi_0^2}\, R(\mathbf{k}_1; \rho) \ldots R(\mathbf{p}_n; \rho), \qquad \text{(A.52)}$$

---

[5]In the constrained instanton formalism, the configuration $\tilde{A}^{\text{inst}}(q)$ and the Green's function (A.51) have poles at correct value $q^2 = -m_v^2$. The residue at the pole for the constrained instanton, which is important to us, agrees with (A.53) modulo corrections which are small in the regime (A.47).

where $R(\mathbf{q}; \rho)$ is the residue of the instanton configuration $\tilde{A}^{\mathrm{inst}}(q, \rho)$ at the pole $q^2 = 0$. An explicit computation gives

$$R(\mathbf{q}; \rho) = \frac{1}{g} \rho^2 |\mathbf{q}|, \tag{A.53}$$

where, as before, we omit the tensor structure, depending on the instanton orientation.

For

$$n \gg 1, \tag{A.54}$$

integration with respect to ρ in (A.52) can be carried out by the saddle-point method. Taking into account the very weak dependence of the measure μ on ρ, we obtain

$$A_{2 \to n}^{\mathrm{inst}} = c_1 e^{-\frac{8\pi^2}{g^2}} (n+2)! \left(\frac{c_2}{g \phi_0^2} \right)^{n+2} |\mathbf{k}_1| |\mathbf{k}_2| |\mathbf{p}_1| \dots |\mathbf{p}_n|, \tag{A.55}$$

where the constant c_1 does not depend on the momenta and depends weakly on the remaining parameters of the problem, and c_2 is a numerical constant. It is clear that the amplitude (A.55) is actually pointlike; the dependence on the momenta in it factorizes.

In the regime (A.47), (A.54), it is not difficult to find the behavior of the instanton cross section

$$\sigma_{2 \to n}^{\mathrm{inst}}(E) = \frac{1}{k_1 \cdot k_2} \frac{1}{n!} \int d\Omega_n |A_{2 \to n}^{\mathrm{inst}}|^2 \sim \frac{1}{n!} \left(\frac{\mathrm{constant} \cdot E^2}{g \phi_0^2 n} \right)^{2n} e^{-\frac{16\pi^2}{g^2}} \tag{A.56}$$

from the expression (A.55). Here, $d\Omega_n$ denotes an element of the n-particle phase space, E is the total energy in the center-of-mass frame. It is clear that the instanton cross section in fact grows quickly (exponentially) with the energy (Ringwald 1990, Espinosa 1990). An even more surprising result is obtained for the total cross section, i.e. the sum of the cross sections (A.56) over n. The dominant contribution to this sum for large E comes from the process with number of final particles of the order of

$$n \sim \left(\frac{E^4}{g^2 \phi_0^4} \right)^{1/3}.$$

The cross section itself grows exponentially with energy[6],

$$\sigma_{\mathrm{tot}}^{\mathrm{inst}}(E) \propto \exp \left[-\frac{16\pi^2}{g^2} + \mathrm{constant} \left(\frac{E^4}{g^2 \phi_0^4} \right)^{1/3} \right].$$

---

[6]The exponential behavior of the instanton cross section with energy was first observed by McLerran *et al.* (1990) for the case of Higgs particles in the final state. These final states give a small contribution to the total cross section in the energy region where the given analysis is legitimate.

These formulae can be written in a more transparent form, by introducing the notation $\alpha_g = g^2/4\pi$ and $E_0 = \sqrt{6}\pi m_v/\alpha_g$. We note that E_0 agrees in order of magnitude with the sphaleron energy; in the electroweak theory $E_0 \approx 15$ TeV (the constant α_g in this case is the coupling constant α_W, corresponding to the group $SU(2)$). In this notation the total cross section has the form

$$\sigma_{\text{tot}}^{\text{inst}}(E) \propto \exp\left[\frac{4\pi}{\alpha_g}\left(-1 + \frac{9}{8}\left(\frac{E}{E_0}\right)^{4/3}\right)\right], \qquad (\text{A.57})$$

where we have included the value of the numerical constant found by Zakharov (1992), Klebnikov *et al.* (1990) and Porrati (1990). The characteristic number of final particles and their energies have the following orders of magnitude

$$n \sim \frac{1}{\alpha_g}\left(\frac{E}{E_0}\right)^{4/3}, \qquad |\mathbf{p}| \sim \frac{E}{n} \sim m_v\left(\frac{E}{E_0}\right)^{-1/3}.$$

The remarkable result (A.57) shows that in the energy region $E \sim E_0$ the exponential suppression of the instanton processes weakens. This region is characterized by a large (of order $1/\alpha_g$) number of particles, born in the scattering process, induced by the instanton and having a small momentum (of order m_v).

At the same time, it is clear that the result (A.57) cannot hold for all energies: an exponentially large cross section would contradict unitarity. This means that the semiclassical formula (A.49) is not reliable for computing the instanton amplitude at high energies. Indeed, corrections to formula (A.57) are also exponential in nature and become significant at $E \sim E_0$. In the general case, the instanton cross section has the exponential form

$$\sigma_{\text{tot}}^{\text{inst}} \propto \exp\left[\frac{4\pi}{\alpha_g}F\left(\frac{E}{E_0}\right)\right],$$

and computations near the instanton enable us to determine the function F for $E \ll E_0$ in the form of a power series in $(E/E_0)^{2/3}$. To compute the exponent in the most interesting region $E \geq E_0$, we need semiclassical techniques, which are far beyond the scope of this book. This computation has not yet been performed, there are only indirect (albeit very serious) arguments in support of the assertion that $F(E/E_0)$ is negative for all energies.

The last statement, if it is true, eliminates the possibility of searching for electroweak instanton processes (accompanied by non-conservation of the baryon and lepton numbers) at accelerators. Nevertheless, there remains a possibility that QCD instanton processes may be detected in collisions of highly energetic particles even at present-day colliders (Balitsky and Braun 1993, Ringwald and Schrempp 1994).

Bibliography

First, we present a list of books and reviews relating to the theme of the book. Where necessary we indicate in brackets the sections to which a particular review relates. This list is intended to provide the reader with an initial orientation and we do not pretend that it is complete.

Then, we list papers referred to within the text. This list is given in alphabetical order. More detailed references can be found in the books and reviews listed in the first part.

1. Books and reviews

For Part I

The topics dealt with in Part I are considered in many texts, including:

1. N.N. Bogolyubov and D.V. Shirkov. Introduction of the Theory of Quantized Fields. New York: Wiley, 1980. Translated from the Russian. Citations in the present book relate to the second Russian edition. Moscow: Enegroatomizdat, 1980.

2. T.P. Cheng and L.F. Li. Gauge Theory of Elementary Particle Physics. Oxford: Clarendon, 1984.

3. L.D. Faddeev and A.A. Slavnov. Gauge Fields. Introduction to Quantum Theory. Reading, MA: Benjamin–Cummings, 1980. Translated from the Russian. Citations in the present book relate to the second Russian edition. Moscow: Nauka, 1988.

4. C. Itzykson and J.B. Zuber. Quantum Field Theory. New York: McGraw-Hill, 1980.

5. L.B. Okun. Leptons and Quarks. Amsterdam: North-Holland, 1982. Translated from the Russian. Citations in the present book relate to the second Russian edition. Moscow: Nauka, 1990.

6. M.E. Peskin and D.V. Schroeder. An Introduction to Quantum Field Theory. Reading, MA: Addison-Wesley, 1995.

7. A.M. Polyakov. Gauge Fields and Strings. Chur: Harwood, 1994.

8. P. Ramond. Field Theory: A Modern Primer. Reading, MA: Addison-Wesley, 1990.

9. L.H. Ryder. Quantum Field Theory. Cambridge: Cambridge University Press, 1985.

10. M.B. Voloshin and K.A. Ter-Martirosyan. Teoriya kalibrovochnykh vzaimodeistvii elementarnykh chastits. Moscow: Enegroatomizdat, 1984.

11. S. Weinberg. The Quantum Theory of Fields. Vol. 1: Foundations. Cambridge: Cambridge University Press, 1995.

As historical surveys, the following may be recommended:

12. T.W.B. Kibble. Genesis of Unified Gauge Theories. In: 'Salamfestschrift' eds. A. Ali and S. Randjbar-Daemi. Singapore: World Scientific, 1994, p. 592.

13. L. O'Raifeartaigh. The Dawning of Gauge Theory. In: 'Salamfestschrift' eds. A. Ali and S. Randjbar-Daemi. Singapore: World Scientific, 1994, p. 577.

14. A.N. Tavkhelidze. Color, Colored Quarks, Quantum Chromodynamics. In: 'Quarks-94', Proc. 8th Int. Seminar, eds. D.Yu. Grigoriev *et al.* Singapore: World Scientific, 1995, p. 3.

We now list references to books and reviews relating to the individual chapters.

Chapters 1 and 2

The gauge invariance of classical electrodynamics is discussed in many texts, for example:

15. L.G. Chambers. An Introduction to the Mathematics of Electricity and Magnetism. London: Chapman and Hall, 1973.

16. L.D. Landau and E.M. Lifshitz. The Classical Theory of Fields (Course of Theoretical Physics, Volume 2), 4th revised ed. New York: Butterworth–Heinemann, 1997. Translated from the Russian. Moscow: Nauka, 1967.

17. I.E. Tamm. Osnovy teorii electrichestva. Moscow: Nauka, 1966.

Chapter 3

Books on group theory, written for physicists:

18. A.P. Balachandran and C.G. Trahern. Lectures on Group Theory for Physicists. Naples: Bibliopolis, 1984.

19. J.Q. Chen. Group Representation Theory for Physicists. Singapore: World Scientific, 1989.

20. C.J. Isham. Lectures on Groups and Vector Spaces for Physicists. Singapore: World Scientific, 1984.

Many concrete results from the theory of simple Lie groups and their representations are contained in the review:

21. R. Slansky. Group Theory for Unified Model Building. Phys. Rep., **79**, p. 1, 1981.

Chapters 4–6

Reviews:

22. E.S. Abers and B.W. Lee. Gauge Theories. Phys. Rep., **9**, p. 1, 1973.

23. S. Coleman. Secret Symmetry: An Introduction to Spontaneous Symmetry Breakdown and Gauge Fields. In: Laws of Hadronic Matter, Proc. 1973 Int. School of Subnuclear Physics, Erice, ed. A. Zichichi. New York: Academic, 1975.

For Part II

Solitons and instantons in field theory are considered in Polyakov [7] and in the books:

24. M.I. Monastyrsky. Topology of Gauge Fields and Condensed Matter. New York: Plenum, 1993. Translated from the Russian. Moscow: PAIMS, 1991.

25. R. Rajaraman. Solitons and Instantons. An Introduction to Solitons and Instantons in Quantum Field Theory. Amsterdam: North-Holland, 1982.

26. A.S. Schwarz. Quantum Field Theory and Topology. Berlin: Springer, 1994. Translated from the Russian. Moscow: Nauka, 1989.

Chapter 7

Reviews:

27. S. Coleman. Classical Lumps and Their Quantum Descendants. In: New Phenomena in Subnuclear Physics, Proc. 1975 Int. School of Subnuclear Physics, Erice, ed. A. Zichichi. New York: Plenum, 1977.

28. L.D. Faddeev. Solitony. In: Nelokal'nye, nelineinye i nerenormiruemye teorii polya, materialy 4 Mezhdunarodnogo soveshaniya po nelokal'nym teriyam polya. Dubna: OIYaI, 1976.

29. L.D. Faddeev and V.E. Korepin. Quantum Theory of Solitons. Phys. Rep., **42**, p. 1, 1978.

30. R. Jackiw. Quantum Meaning of Classical Field Theory. Rev. Mod. Phys., **49**, p. 681, 1977.

31. A. Vilenkin. Cosmic Strings and Domain Walls. Phys. Rep., **121**, p. 262, 1985.

Vortices in superconductors are considered in physics texts on condensed matter, for example, in the book:

32. E.M. Lifshitz and L.P. Pitaevskii. Statisticheskaya fizika. Chast' 2. Moscow: Nauka, 1978.

Chapter 8

A study of homotopy theory for physicists is contained in Schwarz [26].

Chapter 9

In addition to the reviews [27, 30], the following review is to be recommended:

33. P. Goddard and D.I. Olive. New Developments in the Theory of Magnetic Monopoles. Rep. Prog. Phys., **41**, p. 1357, 1978.

Chapter 10

Review:

34. T.D. Lee and Y. Pang. Non-Topological Solitons. Phys. Rep., **221**, p. 251, 1992.

Chapters 11 and 12

Review:

35. S. Coleman. The Uses of Instantons. In: The Whys of Subnuclear Physics, Proc. 1977 Int. School of Subnuclear Physics, Erice, ed. A. Zichichi. New York: Plenum, 1979.

The theory of the decay of a false vacuum at zero and non-zero temperatures, and its applications in cosmology are considered in the following books:

36. E.W. Kolb and M.S. Turner. The Early Universe. New York: Addison-Wesley, 1990.

37. A.D. Linde. Particle Physics and Inflationary Cosmology. Chur: Harwood, 1990. Translated from the Russian. Moscow: Nauka, 1990.

Chapter 13

The review by Coleman [35] and the reviews:

38. R. Jackiw. Introduction to Yang–Mills Quantum Theory. Rev. Mod. Phys., **52**, p. 661, 1980.

39. A.I. Vainshtein, V.I. Zakharov, V.A. Novikov and M.A. Shifman. ABC of Instantons. Sov. Phys.–Usp., **24**, p. 195, 1982.

For Part III and the Appendix

Fermions in gauge theories are considered in the books [1–11] and in the reviews [22, 23]. The interaction of fermions with topological scalars and gauge fields is discussed in the books [25, 26] and in the reviews [30, 35, 39]. In addition, see the reviews:

40. A.G. Cohen, D.B. Kaplan and A. Nelson. Progress in Electroweak Baryogenesis. Ann. Rev. Nucl. Part. Sci., **43**, p. 27, 1993 (Section 17.4).

41. J.E. Kim. Light Pseudoscalars, Particle Physics and Cosmology. Phys. Rep., **150**, p. 1, 1987 (Section A.2).

42. V.A. Matveev, V.A. Rubakov, A.N. Tavkhelidze and M.E. Shaposhnikov. Nonconservation of baryon number under extremal conditions. Sov. Phys.–Usp., **31**, p. 916, 1988 (Section 17.4).

43. M.P. Mattis. The Riddle of High-Energy Baryon Number Violation. Phys. Rep., **214**, p. 159, 1992 (Section A.4).

44. V.A. Rubakov. Monopole Catalysis of Proton Decay. Rep. Prog. Phys., **41**, p. 1357, 1978 (Section 16.2).

45. V.A. Rubakov and M.E. Shaposhnikov. Electroweak Baryon Number Nonconservation in the Early Universe and in High-Energy Collisions. Phys.–Usp., **39**, p. 461, 1996 (Sections 17.4, A.4).

46. P.G. Tinyakov. Instanton-like Transitions in High-Energy Collisions. Int. J. Mod. Phys., **A8**, p. 1823, 1993 (Section A.4).

2. Papers

1. A.A. Abrikosov (1957) Zh. Eksp. Teor. Fiz., **32**, p. 1442.

2. S.L. Adler (1969) Phys. Rev., **177**, p. 2426.

3. I. Affleck (1981) Nucl. Phys., **B191**, p. 429.

4. M.G. Alford and F. Wilczek (1989) Phys. Rev. Lett., **62**, p. 1071.

5. J. Ambjorn, J. Greensite and C. Peterson (1983) Nucl. Phys., **B221**, p. 381.

6. J. Ambjorn and V.A. Rubakov (1985) Nucl. Phys., **B256**, p. 434.

7. P.W. Anderson (1963) Phys. Rev., **130**, p. 439.

8. M. Ansourian (1977) Phys. Lett., **70B**, p. 301.

9. J. Arafune, P.G.O. Freund and C.J. Goebel (1975) J. Math. Phys., **16**, p. 433.

10. I.I. Balitsky and V.M. Braun (1993) Phys. Rev., **D47** p. 1879.

11. T. Banks and C.M. Bender (1973) Phys. Rev., **D8**, p. 3366.

12. T. Banks, C.M. Bender and T.T. Wu (1973) Phys. Rev., **D8**, p. 3346.

13. T. Banks, G. Farrar, M. Dine, D. Karabali and B. Sakita (1990) Nucl. Phys., **B347**, p. 581.

14. T. Banks and E. Rabinovici (1979) Nucl. Phys., **B160**, p. 349.

15. A.A. Belavin and A.M. Polyakov (1975) Pis. Zh. Eksp. Teor. Fiz., **22**, p. 503.

16. A.A. Belavin, A.M. Polyakov, A.S. Schwarz and Yu.S. Tyupkin (1975) Phys. Lett., **59B**, p. 85.

17. J.S. Bell and R. Jackiw (1969) Nuovo Cimento, **A60**, p. 47.

18. A.I. Bochkarev and M.E. Shaposhnikov (1987) Mod. Phys. Lett., **A2**, p. 991.

19. N.N. Bogoliubov, B.V. Struminskii and A.N. Tavkhelidze (1965) Preprint JINR D 1968.

20. E.B. Bogomolny (1976) Yad. Fiz., **24**, p. 861.

21. J. Boguta (1983) Phys. Rev. Lett., **50**, p. 148.

22. L.S. Brown, R.D. Carlitz and C. Lee (1977) Phys. Rev., **D16**, p. 417.

23. N. Cabbibo (1963) Phys. Rev. Lett., **10**, p. 531.

24. C.G. Callan (1982) Phys. Rev., **D25**, p. 2141;
    — (1982) Phys. Rev., **D26**, p. 2058;
    — (1982) Nucl. Phys., **B212**, p. 391.

25. C.G. Callan and S. Coleman (1977) Phys. Rev., **D16**, p. 1762.

26. C.G. Callan, R.F. Dashen and D.J. Gross (1976) Phys. Lett., **63B**, p. 334.

27. — (1978) Phys. Rev., **D17**, p. 2717.

28. S. Coleman (1977) Phys. Rev., **D15**, p. 2929.

29. — (1985) Nucl. Phys., **B262**, p. 263.

30. S. Coleman, V. Glaser and A. Martin (1978) Commun. Math. Phys., **58**, p. 211.

31. S. Coleman and E. Weinberg (1973) Phys. Rev., **D7**, p. 1888.

32. R. Dashen, B. Hasslacher and A. Neveu (1974a) Phys. Rev., **D10**, p. 4130.

33. — (1974b) Phys. Rev., **D10**, p. 4138.

34. T. Dereli, J.H. Swank and L.J. Swank (1975) Phys. Rev., **D12**, p. 3541.

35. G.H. Derrick (1964) J. Math. Phys., **5**, p. 1252.

36. S. Deser, R. Jackiw and S. Templeton (1982) Ann. Phys., **140**, p. 372.

37. L. Dolan and R. Jackiw (1974) Phys. Rev., **D9**, p. 3320.

38. F. Englert and R. Brout (1964) Phys. Rev. Lett., **13**, p. 321.

39. O. Espinosa (1990) Nucl. Phys., **B343**, p. 310.

40. E. Fradkin and S. Shenker (1979) Phys. Rev., **D19**, p. 3682.

41. R. Friedberg and T.D. Lee (1977) Phys. Rev., **D15**, p. 1694;
    — (1977) Phys. Rev., **D16**, p. 1096.

42. R. Friedberg, T.D. Lee and A. Sirlin (1976) Phys. Rev., **D13**, p. 2739;
    — (1976) Nucl. Phys., **B115**, p. 1;
    — (1976) Nucl. Phys., **B115**, p. 32.

43. H. Fritzsch, M. Gell-Mann and H. Leutwyler (1973) Phys. Lett., **47B**,
    p. 365.

44. S. Fubini (1976) Nuovo Cimento, **34A**, p. 521.

45. H. Georgi and S.L. Glashow (1972) Phys. Rev. Lett., **28**, p. 1494.

46. — (1974) Phys. Rev. Lett., **32**, p. 438.

47. S.L. Glashow (1961) Nucl. Phys., **22**, p. 579.

48. J. Goldstone (1961) Nuovo Cimento, **19**, p. 154.

49. V.N. Gribov (1976) Unpublished.

50. D.Yu. Grigoriev and V.A. Rubakov (1988) Nucl. Phys., **B299**, p. 67.

51. D.J. Gross and F. Wilczek (1973) Phys. Rev. Lett., **30**, p. 1343.

52. G.S. Guralnik, C.R. Hagen and T.W.B. Kibble (1964) Phys. Rev.
    Lett., **13**, p. 585.

53. M.Y. Han and Y. Nambu (1965) Phys. Rev., **139B**, p. 1006.

54. P. Hasenfratz and G. 't Hooft (1976) Phys. Rev. Lett., **36**, p. 1119.

55. P.W. Higgs (1964) Phys. Rev. Lett., **13**, p. 508.

56. R. Jackiw, C. Nohl and C. Rebbi (1977) Phys. Rev., **D15**, p. 1642.

57. R. Jackiw and C. Rebbi (1976a) Phys. Rev. Lett., **37**, p. 172.

58. — (1976b) Phys. Rev., **D13**, p. 3398.

59. — (1977) Phys. Rev., **D6**, p. 1052.

60. R. Jackiw and P. Rossi (1981) Nucl. Phys., **B190**, p. 681.

61. B. Julia and A. Zee (1975) Phys. Rev., **D11**, p. 2227.

62. K. Kajantie, M. Laine, K. Rummukainen and M.E. Shaposhnikov (1996) Phys. Rev. Lett., **77**, p. 2887.

63. S.Yu. Khlebnikov, V.A. Rubakov and P.G. Tinyakov (1990) Mod. Phys. Lett., **A5**, p. 1983.

64. T.W. Kibble, G. Lazarides and Q. Shafi (1982) Phys. Rev., **D26**, p. 435.

65. D.A. Kirzhnits (1972) Pis. Zh. Eksp. Teor. Fiz., **15**, p. 745.

66. D.A. Kirzhnits and A.D. Linde (1972) Phys. Lett., **72B**, p. 741.

67. — (1976) Ann. Phys., **101**, p. 195.

68. J. Kiskis (1977) Phys. Rev., **D15**, p. 2329.

69. — (1978) Phys. Rev., **D18**, p. 3690.

70. F.R. Klinkhamer and N.S. Manton (1984) Phys. Rev., **D30**, p. 2212.

71. M. Kobayashi and K. Maskawa (1973) Prog. Theor. Phys., **46**, p. 652.

72. N.V. Krasnikov, V.A. Rubakov and V.F. Tokarev (1979) J. Phys. A: Math. Gen., **12**, p. L343.

73. V.A. Kuzmin (1970) Pis. Zh. Eksp. Teor. Fiz., **13**, p. 335.

74. V.A. Kuzmin, V.A. Rubakov and M.E. Shaposhnikov (1985) Phys. Lett., **155B**, p. 36.

75. A.N. Kuznetsov and P.G. Tinyakov (1997) Phys. Rev., **D56**, p. 1156.

76. L.N. Lipatov (1977) Zh. Eksp. Teor. Fiz., **72**, p. 411.

77. J.H. Lowenstein and J.A. Swieca (1971) Ann. Phys., **68**, p. 172.

78. M. Maggiore and M. Shifman (1992) Nucl. Phys., **B365**, p. 161.

79. N.S. Manton (1983) Phys. Rev., **D28**, p. 2019.

80. V.A. Matveev, V.A. Rubakov, A.N. Tavkhelidze and V.F. Tokarev (1987) Nucl. Phys., **B282**, p. 700.

81. L. McLerran, A. Vainshtein and M. Voloshin (1990) Phys. Rev., **D42**, p. 171.

82. Y. Miyamoto (1965) Prog. Theor. Phys. Suppl., Extra No., p. 187.

83. M.I. Monastyrsky and A.M. Perelomov (1975) Pis. Zh. Eksp. Teor. Fiz., **21**, p. 294.

84. E. Mottola and A. Wipf (1989) Phys. Rev., **D39**, p. 588.

85. Y. Nambu (1960) Phys. Rev. Lett., **4**, p. 380;
    — (1960) Phys. Rev., **117**, p. 648.

86. H.B. Nielsen and M. Ninomiya (1983) Phys. Lett., **130B**, p. 389.

87. H.B. Nielsen and P. Olesen (1973) Nucl. Phys., **B61**, p. 45.

88. N.K. Nielsen and B. Schroer (1977a) Nucl. Phys., **B120**, p. 62.

89. — (1977b) Nucl. Phys., **B127**, p. 493.

90. A. Patrascioiu (1979) Phys. Rev., **D20**, p. 491.

91. R.D. Peccei and H. Quinn (1977) Phys. Rev. Lett., **38**, p. 1440.

92. H.D. Politzer (1973) Phys. Rev. Lett., **30**, p. 1346.

93. A.M. Polyakov (1974) Pis. Zh. Eksp. Teor. Fiz., **20**, p. 430.

94. M. Porrati (1990) Nucl. Phys., **B347**, p. 371.

95. M.K. Prasad and C.M. Sommerfield (1975) Phys. Rev. Lett., **35**,
    p. 760.

96. C. Rebbi and R. Singleton (1996) Phys. Rev., **D54**, p. 1020.

97. A. Ringwald (1988) Phys. Lett., **213B**, p. 61.

98. — (1990) Nucl. Phys., **B330**, p. 1.

99. A. Ringwald and F. Schrempp (1994) In: Proc. Int. Seminar 'Quarks-
    94', eds. D.Yu. Grigoriev *et al.* Singapore: World Scientific, 1995.

100. V.A. Rubakov (1981) Pis. Zh. Eksp. Teor. Fiz., **33**, p. 658.

101. — (1982) Nucl. Phys., **B203**, p. 311.

102. — (1986) Prog. Theor. Phys., **75**, p. 366.

103. A.D. Sakharov (1967) Pis. Zh. Eksp. Teor. Fiz., **5**, p. 32.

104. A. Salam (1968) In: Elementary Particle Theory. Proc. 1968 Nobel
     Symposium, ed. A. Svartholm. Stockholm: Lerum.

105. A.S. Schwarz (1977) Phys. Lett., **67B**, p. 172.

106. J. Schwinger (1962) Phys. Rev., **125**, p. 397.

107. T.H.R. Skyrme (1961) Proc. R. Soc., **A260**, p. 127.

108. V. Soni (1980) Phys. Lett., **93B**, p. 101.

109. G. 't Hooft (1974) Nucl. Phys., **79**, p. 276.

110. — (1976a) Phys. Rev. Lett., **37**, p. 8.

111. — (1976b) Phys. Rev., **D14**, p. 3432.

112. Yu.S. Tyupkin, V.A. Fateev and A.S. Schwarz (1975) Pis. Zh. Eksp. Teor. Fiz., **21**, p. 91.

113. V.G. Vaks and A.I. Larkin (1961) Zh. Eksp. Teor. Fiz., **40**, p. 282.

114. A. Vilenkin and A.E. Everett (1982) Phys. Rev. Lett., **48**, p. 1867.

115. M.B. Voloshin, I.Yu. Kobzarev and L.B. Okun (1974) Yad. Fiz., **20**, p. 1229.

116. S. Weinberg (1967) Phys. Rev. Lett., **19**, p. 1264.

117. — (1974) Phys. Rev., **D9**, p. 3357.

118. — (1978) Phys. Rev. Lett., **40**, p. 223.

119. F. Wilczek (1978) Phys. Rev. Lett., **40**, p. 279.

120. E. Witten (1982) Phys. Lett., **117B**, p. 324.

121. — (1985) Nucl. Phys., **249**, p. 557.

122. T.T. Wu and C.N. Yang (1975) Phys. Rev., **D12**, p. 3845.

123. L. Yaffe (1989) Phys. Rev., **D40**, p. 3463.

124. C.N. Yang and R.L. Mills (1954) Phys. Rev., **96**, p. 1.

125. V.I. Zakharov (1991) Phys. Rev. Lett., **67**, p. 3650.

126. — (1992) Nucl. Phys., **B371**, p. 637.

127. Ya.B. Zel'dovich, I.Yu. Kobzarev and L.B. Okun (1974) Zh. Eksp. Teor. Fiz., **67**, p. 3.

Index[1]

---

[1]This Index supplements the Contents list, without repeating it. The Index includes terms and contents not directly reflected in the Contents list.